NEURAL ASPECTS
OF TACTILE SENSATION

ADVANCES IN PSYCHOLOGY

127

Editor:

G. E. STELMACH

ELSEVIER
Amsterdam – Lausanne – New York – Oxford – Shannon – Singapore – Tokyo

NEURAL ASPECTS
IN
TACTILE SENSATION

Edited by

J.W. MORLEY

School of Physiology and Pharmacology
University of New South Wales,
Sydney, Australia

1998

ELSEVIER
Amsterdam – Lausanne – New York – Oxford – Shannon – Singapore – Tokyo

NORTH-HOLLAND
ELSEVIER SCIENCE B.V.
Sara Burgerhartstraat 25
P.O. Box 211, 1000 AE Amsterdam, The Netherlands

Library of Congress Cataloging in Publication Data.
A catalog record from the Library of Congress has been applied for.

ISBN: 0 444 822 828

♾ The paper used in this publication meets the requirements of ANSI/NISO Z39.48-1992 (Permanence of Paper).

Printed in The Netherlands

754958
R

Preface

Touch is an important part of everyday life. It affords us the ability to accurately determine the surface properties of objects, as well as facilitating many functions of the motor system. Research into touch has a long history, however it is only in the last 30 to 40 years that the neural mechanisms underlying tactile sensation have begun to be unravelled. This book consists of nine chapters in which investigators active in the field of tactile reseach discuss recent advances in our understanding of the old and intimate sense of touch.

The first chapter, by G.K. Essick, deals with our ability to detect and discriminate tactile stimuli moving over the skin, and the neural mechanisms that underlie this ability. In the second chapter A.W. Goodwin discusses the discrimination of curved surfaces on the skin, and the way in which information about curvature is coded by mechanoreceptive afferents. Chapter three by V.G. Macefield provides an account of the responses of peripheral afferents in human subjects that signal information about touch, finger movement and cutaneous force. S.S. Hsiao in chapter four compares and contrasts neural coding in the tactile system with that in the visual system. In chapter five A.B. Turman and colleages detail the functional anatomy of the somatosensory region of the cerebral cortex, and provide discussion on information processing in the somatosensory system. This theme is continued in chapter six by Y. Iwamura who details the result of his studies on the response properties of neurons in the somatosensory cortex, and extended in chapter seven by R. Romo and colleagues in which the processing of tactile stimuli by neurons in the sensory-motor cortex is discussed. Chapter eight by C.E. Chapman provides an account of perceptual constancy in the tactile system, specifically with reference to the neural coding of surface texture. The final chapter, chapter nine, by M. B. Calford and colleagues details the conditions under which short term plasticity is evident in the somatosensory systems.

CONTENTS

CONTRIBUTORS

G.K. Essick, School of Dentistry and Curriculum in Neurobiology, University of North Carolina, Chapel Hill, NC 272599-7455 USA.

M.B. Calford, Psychobiology Laboratory, Division of Psychology, The Australian National University, Canberra City, ACT 0200, Australia.

C.E. Chapman, Centre de recherche en sciences neurologiques, Université de Montréal, Montréal H3C 3J7, Canada.

J.C. Clarey, Vision, Touch and Hearing Research Centre, The University of Queensland, Brisbane, Qld 4072, Australia.

W. García, Insituto de Fisiología Celular, Universidad Nacional Autónoma de México, 04510 México, D.F., Mexico.

A.W. Goodwin, Department of Anatomy and Cell Biology, University of Melbourne, Parville, Victoria 3052 Australia.

A. Hernández, Insituto de Fisiología Celular, Universidad Nacional Autónoma de México, 04510 México, D.F., Mexico.

S.S Hsiao, The Krieger Mind/Brain Institute, Johns Hopkins University, Baltimore MD, USA.

Y. Iwamura, Department of Physiology, Toho University School of Medicine, 21-16 Omori-Nishi 5chome, Otaku, Tokyo, Japan 143.

V.G. Macefield, Prince of Wales Medical Research Institute, Randwick, Sydney NSW 2031, Australia.

H. Merchant, Insituto de Fisiología Celular, Universidad Nacional Autónoma de México, 04510 México, D.F., Mexico.

J.W. Morley, School of Physiology and Pharmacology, University of New South Wales, Sydney 2052, Australia.

M.J. Rowe, School of Physiology and Pharmacology, University of New South Wales, Sydney 2052, Australia.

R. Romo, Insituto de Fisiología Celular,Universidad Nacional Autónoma de México, 04510 México, D.F., Mexico.

A.B. Turman, Department of Biomedical Sciences, Faculty of Health Sciences, University of Sydney, NSW 2141, Australia.

R.Tweedale, Vision, Touch and Hearing Research Centre, The University of Queensland, Brisbane, Qld 4072, Australia.

A. Zainos, Insituto de Fisiología Celular,Universidad Nacional Autónoma de México, 04510 México, D.F., Mexico.

Neural Aspects of Tactile Sensation
J.W. Morley (Editor)
1998 Elsevier Science B.V.

Factors Affecting Direction Discrimination of Moving Tactile Stimuli

G. K. Essick

A tactile stimulus moving across the skin evokes a rich perceptual experience and one that has been of interest to psychologists for over a hundred years (cf. Hall and Donaldson, 1885). No other attribute of the percept has received greater attention than that of the direction of motion, a characteristic that importantly distinguishes moving from non-moving stimuli. As a result of extensive investigation, it has been established that an individual's capacity to distinguish opposing directions of movement is not simply characterized: The capacity varies, for example, as a function of stimulus velocity, the length of skin traversed, and the body site over which the stimulus moves. Moreover, a most influential determinant of the capacity is the degree to which the skin is stretched laterally. To illustrate, stimuli that appear comparable to the naive observer can result in estimates of direction discrimination that differ more than 10-fold due to differences in the extent to which the surrounding skin is stretched.

In this chapter, factors that affect normal human capacity to discriminate direction are explored. The chapter begins with a detailed description of the diverse modes of moving stimulus delivery that have been employed to study direction discrimination (section 1.). The impact of variation in the parameters of stimulation (e.g., velocity of motion) is then reviewed (section 2.). Important to the goal of this publication, both topics are discussed in relation to the hypothetical neural mechanisms by which information about direction is encoded. The text is organized to serve as a comprehensive reference to basic and clinical somatosensory researchers who conduct psychophysical studies on direction discrimination.

1. Stimuli Employed To Assess Direction Discrimination

Direction discrimination is typically defined as the perceptual capacity to distinguish opposing directions of stimuli moving across the skin in an otherwise identical or very similar manner. In the first half of this chapter it will be shown that this capacity depends highly upon the exact manner in which the skin is contacted. The precise relationship among a stimulus' mechanical action, the peripheral neural encoding of information about direction, and psychophysical performance is poorly understood. Progress in this area has been slow for a number of reasons.

First, although psychophysical investigators have generally agreed that direction discrimination varies with the magnitude of lateral skin stretch, they have rarely measured the extent to which the stimuli employed in their experiments stretched the skin (but see Gould et al., 1979 pp. 66, 69; Olausson and Norrsell, 1993 pp. 552-553). Different degrees of lateral skin stretch have been reported based on gross visual inspection (Hall and Donaldson, 1885 p. 567; Essick and Whitsel, 1985a p. 190; Srinivasan et al., 1990 p. 1325), by inference from the nature of the stimulation (Loomis and Collins, 1978 p. 487; Gould et al., 1979 pp. 66-68; Srinivasan et al., 1990 pp. 1325-1326; Norrsell and Olausson, 1992 p. 158; Norrsell and Olausson, 1994 p. 534; Gardner and Sklar, 1994 p. 2415; Olausson, 1994 pp. 305-306; Essick et al., 1996) or from the magnitude of the forces applied to the skin (Gould et al., 1979 pp. 66-68; Norrsell and Olausson, 1992 p. 158; Olausson and Norrsell, 1993 p. 555). Second, only a few studies have directly addressed the peripheral neural encoding of information about the direction of stimuli that do (Greenspan, 1992; Edin et al., 1995; Essick and Edin, 1995; see also Goodwin and Morley, 1987; LaMotte and Srinivasan, 1987a,b, 1991; Srinivasan et al., 1990), do not (Ray et al., 1985; Gardner, 1988; Gardner and Palmer, 1989; Gardner et al., 1989), and only (Edin and Abbs, 1991; Edin, 1992; Srinivasan et al., 1990) stretch the skin laterally. Third, psychophysical and clinical investigators have often employed 'natural' stimuli (e.g., a moving brush, stylus or probe) to which more than one peripheral direction-encoding mechanism may be sensitive (cf. section 1.3.3.). With natural moving stimuli, it could not be determined whether perceptual processes relied on one particular source or all available peripheral neural sources of information about direction of motion.

1.1. Stimuli that translate without stretching

With the exception of an occasional blast of wind or crawl of an insect, stimuli that translate across the skin without lateral stretch are seldom encountered in nature. Such stimuli, however, are unquestionably the simplest of laterally moving stimuli and serve as an important experimental tool. This is because the only skin receptors that are stimulated are necessarily those directly beneath or in contact with the moving object. As such, the total skin field from which information about direction is provided is easily defined and can be systematically varied and studied. Second, as will be described in section 1.1.5., the responses evoked in individual mechanoreceptors are much simpler than those evoked by stimuli that stretch the skin. 'Frictionless' stimuli used in the sensory laboratory have included a moving stream of air, an edge sequentially simulated along adjacent positions on the skin, a rolling wheel, and a moving jet of water.

1.1.1. Moving air-stream stimuli

One of the first systematic studies of direction discrimination conducted with an air-stream stimulus was reported by Gould et al. (1979). The rationale was "to utilize a stimulus that produced an indentation that could be moved along the skin without a component of friction that would stretch the skin" (p. 67). To summarize, air was heated to the thermoneutral zone of the forearm skin and forced at 20 lb/in^2 through a 24-gauge needle. The tip of the needle was filed to a flat surface and kept as close to the skin surface as possible without contact. Approximately 0.7 gm-wgt of force was applied. Movement of the needle across the skin was accomplished by a modified rack and pinion manipulator (cf. section 1.3.1.).

Norrsell and Olausson (1992, 1994) more recently studied direction discrimination with an air-stream stimulus. The stream was moved across the skin by the same mechanical devices employed for metal-rod stimuli (cf. section 1.3.1.). The stream was emitted from a nozzle with 0.6 mm (Norrsell and Olausson, 1992 pp. 156-157) or 2 mm internal diameter (1994 p. 534) held either 6 mm (Norrsell and Olausson, 1992) or 3 mm (1994) from the skin surface. In the earlier study, the impact on direction discrimination of variations in expulsion pressure from 1 to 6 bar was evaluated; in Norrsell and Olausson (1994), in expulsion force from 0.5 to 8 gm-wgt. In later experiments the air was heated to 37°C, and its temperature and pressure were monitored continuously.

1.1.2. Simulated moving stimuli

The use of a novel dense-array tactile stimulator to study human direction discrimination was reported by the author in 1992 (Essick, 1992 pp. 520-521; see also Gardner and Sklar, 1986). Briefly summarized, the stimulator was composed of a transducer module, driving electronics, and host PC computer. The transducer consisted of 288 0.01-inch-diameter probes arranged in a matrix. The probe center-to-center was nominally 0.12 across the 24 rows and 12 columns so that the entire array was only 2.7 cm x 1.3 cm in dimension. The probes extended through an adjustable, flat display plate. The plate was secured to the cutaneous test site by thin, double-sided tape that surrounded the array of probes (see Essick et al., 1996 for additional details regarding the stimulator).

Each probe was attached to a piezoelectric bender/bimorph. In response to an imposed voltage, the bender/bimorph deflected and pushed the probe lightly into the skin. In most experiments, square wave pulses at a frequency of 204 Hz (50% duty cycle) were applied to sequences of bender/bimorphs to generate vibrating edge-like patterns that translated in close spatial proximity across the skin. The velocity of movement could be increased five octaves in magnitude (viz., from 0.75 to 1.5, 3, 6, 12 and 24 cm/sec) by successively halving the number of cycles, c, between the activation of spatially adjacent rows (from 32 to 16, 8, 4, 2 and 1, respectively). Percepts of robust, smooth, continuous motion were evoked except at the slowest velocity of motion (Szaniszlo et al., 1996). The number of pulses applied to each row determined the width of the stimulus in the direction of motion: Movement of a stimulus k rows wide onto, across, then off of the skin was generated by applying k x c pulses to each row. The number of rows sequentially activated determined the length of skin traversed in increments of 0.12 cm. The perceived intensity or loudness of the stimuli was controlled by the magnitude of the voltage pulse from 0 to

40 v. Although the amplitude of the probes has not been measured, the displacement of the probes in a similar device with different controlling electronics, the OPTACON, has been estimated to attain 65μ when lightly loaded by the skin (Bliss et al., 1970).

All testing completed to date has been conducted on the facial or perioral skin. It has been the author's observation that this site represents an ideal recipient for the vibrotactile stimulation provided by the dense-array stimulator. The tissue is remarkably compliant, and unpredictable alterations in the pressure exerted by the subject do not result in critical changes in stimulus loudness. (Well known to users of the OPTACON, significant alterations in tactile loudness are commonly observed with the less compliant tissues of the finger.) Second, the detection sensitivity of the facial skin is such that a wide spectrum of stimulus intensities can be attained: The threshold voltage for detecting most moving stimuli is about 5 v. Yet, an uncomfortable intensity of stimulation can be achieved at 40 v. Third, the rather diffuse 'hum' noted on digital skin is less of a problem on the face, at least for many subjects. This may be due to the lack of Pacinian corpuscles in the facial skin (Barlow, 1987).

Recently, Gardner and Sklar (1994; 1986) reported use of the OPTACON to study direction discrimination. A four-msec-long square wave pulse was applied to every, every other, every fourth or every fifth row to generate edge-like patterns that translated either toward or away from the tip of the index finger. The time between the activation of successive rows was always 10, 20 or 40 msec. Velocity of apparent motion varied from 3.0 to 48 cm/sec as dictated by the spatial and temporal intervals. Percepts of smooth continuous motion were always, inconsistently, and never evoked by the 100-, 50- and 25-Hz stimulation, respectively.

Although the dense-array tactile stimulator grossly resembles the OPTACON, the stimuli employed by Essick and colleagues differed significantly from those employed by Gardner and Sklar. First, whereas the separation between successively activated rows was kept as short as possible in all experiments of Essick, it was systematically varied and studied by Gardner. Second, in the studies of Essick, different velocities of motion were attained by varying only the time between the activation of adjacent rows. In contrast, in the studies of Gardner, different velocities of motion were attained by varying the temporal frequency of stimulation and the separation between successively activated rows. Third, unlike in the studies of Essick, only one pulse was delivered to each row in the studies of Gardner. Fourth, the delay between the offset of activation of one row and the onset of the adjacent row always approximated 2.5 msec in the studies of Essick. In contrast, the delay varied from 6 to 36 msec in the studies of Gardner depending on the temporal frequency. Fifth, since the separation between successively activated rows was held constant in the studies of Essick, the length of skin traversed was directly proportional to the number of rows activated. In contrast, the path length was determined by both the number of and separation between successively activated rows in the studies of Gardner. Sixth, the within-row density of probes for the dense-array tactile stimulator was twice that of the OPTACON.

These differences likely explain a number of discrepancies in the psychophysical relations obtained with the two tactile stimulators. For example, Essick and colleagues have consistently found that directional sensitivity in d' units varies as an inverted 'U-shaped' function of stimulus velocity over the range 0.5 to 24 cm/sec (Essick et al., 1996 *Application 1*). In contrast, Gardner and Sklar (1994) found that directional d' is unaffected by apparent velocity over the range 4 to 35 cm/sec. Moreover, direction could not be discriminated when only two rows were sequentially activated. Preliminary experimentation with the dense-array tactile stimulator, however, has shown that subjects can discriminate opposing orders in which two rows are successively activated (unpublished). Moreover, threshold distances as short as 0.12 cm (viz., that between two adjacent rows) have been observed on highly innervated skin sites.

1.1.3. Rolling wheel stimuli

Perhaps the most intuitive mode of moving stimulus delivery, yet one of the least studied, is that provided by a frictionless wheel rolled across the skin. Only one study to the author's knowledge has employed such a device to investigate direction discrimination (but see Gould et al., 1979 p. 65). In Olausson (1994), direction discrimination of a 1-mm wide and 15-mm wide Plexiglas rim, rolled across the skin, was studied. The wheels, 40 mm in diameter, were attached to ball bearings and suspended in a manner to assure maintenance of a constant load independently of the skin's curvature. The load exerted by the narrow wheel was changed from 1 to 5.5 to 15 gm; by the wide wheel, from 5.5 to 15 gm. Translation of the wheels across the skin was accomplished by a computer-interfaced, modified X-Y plotter (cf. section 1.3.1.).

1.1.4. Moving water-jet stimuli

One study has reported the use of a water-jet stimulus to systematically investigate direction discrimination (Loomis and Collins, 1978). Briefly summarized, tap water was pumped at 20 psi through a nozzle with 1.3 mm internal diameter to reach a thin, flexible rubber membrane (dental dam material) at a distance of 15 cm. One side of the membrane was exposed to the water jet and the other side covered the test site. The subject was further isolated from the water spray by a open-end containment box, one end of which was sealed to the membrane so as to expose a 3" x 3" window. Lateral movement of the water jet was effected by a microscope stage positioner to which the nozzle was connected. Stimulus onset and offset were controlled by a solenoid-operated Y valve that diverted the continuous flow of tap water (13.8 cc/sec) away from and to a runoff reservoir.

Loomis and Collins contended that the point stimulus moved completely without friction across the skin. Yet the exquisite degree of direction discrimination observed with this mode of stimulation approximated that observed with stimuli that only stretched the skin. Loomis and Collins noted that the skin was indented for 2 or more mm over a circular area about 1.8 mm in diameter and received a force of ca 6.5 gm-wgt. Thus, it is plausible that the stimulus, although frictionless, so deformed the skin at and around the point of application that its mechanical action resembled that of a stimulus that significantly stretched the skin.

1.1.5. Peripheral neural mechanisms for encoding direction

The peripheral neural responses to stimuli that translate across the skin without lateral stretch are unequivocally simpler than those evoked by stimuli that stretch the skin (Ray et al., 1985; Gardner, 1988; Gardner and Palmer, 1989; Gardner et al., 1989). Ray et al. (1985) studied the response of single guard and down hair afferents supplying the skin of the feline hindlimb (sural and posterior tibial nerves) to a fine air-jet stimulus. The hair was cut to a short uniform length above the skin. The moving stream of air was provided by an EEG pen whose velocity and position were precisely controlled by a galvanometer and programmable wave-form generator. The tip of the pen was 0.5 mm in diameter and was positioned 2-4 mm above the clipped hairs. Stimulus force was typically maintained at 2-4 mN.

Through detailed mapping of the receptive fields (RFs) of the individual guard hair afferents, it was convincingly shown that the pattern of discharge ("impulse clustering") reflected the spatial distribution of receptor terminals in the skin. As a result, opposing directions of stimulus motion evoked discharge patterns that were, to a first approximation, mirror reversals (e.g., see Ray et al., *Fig. 6.* on p. 339). The receptive fields of individual down hair afferents were more homogeneous than those of guard hair afferents, but the responses to opposing directions of the frictionless stimuli were remarkably similar (e.g., see *Fig. 12.* on p. 343). It was suggested that a detailed RF organization could be demonstrated for the down hair afferents with use of a more discrete, near threshold stimulus. Consistent with this suggestion, the RF structure of even guard hair afferents is obscured when mapped with a small brush that applies only ca ten times the force of the air jet (Whitsel et al., 1972). Similar to the findings of Ray et al. (1985), a detailed receptive field organization has been demonstrated with small discrete moving stimuli for fast and slowly adapting type I afferents innervating monkey (e.g., Connor et al., 1990 *Figure 4.* on p. 3826) and human (e.g., Phillips et al., 1990 *Fig. 1.* on p. 591) glabrous skin, and for low-threshold Aβ-fiber mechanoreceptive (Vallbo et al., 1995) and C-fiber mechanoreceptive (Vallbo et al., 1993 *Fig. 3.* on p. 303) afferents supplying the hairy skin of the human forearm.

Gardner and colleagues (Gardner, 1988; Gardner and Palmer, 1989; Gardner et al., 1989) studied the responses of single mechanoreceptive afferents in the monkey median and ulnar nerves to the OPTACON stimuli described in section 1.1.2. It was discovered that the slowly adapting afferents did not respond to the stimuli. In contrast, fast adapting afferents supplying the digits exhibited uniform sensitivity over the entire receptive field: Each row of probes pulsed within the RF evoked one impulse in about 75% of the mechanoreceptors tested. For 25% of the afferents, two impulses were evoked at the RF center in one of the two opposing directions of motion. Moreover, for all fast adapting afferents, the RF boundaries shifted one row (ca 1.2 mm) in the direction of motion. Thus, the last row to elicit a response in each direction was ineffective in evoking a response in the opposing direction. These relatively subtle directional asymmetries in the number of spikes and RF boundaries were interpreted to reflect the operation of lateral facilitation and postspike facilitation, respectively (Gardner and Palmer, 1989 p. 1432).

Most recently, Vallbo et al. (1995) mapped the RF structure of mechanoreceptive afferents in the human lateral antebrachial cutaneous nerve which supplies the radial forearm. The mapping procedure was similar to that employed by Ray et al. (1985) but the stimulus consisted of a probe, 1 mm in diameter, that exerted only 2.2 or 2.5 mN of force. Comparable maps were obtained with opposing scanning directions (proximal-to-distal and distal-to-proximal) for slowly adapting type I and type II afferents and for rapidly adapting field afferents. For rapidly adapting hair units, scanning movements evoked more spikes in the distal-to-proximal direction than in the proximal-to-distal direction, though the detailed RF structure appeared the same.

In summary, the studies reviewed above demonstrate or suggest the following. First, only those afferents whose receptive fields are directly contacted are activated by light frictionless stimuli. Second, compared to stimuli that significantly stretch the skin (cf. section 1.3.3.), the pattern of discharge is notably simpler, and directional asymmetries in the pattern and intensity of discharge are relatively minor. Accordingly, the mechanism by which information about stimulus direction is provided to perceptual processes is limited, in large part, to the times at which spatially discrete sets of peripheral mechanoreceptive afferents are sequentially activated.

1.2. Stimuli that stretch without translating

Stimulation that laterally stretches the skin without translation is most commonly encountered in nature. To illustrate, even the smallest purposeful movement of the body alters stresses and strains in the skin over a relatively large area (e.g., see Edin and Abbs, 1991). Although patients' capacity to report the direction of skin stretch effected by an externally applied stimulus has been described in the clinical literature (Halpern, 1949; Bender et al., 1982), the capacity of neurologically normal subjects has received less attention. For psychophysical studies of direction discrimination, a small probe glued to the skin and a sliding glass plate have each been used as a stimulus device.

1.2.1. Moving probe glued to skin

Gould et al. (1979) conducted three series of experiments during which a rubber-tipped probe, 2 mm in diameter, was glued to the skin of the test site. The probe's initial position was carefully adjusted to assure that the skin was neither depressed or elevated with regard to its resting position. In the first and second series of experiments, the subjects' capacity to discriminate the direction of 1- and 2-mm excursions of the probe, respectively, was studied. The movements were generated presumably by the rack and pinion manipulator that the investigators used to study floating-probe and air-stream stimuli (cf. section 1.3.1.). In the third series of experiments, the probe was attached to a servo-controlled Ling vibrator mounted to a stereotaxic manipulator. For the experiments conducted with this apparatus, excursions were specified to a resolution of 0.1 mm for a ramp velocity of 1 cm/sec, and to 0.5 mm for a ramp velocity of 1 mm/sec.

1.2.2. Sliding glass plate

Srinivasan et al. (1990) used a servo-controlled mechanical stimulator driven by a hydraulic system (LaMotte et al., 1983) to slide a smooth glass plate tangentially to the finger tip. The position, velocity of movement, extent of excursion, and load exerted by the plate were highly controlled and monitored. Moreover, the mechanical action of the stimulus on the skin was observed and recorded by videocamera. By limiting the extent of the excursions, the skin was subjected to identical displacements (stretch without translation) in opposing direction. With longer excursions, the stretch was followed by a continuous translation (slip) without detectable mechanical transients (i.e., without "stick-slip"; Srinivasan et al., 1990 p. 1330).

1.2.3. Peripheral neural mechanisms for encoding direction

The peripheral neural responses to stimuli that laterally stretch the skin without translation have been investigated during three groups of studies. In one group, the primary purpose was to evaluate the peripheral neural encoding of textured surfaces stroked across the monkey finger pad. During some experiments, however, the response to a smooth, untextured, flat surface was also obtained. Although the stimulus surface translated across the receptive field, the mechanical action was equivalent to simple stretch since there were no textural elements which could successively activate mechanoreceptors or their individual terminal endings.

LaMotte and Srinivasan (1987a p. 1662, 1987b pp. 1680-1681; see also LaMotte and Whitehouse, 1986 p. 1118) evaluated the response of mechanoreceptive afferents supplying the monkey finger tip to a smooth surface that moved 18 mm across the RF at a velocity of 10 mm/sec and compressional force of 20 gm-wgt. In response to the initial skin stretch, fast adapting afferents exhibited a few impulses for both opposing directions of motion. Many slowly adapting afferents, however, exhibited a sustained response and a pronounced directional preference that could not be attributed to differences in the extent of lateral skin displacement (medial-to-lateral or lateral-to-medial). In subsequent experiments, the excursion was reduced to 5.5 mm to prevent translation (slip) of the surface with respect to the skin (Srinivasan et al., 1990 pp. 1327-1330; LaMotte and Srinivasan, 1991 p. 54). Both fast adapting type I and type II afferents responded only at stimulus onset during the dynamic phase of skin stretch. Consistent with the earlier studies, over 60% (11 of 16) of the slowly adapting afferents responded continuously to the stimuli in a directionally selective manner. Moreover, discharge rates in the two opposing directions differed up to ten-fold. Importantly, in all three classes of afferents, the discharge patterns evoked by stimuli whose excursions approximated 14.0 mm were identical to those evoked by the shorter excursion. In addition to stretching, the stimuli whose excursions approximated 14.0 mm translated smoothly across the skin. Accordingly, it was concluded that the afferents were not sensitive to translation of the featureless surfaces.

Comparable findings were reported by Goodwin and colleagues using a smooth surface, sinusoidally moving over the skin with a peak-to-peak excursion of 80 mm (Goodwin et al., 1989 pp. 1291-1292; Sathian et al., 1989 p. 1276). In contrast to the work of LaMotte and of Srinivasan, responses were also evoked throughout the translation in monkey fast adapting type II afferents supplying the digits. These responses were thought to be evoked by low-amplitude, high-frequency vibrations that accompanied the sinusoidal movement.

In a second group of studies, the responses of individual afferents in the human radial nerve were recorded to movements of one or more digits (Edin and Abbs, 1991). Fast adapting type I afferents whose RFs were located near the joints signaled changes in joint position independently of their direction. In contrast, individual slowly adapting type I and II afferents with RFs up to 8 cm from the joints were stimulated in complex directionally selective manners (the effect of movement of one finger in a particular direction could be reversed by movements of the other fingers).

In a third group of studies, the strain sensitivity of individual afferents was directly measured. Edin (1992) studied the slowly adapting mechanoreceptive afferents supplying the back of the hand and dorsal forearm. Highly controlled degrees of skin stretch were applied across the RF using a hand-held device consisting of a handle, two parallel stiff beams and a stepping motor. The beams were positioned on opposite sides of the RF. The tips of the beams were supplied with hooks to prevent skin slippage during stimulation. Stretch excursions approximated 250, 500, 1000, 2000 or 4000μ and thus encompassed the range of stimulation that normally accompanies physiological movements at the metacarpophalangeal joints. Stretch velocity was always 0.624 cm/sec. Both type I and type II slowly adapting afferents exhibited exquisite dynamic and static sensitivity to skin stretch (the levels rivaled those reported for muscle spindle afferents). In addition, of eight afferents tested, strain sensitivity was higher in each of two opposing directions along one particular orientation, thus defining a preferred axis of stimulus orientation.

In summary, the studies reviewed above demonstrate the following. First, minute degrees of lateral skin strain evoke direction- and/or orientation-specific patterns of activity in slowly adapting afferents supplying the skin. Second, given the distensibility of the skin, sufficient strains for mechanoreceptor activation are generated at appreciable distances from the site of stimulus application. Accordingly, information about the direction of skin stretch is made available by at least two hypothetical mechanisms (Srinivasan et al., 1990 p. 1330): (i) the discharge intensities of the population of direction- and/or orientation-selective afferents that are activated by the stimulus (Edin and Abbs, 1991 pp. 667-668; Edin, 1992 p. 1112), and (ii) the spatial distribution of activated afferents. The distribution is determined by the complex spatial pattern of strain generated by a stimulus and the strain sensitivities of the afferents whose RFs fall within the pattern. Both hypothetical encoding mechanisms require central neural processes that can 'decipher' the complex intensive and spatial distributions of peripheral activity.

1.3. Stimuli that translate and laterally stretch the skin

Most objects in nature that move over the skin not only translate across but possess the potential to laterally stretch the skin. As such, these stimuli are more complex than those that only translate or stretch. In psychophysical studies of direction discrimination, natural moving stimuli have included a moving probe (point, contactor, rod or stylus) and a moving brush.

1.3.1. Moving probe stimuli

Hall and Donaldson (1885 pp. 557-558) were among the first investigators to report the use of a natural moving stimulus, a metallic point/contactor, to assess human direction discrimination. The stimulus object was secured to the bottom of a small cup counterbalanced to a second cup at the opposite end of a "swinging arm". The object was drawn over the skin by translation of the arm. Movement was generated by a mechanical apparatus (of remarkable sophistication for its day) that assured delivery of the stimulus along the experimenter-specified chord of skin at a constant velocity. Stimulus force, area of contact, and pressure were varied by altering the weights placed in the cup or the size of the stimulus object. Contactors of diameter 2, 8 and 12 mm and weights of 15, 45 and 75 gm were used in the experiments reported.

A contemporary version of Hall and Donaldson's device was employed by Gould et al. (1979 p. 66). A "floating probe" was fitted into a plastic sleeve attached to a rack and pinion manipulator. Adjustable stops permitted accurate movement of the probe along set excursions over the skin with gross control of velocity. The steel probe initially described by Gould et al. exerted 5 gm-wgt of force perpendicular to the skin surface (the tip diameter was not specified). In a second series of experiments, a probe weighing only 0.7 gm and terminating in a spherical tip 1 mm in diameter was employed to minimize lateral skin stretch.

Norrsell and Olausson (1992 pp. 156-157; Olausson and Norrsell, 1993 p. 547) used a similar stimulus object in their work. The tip of the stimulus rod was either "dull" with diameter of 1 mm (Norrsell and Olausson, 1992) or ball-shaped with diameter of 2 mm (Olausson, 1993). The vertical load exerted by the rod was varied from 1 to 6 gm (Norrsell and Olausson, 1992) and from 0.5 to 8 gm (Olausson and Norrsell, 1993) by the addition of weights to its shaft. In the earlier study, the metal rod was moved by hand with the help of a mechanical guide. In the later study, the rod was mounted to a holder attached to a computer-interfaced, modified X-Y plotter. In both studies it was determined that the velocity of the rod increased with the extent of movement, but was always within the range 6 to 22 mm/sec (Norrsell and Olausson, 1992) or 29 to 35 mm/sec (Olausson and Norrsell, 1993).

1.3.2. Moving brush stimuli

Moving brush stimuli have been extensively employed in the psychophysical laboratories of the author and B.L. Whitsel for assessment of human direction discrimination (Dreyer et al., 1978, 1979b; Whitsel et al., 1978, 1979; Essick, 1983, 1991, 1992; Essick and Whitsel, 1985a,b; Essick et al., 1988, 1989, 1990, 1991, 1992, 1996; see also Murray et al., 1994). In all experiments, a servomotor and controlling electronic circuitry were used to apply constant velocity stimuli to a restricted portion of the skin at a preselected orientation. Stimuli at experimenter-specified velocities were delivered in clockwise and counter-clockwise directions with an error <1% at velocities greater than 10 cm/sec, <5% at velocities between 2.5 and 10 cm/sec, and <10% at velocities less than 2.5 cm/sec.

The test site was exposed to the bristles through an aperture in a thin acrylic or Teflon plate positioned on, and in recent years taped to, the skin. The dimensions of the aperture defined the length and width of skin stimulated. The edges of the plate adjacent to the aperture were beveled toward the skin. As a result, the bristles moved smoothly as a coherent edge-like bundle onto, across, then off of the skin. It was assumed that the skin surrounding the aperture (and thus covered by the plate) was not stretched significantly by the stimuli.

The brush used in most psychophysical experiments of direction discrimination was recently shown to apply a force perpendicular and parallel to the skin surface of about 38.4 mN and 31.7 mN, respectively (Edin et al., 1995 pp. 831-833). The standard deviation of these forces upon replication of the same stimuli averaged 6% of the mean, indicating a high degree of stimulus-force reliability. The applied stress was calculated to approximate 103 mN/cm^2 at 46 degrees to the skin surface. Whereas the magnitude of the applied stress was relatively unaffected by increases in the velocity of motion, the angle of stress application decreased slightly (a greater proportion of the resultant force was directed parallel to the skin surface). On average, the dynamic friction between the skin and the brush approximated 0.72.

1.3.3. Peripheral neural mechanisms for encoding direction

Peripheral neural responses in monkey, cat and man to stimuli that translate across and stretch the skin have been extensively studied by investigators in the past. The purpose of most of these studies, however, was to determine how information about surface features and textures was encoded (e.g., see Greenspan 1992 p. 885 and Edin et al.,

1995 pp. 841-842). The influence of the velocity and direction of movement was usually of passing interest. Moreover, findings regarding the parameters of stimulus motion mostly confirm those of studies whose primary purpose was to evaluate their impact. The literature selected for review below is by no means exhaustive.

The neural response evoked by minute degrees of movement within the receptive field was described by Gardner and Spencer (1972). These investigators applied small shearing movements within the RF of 203 single afferents in the superficial radial nerve of the cat. The stimulus object was a small brass rod, 2.5 mm in diameter and weighing 3.5 gm. The excursion of the rod resembled one-half cycle of a 50-Hz sinusoid with a peak displacement of 16 to 620μ (and thus with average velocities approximating 0.32 to 12.4 cm/sec). All major classes of low-threshold cutaneous mechanoreceptive afferents were isolated and studied. Although no information was obtained about the encoding of direction, a number of important results were found about the afferents' sensitivity to threshold moving stimuli within the RF. Specifically, with increases in probe excursion, greater numbers of spikes were evoked, response latencies decreased, and the neural response became more consistent from trial to trial. Moreover, the response evoked by these minimal stimuli was not uniform, but decremented with distance from the RF center.

During the same year, Whitsel et al. (1972) described the response of 58 mechanoreceptive afferents supplying the monkey hairy hindlimb to brushing stimuli delivered across the entire RF. Compared to the high degree of directional response asymmetry exhibited by single somatosensory cortical neurons, that exhibited by the afferents to the same stimuli appeared negligible and received no quantitative analysis. More recently, the response to brushing stimuli was evaluated for mechanoreceptors supplying the feline hairy skin (Greenspan, 1992) and human glabrous and hairy skin (Edin et al., 1995; Essick and Edin, 1995).

Using rigorous quantitative techniques, Greenspan (1992) found that most mechanoreceptive afferents exhibited different response rates to stimuli moving in opposing directions along the proximo-distal axis of the feline hindlimb. No particular direction, however, was consistently preferred by any class of receptors. The moving brush stimulus was smaller than the RF.

The author and his colleagues in the Department of Physiology, University of Umeâ extensively studied the responses of 70 low-threshold mechanoreceptive afferents in the human median, radial and inferior alveolar nerves to well controlled brush stimuli moving across the RF (Edin et al., 1995; Essick and Edin, 1995). The brush was equipped with transducers to sample the normal (indenting) and tangential (stretching and compressing) forces applied across the skin. Similar to the author's psychophysical experiments, the stimuli were delivered to the skin through an aperture in a Teflon plate. The aperture was typically 1.4 cm x 1.4 cm and was always larger than the dimensions of the RF. Since the brush was wider than the receptive fields, it resembled a broad edge that moved onto, across, then off of the RF. Opposing directions of motion were typically delivered parallel to the long axis of the hand or transversely across the face.

Over 70% of the afferents exhibited statistically significant differences in the mean-firing-rate responses evoked by opposing direction of motion (Essick and Edin, 1995; e.g., see Figure 1). The directional differences in the firing rate could not be attributed to unavoidable directional differences in the forces applied across the receptive field by the stimuli. Similar to Greenspan's finding for the feline hindlimb, no direction of motion on the human hand or face consistently led to higher response rates so as to suggest a preferred regional direction of motion.

Detailed analyses of the temporal patterns of discharge of individual afferents indicated that the discharge reflected up to three partially overlapping phases of the moving brush stimulus: skin compression, indentation and stretch (Edin et al., 1995). Although the details of the discharge patterns were preserved to a high degree with variation in stimulus velocity, they differed for opposing and orthogonal directions of motion, suggesting asymmetrical sensitivity to stretch/compression with respect to the orientation of stimulation. The patterns of discharge were not assigned a functional role, but were considered an epiphenomenon: "In order to reliably respond with a certain number of spikes or a certain mean firing rate to a particular

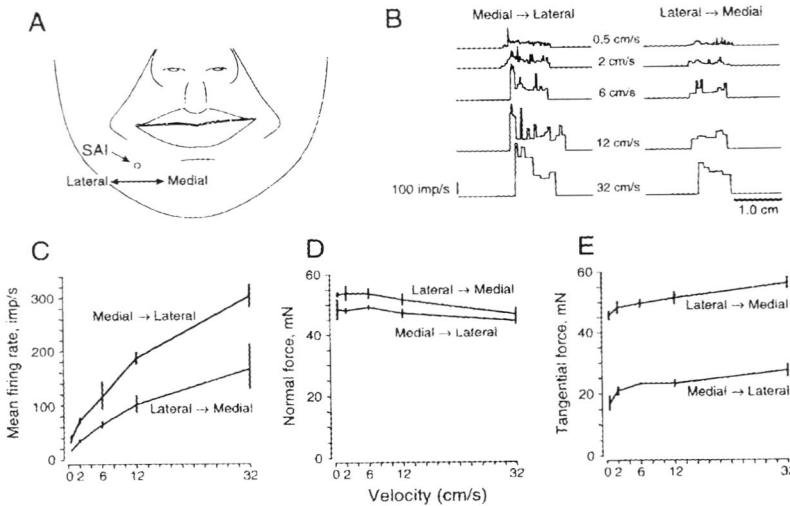

Figure 1. Response (**B**) of typical SAI unit to different velocities and directions of brushing stimuli moving transversely across the face. **A**. Receptive field was located below the corner of the mouth within the distribution of the right mental nerve. Note that medial-to-lateral movement evoked twice the mean firing rate observed for lateral-to-medial movement (**C**), even though the normal force (**D**) and tangential force (**E**) applied across the receptive field were less.

stimulus (i.e., with a certain intensity), the receptor is endowed with certain response characteristics that reflect the sequential compression, indentation and stretch of the skin" (Edin et al., 1995 p. 846). Directional asymmetries in the pattern were hypothesized to reflect nonisotropic properties of the tissues and asymmetric positioning of receptor terminal endings in the skin. Furthermore, it was shown that patterns evoked in the same afferents by stimuli moving precisely in the same direction could differ as a result of nonobvious differences in the manner the skin was stroked (e.g., as a result of subtle differences in skin tension).

Restricting the area of skin surrounding the RF by reducing the size of the aperture had either little effect on the discharge pattern or altered those components evoked by the brush when in contact only with the surrounding skin. Moreover, the extent to which the stimulation extended beyond the RF boundaries has a negligible effect on the individual afferent's capacity to encode information about direction.

In the studies of Greenspan (1992), Edin et al. (1995), and Essick and Edin (1995), care was taken to assure that the stimuli were delivered directly across the geometric center of the receptive field. It would have been informative to know if the observed directional asymmetries were present for stimuli delivered only to the periphery of the RF and to the surrounding skin. That this would be the case was suggested by Goodwin and Morley (1987). These co-workers showed that the directional differences in the response of mechanoreceptors supplying the monkey digital skin were not generated by asymmetries in the position of the stimuli with respect to the RF center. The stimulus consisted of a sinusoidally moving grating whose position of contact within the RF was systematically varied by rotating the support plate on which the finger was secured. The orientation of motion was perpendicular to the long axis of the finger. It was found that neither the presence of direction asymmetry nor the preferred direction of motion was determined by the position of the stimulus in relation to the RF center. Moreover, maximum response magnitude did not necessarily occur at the RF center. Approximately 60% of the slowly adapting and 60% of the fast adapting afferents exhibited directional response asymmetry. However, only 3 of 30 afferents exhibited a reversal in directional sensitivity as the position of the stimulus was shifted from the radial to the ulnar side of the RF.

In summary, the two mechanical actions of natural moving stimuli, i.e., translation and stretch, each generate stresses and strains to which peripheral neural encoding mechanisms are differentially sensitive. As a result, direction of natural stimulus motion is encoded by more than one hypothetical peripheral mechanism:(i) by the times at which spatially discrete sets of peripheral mechanoreceptive afferents are sequentially activated,

   (ii) by the population of directionally specific discharge patterns evoked in individual afferents,
   (iii) by directionally specific intensities at which the population of afferents activated by the stimulus discharge, and
   (iv) by the spatial distribution of afferents activated.

Mechanism (i) requires only physical translation across the skin, but to a spatial extent and temporal frequency for which central processes can determine direction from sequentially received positional information (cf. section 2.2.2.). Mechanisms (iii) - (iv) require lateral skin stretch: That is, little if any information about direction is encoded by these mechanisms for frictionless stimuli. Mechanism (ii) provides the least reliable information about direction since the discharge patterns reflect mechanical events that accompanies skin compression, indentation and stretch, rather than direction per se.

2. Impact of Variations in the Parameters of Stimulation

2.1. Spatial parameters

Spatial parameters include all aspects of the physical stimulation that affect the number, density or distribution of afferents activated by a moving tactile stimulus. Also included are those mechanical features of stimulus delivery that determine the pattern of discharge evoked in individual afferents. As will be made evident below, the capacity to discriminate direction improves with the number of mechanoreceptors activated by a moving tactile stimulus. The number can be increased by increasing the extent to which the skin is stretched laterally, increasing the length or width of skin traversed, testing a site with a higher innervation density, traversing the test site at an orientation that maximizes the number of RFs contacted, or by using a stimulus that moves smoothly, rather than discontinuously, across the skin.

2.1.1. Extent to which the skin is stretched laterally

The impact of variations in skin stretch on direction discrimination has been studied by varying the load applied by a moving-probe stimulus, varying the extent to which the skin is stretched by the same moving-probe stimulus, and by employing diverse modes of stimulus delivery for which the dynamic friction at the stimulus-skin interface differed.

2.1.1a. Effects of variation in stimulus load

A number of investigators have evaluated the impact of variations in the load applied to the skin by a moving probe on direction discrimination. It was assumed that variations in the load would result in changes in both the normal and tangential forces applied to the skin. Increases (decreases) in the tangential force were anticipated to result in more (less) skin stretch. The lowest forces employed always exceeded the absolute detection threshold to assure that direction discrimination would not be compromised by the subjects' inability to feel the stimuli (for discussion of this topic, see Murray et al., 1994).

Hall and Donaldson (1885) estimated the distance required of four subjects to report in which of two opposing directions a metallic contactor moved (presumably along the forearm, p. 567). Velocity of motion was 2 mm/sec. The contactor, 12 mm in diameter, was applied with weights of 15, 45 and 75 gm. It was found that the distance decreased an average of 63% (from 0.76 cm to 0.28 cm) with an increase in weight from 15 to 75 gm. Three-quarters of the total change was observed for the increase from 15 to 45 gm. Hall and Donaldson noted that heavier weights "indent and stretch the skin". Moreover, the impact of the greater mechanical action exerted by heavier objects was succinctly noted: "If there be such a thing as a simple motor sense in the skin at the root of our inference, it is in such cases not independent of aid from other sources" (p. 567).

Using a floating-probe stimulus, Gould et al. (1979) obtained estimates of the threshold distance for direction discrimination across the right forearm of ten subjects. It was estimated that the direction of a probe weighing 5 gm could be discriminated correctly 89.3 percent of the time when moved over 4.3 mm of skin. Four subjects were tested in a second series of experiment with a probe weighing only 0.7 gm. For the lighter stimulus, the threshold distance for direction discrimination was 14 mm, a distance 226% longer than that for the heavier stimulus. The difference was attributed to the lesser degree that the skin was stretched laterally by the lighter stimulus.

The impact of stimulus load was systematically investigated in two series of experiments by Norrsell and Olausson. In the first series (Norrsell and Olausson, 1992), threshold distances for direction discrimination on the forearm were obtained for four subjects with a dull metal-tip stimulus. With an increase in force over the range 1 to 6 gm, the threshold distance fell roughly ten-fold (viz., from ca 30 mm to

3 mm, p. 157). In a second series of experiments (Olausson and Norrsell, 1993), similar stimuli were delivered with each of five weights, equally spaced on a logarithmic scale from 0.5 to 8 gm. For seven subjects percent correct direction discrimination, P(c), increased linearly with the logarithm of load from an average of ca 60% at 0.5 gm to 90% at 8 gm. Both the fall in the threshold distance for direction discrimination (Norrsell and Olausson, 1992) and the increase in P(c) (Olausson and Norrsell, 1993) for stimuli that exerted greater loads were attributed to increased lateral skin stretch.

2.1.1b. Effects of limitations in skin distensibility

In the work described above, it was demonstrated that subjects' directional sensitivity increased for stimuli that applied greater forces to the skin. In response to greater skin stretch, more information about direction was provided by peripheral neural mechanisms sensitive to lateral skin strain. As a further test, Norrsell and Olausson hypothesized that the increase in psychophysical performance with load could be prevented by "bracing the skin" to minimize the degree to which it was stretched.

Two procedures were employed to diminish the extent that the skin was deformed laterally. During some sessions of the first series of experiments (cf. section II.A.1.a.), the chord of skin traversed by the stimulus tip was surrounded by surgical sticky plaster, leaving only a window 4 mm wide and 50 mm long (Norrsell and Olausson, 1992 pp. 156, 158-159). As such, the spatial extent over which the stimulation could exert an effect was greatly reduced. Consistent with the investigators' hypothesis, thresholds remained high with the plaster mask. For example, whereas the thresholds obtained with the heaviest (6-gm) probe approximated 2-3 mm without the mask, they commonly exceeded 20 mm with the mask.

A different procedure was employed in the second series of experiments to minimize the extent of skin stretch (Olausson and Norrsell, 1993 pp. 552-553). In half of the sessions, the arm was straightened at the elbow to brace the skin of the test site. In the other sessions, the subject's elbow remained at a right angle ("bent"). It was discovered that the slope of the P(c)-versus-load relationship was unaffected by the position of the arm. In contrast, the intercept differed significantly. P(c) averaged 8% higher (range among the seven subjects = 1 to 13% higher) when the arm was bent rather than straight. In separate sessions, the lateral movement of the skin in response to a 5-gm load was measured with the arm bent and with the arm straight. As predicted, the distensibility of the skin was greater when the arm was bent for all subjects. None of the appreciable among-subject variability in P(c), however, could be attributed to among-subject variability in the skin's distensibility.

2.1.1c. Effects of variation in mode of stimulus delivery

Perhaps the most decisive studies to evaluate the impact of skin stretch have been those that have employed stimuli that do not stretch, stretch, and only stretch the skin. Gould et al. (1979) delivered all three types of stimuli to the forearm. It was found that distance thresholds for direction discrimination were unequivocally less for stimuli that stretched the skin. For example, thresholds obtained with an air stream that exerted 0.7 gm-wgt averaged 11.3 mm and approximated those obtained with a probe (viz., 14.0 mm) that exerted the same light normal force to the skin. In contrast, thresholds estimated with a probe glued to the skin were <2 mm for the same four subjects.

In separate experiments conducted on ten subjects, estimates of the threshold obtained with a 5 gm-wgt probe (4.3 mm) and with a probe glued to the skin (< 1 mm) were found to differ to a lesser extent. This was attributed to greater skin stretch produced by the heavier probe (i.e., the mechanical action exerted by the 5-gm-wgt probe more closely resembled that exerted by the probe glued to the skin). In a third series of direction-discrimination experiments, a probe glued to the skin was attached to a servo-controlled Ling vibrator to allow precise determination of the threshold excursion for direction discrimination. For three subjects the thresholds averaged 0.6 mm for stretch at 1 cm/sec.

In summary, Gould et al. (1979) demonstrated that at the very same skin site the threshold distance or excursion for direction discrimination could vary >20-fold as a result of differences in the extent to which the skin was stretched laterally. Thresholds were highest for light floating-probe and frictionless air-stream stimuli that minimally stretched the skin. In contrast, threshold were least for a probe glued to the skin, a stimulus with infinitely high friction that maximally stretched the skin.

Norrsell and Olausson systematically investigated the impact of the load exerted by a moving air-stream stimulus in two series of experiments. The experiments most closely paralleled those conducted on the same subjects with a metal rod (cf. sections 2.1.1a. and 2.1.1b.). In their first series of experiments (Norrsell and Olausson, 1992), threshold distances for direction discrimination were obtained over the range 1 to 6 bar, corresponding to 1.6-9.6 gm-wgt. In contrast to the metal-rod stimulus, the threshold distances obtained with the air stream were unaffected by force and never fell below 17.5 mm (Norrsell and Olausson, 1992 *Fig. 3.* on p. 158). Thresholds with the air-stream stimulus typically approximated the highest observed with the rod, i.e., the thresholds obtained with the lowest forces.

In the second series of experiments with seven subjects (Norrsell and Olausson, 1994), air-stream stimuli were delivered at each of five weights, equally spaced on a logarithmic scale from 0.5 to 8 gm. In contrast to a rod stimulus (Olausson and Norrsell, 1993), no correlation was observed between percent correct direction discrimination, P(c), and the logarithm of the load. The same results were found regardless of whether the arm was bent or straight. The insensitivity of the threshold distance for direction discrimination (Norrsell and Olausson, 1992) and of P(c) (Norrsell and Olausson, 1994) to greater vertical loads was attributed to the failure of the air-stream stimuli to laterally stretch the skin. Insensitivity of P(c) to greater loads of a frictionless rolling-wheel stimulus was also reported (Olausson, 1994 p. 309).

The findings reviewed above bring into question the relative sensitivities of the mechanisms which rely on the responsiveness of mechanoreceptive afferents to lateral skin strain to those that rely on the spatiotemporal sequence of mechanoreceptor activation (e.g., Olausson and Norrsell, 1993 p. 556). To address this issue, the author and his colleagues studied direction discrimination of both stimuli that did and stimuli that did not stretch the skin (Essick et al., 1996

Application I). In contrast to previous studies, the mechanical actions exerted by the two modes of stimulation were restricted to the same skin area. Six neurologically normal young adults were recruited to participate in six sensory testing sessions. In three sessions for each subject, the stimuli were provided by a brushing stimulator (cf. section 1.3.2.); and in three, by the dense-array tactile stimulator (cf. section 1.1.2.). During all testing, the same cutaneous region on the chin was traversed by stimuli moving at each of five velocities between 0.5 and 24 cm/sec. The brushing stimuli were delivered through an aperture in a Teflon plate. The plate was positioned securely to the skin to restrict stretch to the confines of the aperture. The dense-array tactile stimulator was used to simulate a moving edge whose width and total excursion matched the dimensions of the aperture over which the brushing stimuli were applied (see Figure 2).

It was found that directional sensitivity in d' units was slightly greater for the simulated frictionless motion (2.63 units) than for the natural brushing motion (1.97 units). This suggested that if the skin area affected is kept the same, central neural mechanisms dependent on the lateral strain sensitivity of mechanoreceptive afferents may not exhibit greater sensitivity than those mechanisms dependent on the spatiotemporal order of activation. In addition, a significant interaction between velocity and mode of stimulation was detected. The velocity at which the subjects exhibited maximum directional sensitivity was less for the simulated motion than for the real motion. It was hypothesized that this was due to differences in the width of the brushing and dense-array stimuli in the direction of motion.

2.1.2. Width of stimulus in the direction of motion

The author and his colleagues studied the impact on direction discrimination of systematic changes in the width of an edge in the direction of motion (Essick et al., 1996 *Application I*). During all sessions, simulated moving stimuli were delivered to the same cutaneous region on the chin at each of 0.45, 1.1, 2.6, 6.1 and 14.7 cm/sec by the dense-array tactile stimulator. In each of two sessions for each of four subjects, the width of the edge attained a width of 1, 2, 4 and 6 rows. For example, the 6-row-wide stimulus resembled an edge that successively changed from 1 to 2 to 3 to ... 6 to 5 to ... 1 row.

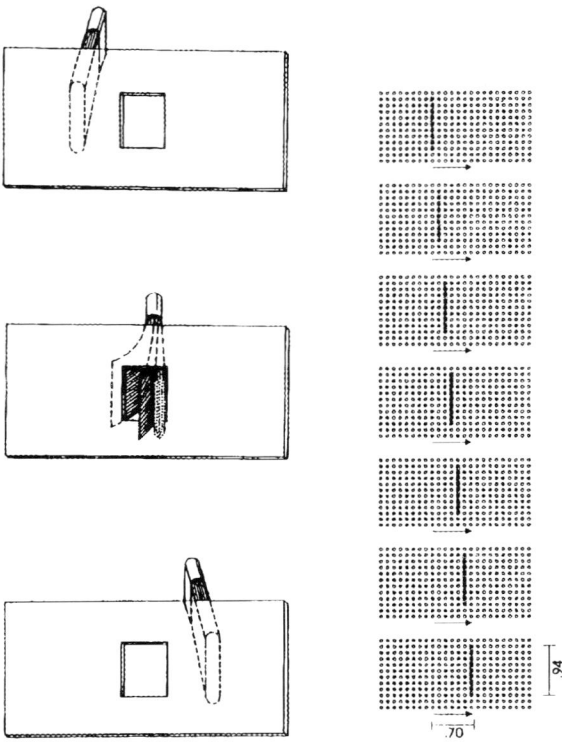

Figure 2. Schematic illustration of delivery of brushing (left panels) and dense-array (right panels) stimuli over the same length and width of skin and in the same direction. Note that the stimuli differed in width perpendicular to the direction of motion. Moreover, the brushing stimuli were feather-like in nature; the dense-array stimuli were vibrotactile in

It was found that directional sensitivity in d' units differed significantly for stimuli of different widths. Mean d' was less for the widest (6-row) stimulus than for stimuli 1, 2 or 4 rows wide. The difference was not the same at all velocities as revealed by a significant velocity-by-width interaction. For stimuli that were consistently 1-row wide, directional sensitivity was greatest at test velocities 1.1 and 2.6 cm/sec. In contrast, for the widest stimulus, directional sensitivity was greatest at test velocities 2.6 and 6.1 cm/sec.

In summary, faster velocities were required of frictionless stimuli that were wider in the direction of motion to obtain about the same amount of information. Thus, faster velocities were needed to discern the spatiotemporal sequence of mechanoreceptor activation. Since information about direction was only provided by those RFs under the leading and trailing edges of the stimulus, the prolonged activation of RFs under the central portion may have detracted from the critical information at the edges. This negative influence on direction discrimination was perhaps minimized by faster velocities of stimulus motion.

2.1.3. Width of stimulus perpendicular to the direction of motion

Olausson (1994) recently evaluated the capacity of 12 subjects to discriminate opposing directions of two frictionless wheels rolled along the forearm. The wheels were 1 mm and 15 mm wide in the dimension perpendicular to the direction of motion. Percent correct direction discrimination, P(c), was estimated for each of five traverse distances, equally spaced on a logarithmic scale from 2 to 32 mm. Velocity of movement was 33 mm/sec. For four subjects, the stimuli were applied with 5.5 gm-wgt of force; for four different subjects, 1 gm-wgt per 1 mm of wheel width; and for the remaining four subjects, 15 gm-wgt of normal force. As such, data were obtained under three load conditions: a constant "low" load, a constant pressure, and a constant "high" load, respectively.

For each of the three load conditions and for each of the narrow and wide wheel, P(c) increased linearly with the logarithm of the distance traversed. For stimuli applied with the constant pressure or constant high load, the slope of the P(c)-versus-distance relationship was greater for the wide wheel than for the narrow wheel. Analyses of the estimates at the individual distances revealed that P(c) differed for the two wheels only when the stimuli were delivered over the longer "suprathreshold" distances, 16 and 32 mm. At no distance did P(c) differ for the narrow and wide wheels applied with the constant low load. This was attributed to less-than-optimal stimulation of the mechanoreceptive afferents by the low force. The results were interpreted to imply that spatial summation effected by the 15-fold increase in width of the stimulus improved suprathreshold performance, but did not lower the threshold distance for direction discrimination.

The author has also observed that directional sensitivity varies with the width of skin stimulated to suggest the operation of spatial summation. As an example, two series of experiments were conducted to assess differences in direction discrimination with a two-interval versus single-interval paradigm. The conditions of stimulation were all but identical in the two studies with the exception of the width of skin contacted: In the second series of experiments (unpublished), the width was one-half that of the first series (Essick et al., 1988 pp. 516-518). For the velocity at which the subjects best discriminated direction of motion, estimates of d' from the second study were found to be roughly one-half those from the first study.

The impact of width was systematically studied by Essick and McGuire (1986, published only in abstract format). Directional d' was assessed at ten velocities from 0.5 to 64 cm/sec at a site overlying the right mental foramen. Brushing stimuli were applied over apertures that exposed equal-length chords of skin 1-cm-wide and 0.5-cm-wide. Nine subjects were tested. Mean d' from use of the wide aperture (viz., 2.45 units) exceeded that from use of the narrow aperture (1.84 units; $p<0.001$). The difference in d', however, was not constant at all velocities ($p<0.023$). Over the range 0.5 to 16 cm/sec, d' from use of the wide aperture exceeded that from use of the narrow aperture by 0.73 units, on average. At the highest velocities, 32 and 64 cm/sec, direction was largely ambiguous with use of both apertures.

2.1.4. Magnitude of stimulus excursion

Most psychophysical investigators have assumed that subjects' capacity to discriminate direction increases monotonically with the magnitude of the stimulus excursion (i.e., with stimulus distance, shift magnitude, length of skin traversed or path length). Such a general assumption proved sufficient for the goal of the studies, i.e., to estimate the magnitude of the excursion anticipated to result in a criterion level of performance or sensitivity (Dreyer et al., 1978, 1979b; Loomis and Collins, 1978; Whitsel et al., 1978, 1979; Gould et al., 1979; Norrsell and Olausson, 1992; Essick et al., 1996 *Application 2*). In contrast, a few investigations have sought to precisely characterize the relationship between directional sensitivity and the magnitude of the excursion. As will be made evident below, the utility of this approach for evaluating the impact of this experimental parameter is unquestionable.

In an early psychophysical study, the author evaluated the capacity of two subjects to discriminate opposing directions of brushing stimuli that moved up and down the skin overlying the deltoid muscle of the left shoulder (Essick and Whitsel, 1985b pp. 215, 217). Stimuli were delivered at 25 cm/sec over four lengths of skin from 1 to 6 cm. A linear relationship was observed between directional sensitivity and the length of skin traversed (slope = 1.1 d' units/cm; Essick, 1991 *shoulder study* pp. 331-332).

Two subsequent studies were undertaken to evaluate the generality of the linear relationship for skin sites thought to differ appreciably in innervation density from that of the shoulder. In the first study, stimuli were applied to a site overlying the right mental foramen on the chin (Essick et al., 1989). The impact of a three-fold change in the length of skin traversed from 0.35 to 1 cm was systematically evaluated for eight subjects (data were obtained at each of six lengths of skin). Five velocities from 0.5 to 12 or 32 cm/sec were employed during each testing session. It was discovered that when the brushing stimuli traversed 0.35 cm of skin, very little information about direction was made available. However, when the stimuli traversed 1.0 cm, the percept of direction was largely unambiguous. Moreover, over the range 0.35 to 1.0 cm, subjects' capacity to discriminate direction could be expressed by a linear relationship between d' and the length of skin traversed, TL:

$$d' = m \times (TL - tl_0)$$

where m is the slope and tl_0 is the x-intercept. The slope was interpreted as the rate at which directional information is made available as the length of skin traversed is increased; and tl_0, the predicted maximum length of skin over which no information about direction is made available (Essick et al., 1990). The parameter m was shown to be relatively independent of the velocity of stimulation. Moreover, in contrast to the shoulder, directional sensitivity was found to increase at a relatively high rate, an average of 5.7 d' units/cm, at the 'best' or optimal velocity (Essick, 1991 *Figure 2.* on p. 331). The parameter tl_0 varied as a U-shaped function of stimulus velocity and approximated 0.0 for stimuli moving at the subjects' optimal velocity for direction discrimination.

In the second study, stimuli were applied to a site on the right lower vermilion of three subjects (Essick, 1991 pp. 331-332, 333-334). Data were obtained at each of three lengths of skin: 0.35, 0.4 and 0.5 cm. Five velocities from 0.5 to 32 cm/sec were employed during each testing session. A linear relationship between d' and TL was again observed. At each subject's optimal velocity for direction discrimination, the slope m approximated 12.5 d' units/cm, and was thus 2.2 and 11.4 times that observed on the chin and shoulder, respectively, for separate groups of subjects. As will be made evident in section 2.1.5., this suggests that the red lip is 2.2 and 11.4 times more directionally sensitive than the hairy chin and shoulder, respectively.

A linear relationship between d' and the length of skin traversed suggests that the relationship between P(c) and length is S-shaped (an ogive) when represented graphically. However, over the range 2 to 32 mm on the forearm, Norrsell and Olausson found that the relationship between P(c) and the logarithm of the distance traversed could be satisfactorily described by a line. This was observed both for stimuli that stretched the skin (i.e., a metal rod; Olausson and Norrsell, 1993) and for stimuli that did not stretch the skin (i.e., an air stream, Norrsell and Olausson, 1994; and a rolling wheel, Olausson, 1994). For the former stimuli, the y-intercept of the relationship, but not the slope, was sensitive to the extent that the skin was braced, confirming the importance of lateral stretch to subjects' discriminative performance (Olausson and Norrsell, 1993; section 2.1.1b.). For the latter stimuli, neither the y-intercept nor slope varied with the extent that the skin was braced, confirming the irrelevance of lateral stretch when frictionless stimuli are employed (Norrsell and Olausson, 1994; section II.A.1.c.). The slope, however, increased with an increase in the width of the wheel stimulus perpendicular to the direction of motion, suggesting spatial summation of information from the receptive fields that were contacted (Olausson, 1994; section 2.1.3.).

Using the OPTACON, Gardner and Sklar (1994; section 1.1.2.) studied direction discrimination of simulated motion on the finger of 16 subjects. A linear relationship between d' and total path length was observed. Moreover, for each path length, d' increased with a reduction in the separation between successively activated rows from 4.8 to 3.6 to 2.4 mm. As a result, the overall height of the line representing the linear relationship increased with a reduction in the separation. Although the average slope of the d'-versus-path length relationship approximated 2 d' units/cm, values varied for the individual separations to suggest that a spatial parameter other than total path length, or separation, determined subjects' discriminative performance. Subsequent analyses revealed that paths varying four-fold in total length resulted in the same directional sensitivity for stimuli composed of the same number of successively activated rows. Moreover, all of the data were well represented by a single linear relationship when replotted as a function of the total number of rows, confirming the importance of this spatial parameter. Similar results were observed for all temporal frequencies of stimulations. These results led Gardner and Sklar to conclude that direction discrimination requires a "minimum total output from the afferent population". As a result, "more stimuli will have to be applied to the less densely innervated skin regions and longer paths will need to be traversed to reach a criterion performance level" (p. 2425).

2.1.5. Test site

That direction discrimination varies appreciably among test sites was noted in the earliest experiments. Using identical conditions of stimulus size, weight and velocity, Hall and Donaldson (1885 p. 566) measured the distance traversed before three subjects reported direction of motion. For each subject, four to seven sites were studied including the forearm, upper arm, back, thigh, shin, palm and forehead. Although the distance varied more than four-fold among sites, the patterns of variation differed notably among the subjects. For all subjects, however, the skin of the back required a longer distance. Given the paucity of data, the investigators concluded only that "there is a difference in the areas needful to discriminate motion upon different parts of the dermal surface" and the "area" (distance) is shorter than that required for two-point discrimination.

Loomis and Collins (1978) measured the "shift magnitude" for discriminating direction of a water-jet stimulus delivered to four body sites. The sites included the distal pad of the right index finger, the middle of the forehead, the middle of the belly above the naval, and the lower back to either side of the midline. Thresholds were obtained for three subjects. Consistent with the work of Hall and Donaldson (1885), subjects were least directionally sensitive on the back: Thresholds on the back were roughly 30 times greater than those on the finger, the most sensitive site. Thresholds on the forehead and belly were about ten times greater than those on the finger. Loomis and Collins noted that the threshold shift magnitudes were notably smaller than published estimates for error of localization and two-point discrimination. The thresholds were also appreciably shorter than those found in other studies of human direction discrimination (cf. section 1.1.4.).

Using a brushing stimulus, Whitsel and colleagues estimated the threshold distance for direction discrimination on the thenar eminence of the left hand, the preaxial left arm 10 cm proximal to the elbow (Dreyer et al., 1978; Whitsel et al., 1978, 1979), the volar surface of left digit 2 (Whitsel et al., 1979), and sites 2 cm above and below the left corner of the mouth (Dreyer et al., 1979b; Whitsel et al., 1979). Estimates were obtained at eight or nine velocities between 0.75 and 250 cm/sec at each site tested. A total of six subjects participated in the studies.

In Dreyer et al. (1978) and Whitsel et al. (1978), it was demonstrated that direction was discriminated optimally when stimulus velocity was in the range 3 to 25 cm/sec and 5 to 30 cm/sec, respectively. However, at all velocities the thenar site exhibited lower threshold lengths than the arm site. In Dreyer et al. (1979b) only the two facial sites were studied. Threshold lengths were lower for the site on the mandibular skin than for the site on the maxillary skin. The lowest thresholds for both sites were observed for velocities between 3 and 25 cm/sec.

Data from the finger site were added to the review published by Whitsel et al. (1979). The finger site proved most sensitive with shortest threshold lengths over the velocity range 3 to 10 cm/sec. The sites were ordinally ranked according to their directional sensitivities in the following sequence: volar digit 2 > facial sites > thenar eminence sites > arm site. Moreover, the investigators concluded that the impact of velocity was similar for the different sites (see also Dreyer et al., 1978; but Dreyer et al., 1979b).

The author and his colleagues subsequently conducted four series of experiments to evaluate the impact of variations in the test site on directional sensitivity. In the first series, three subjects were tested at each of five positions on the thenar eminence of the left hand (Essick and Whitsel, 1985b pp. 215-216, 218-220). Brushing stimuli were delivered at 25 cm/sec over the same length of skin at each site for each subject (either 1.0 or 1.5 cm). The distal tip of the thenar was found to be most sensitive; the distoradial edge and base of the thenar were least sensitive. It was noted that the detailed topographic pattern of variation in directional sensitivity paralleled the corresponding volume of representation in primate somatosensory cortex.

In the second series of experiments, three subjects were tested at each of five positions on the left upper limb from the midvolar surface of digit 2 (hereafter referred to as the 'finger pad') to the skin overlying the midportion of the 'shoulder' (Essick and Whitsel, 1985b). As in the first study, brushing stimuli were delivered at 25 cm/sec over the same length of skin at each site for each subject (either 1.0 or 1.5 cm). The mean estimate of d' was greatest on the finger pad, and least on the postaxial proximal forearm ('proximal forearm') and shoulder. Estimates from the radial inter-digital pad ('knuckle'), the base of the thenar ('wrist'), and proximal forearm were not statistically different, but were strikingly less than those from the finger pad.

In a third series of experiments, the same five sites were evaluated on the right side of the body of four new subjects (Essick et al., 1991 *Constant-Traverse-Length Experiments*). The experimental protocol differed from that of the second series as follows. First, a two-interval, rather than a single-interval, paradigm was employed. Second, the width of skin contacted was 1.0 cm, rather than 0.5 cm. Third, the length of skin traversed was 0.75 cm, rather than 1.0 or 1.5 cm. Fourth, estimates were obtained at each of five velocities (0.5, 2, 6, 12 and 32 cm/sec) rather than at only 25 cm/sec. Fifth, a site overlying the right mental foramen on the face ('chin') was also studied.

On the finger pad, direction of motion was largely unambiguous at all velocities (mean d' exceeded 3.5 units). In contrast, on the shoulder, direction of motion was largely ambiguous at all velocities (mean d' was less the 0.65 units). Subsequent analyses revealed that subjects' peak directional sensitivity (cf. section 2.2.1a.) was higher and comparable for the chin, finger pad and knuckle. Subjects' peak directional sensitivity was lower and comparable for the wrist, proximal forearm, and shoulder. In contrast to the work of Whitsel and colleagues, a highly significant interaction between test site and stimulus velocity was detected. This suggested that the impact of velocity differed for the different test sites.

In response to the need for a quantitative measure of the relative directional sensitivities of skin sites, the author and his colleagues proposed use of the rates at which directional information is made available per unit length of skin traversed (Essick et al., 1991). This was deemed appropriate since the rates must be limited to a first approximation by the innervation densities of the sites. Accordingly, the relative directional sensitivity of site 1 to site 2, $rd'_{1:2}$, is estimated as

$$m_1/m_2$$

where m_1 and m_2 are the respective rates at which d' increases with the length of skin traversed, TL, for $TL > tl_0$ (cf. section II.A.4.). As such, given TL_1 and TL_2 at two sites over which moving stimuli result in the same peak directional sensitivity, $rd'_{1:2}$ approximates

$$(d'/TL_1) / (d'/TL_2)$$
$$= TL_2 / TL_1.$$

Use of this ratio as a measure of the relative directional sensitivity of two sites is also parsimonious with the work of other investigators (Gardner and Sklar, 1994; section 2.1.4.).

A fourth series of experiments were conducted on the subjects that participated in the third series to quantitatively characterize the relative sensitivities of the sites along the right upper limb and the site-dependent manner in which velocity impacts direction discrimination (Essick et al., 1991 *Variable-Traverse-Length, Constant-Performance Experiments*). During preliminary testing, the length of skin predicted to result in a maximum of 3' units of directional sensitivity was estimated for each site for each subject. The protocol employed for the third series of experiments was then repeated with use of these lengths.

Analysis confirmed that the traverse lengths employed in the experiments succeeded in matching peak sensitivity among the sites. Longer lengths were required on more proximal test sites to attain 3 d' units of directional sensitivity. Based on the model described above, relative directional sensitivity was found to decrease 6-fold, on average, from the finger pad to the proximal forearm. For two of the subjects, rd' was similar for the shoulder and proximal forearm. For the other two subject, rd' on the shoulder approximated one-half that on the proximal forearm. It was argued that the proximo-distal gradient of directional sensitivity along the upper limb paralleled the innervation density of slowly and rapidly adapting type I afferents.

In addition to the pronounced spatial gradient in relative directional sensitivity, the velocity of stimulus motion at which directional sensitivity was highest increased proximally. This is described in detail below (cf. section 2.2.1a.).

2.1.6. Orientation of stimulus delivery

In early experiments, the author and B.L. Whitsel evaluated the impact of stimulus orientation on directional sensitivity (Essick and Whitsel, 1985b pp. 215, 218-220). Brushing stimuli were delivered in two opposing directions at each of four intersecting orientations on the left thenar eminence. The orientations were (i) parallel to the long axis of the thumb, (ii) perpendicular to the radial edge of the thumb, and those bisecting (i) and (ii). Stimuli were delivered at 25 cm/sec over a length of skin (either 0.75, 1.0 or 1.5 cm) that was kept the same for each of four subjects. It was found that no orientation of motion consistently led to higher estimates of directional sensitivity for all subjects. Moreover, only one subject exhibited a systematic, albeit weak, preference: Values of d' were higher for stimuli delivered parallel to the long axis of the thumb.

Although no orientation preference was detected on the thenar, one cannot conclude that preferences do not exist at other body sites. In support of this conclusion, Gould et al. (1979) described subjects' performance in discriminating direction of stimuli moving transversely and longitudinally across the forearm. Stimulation was provided by a steel probe weighing 5.0 gm. Ten subjects were tested. At the threshold excursion for discriminating the orientation of motion, subjects could also discriminate direction. However, whereas direction was discriminated correctly 73.6 percent of the time in the longitudinal orientation, it was discriminated correctly 89.3 percent of the time in the transverse (medial-lateral) orientation. Although Gould and colleagues did not comment on this difference of 16%, directional sensitivity may have been greater across the forearm.

2.1.7. Continuity of stimulus excursion

At least three studies have systematically evaluated alterations in direction discrimination effected by variations in the continuity of motion across the skin. These studies were conducted, in part, to determine to what extent direction is inferred from an overall change in stimulus position.

Dreyer et al. (1979a; study published in abstract format only) hypothesized that the increase in performance on direction discrimination tasks with an increase in the length of skin traversed could be attributed to either of two spatial factors: the increased "separation of the extreme points of stimulus contact" or the increased "surface area of stimulus contact". To address this issue, the investigators evaluated P(c) at nine velocities of brushing stimuli from 0.5 to 250 cm/sec delivered to the ventral forearm. The apertures through which the skin was exposed were 1, 2, 3, 4, 5, and 6 cm long. For each length, one aperture consisted of a single continuous opening; and a second, of two 0.25-cm-long slit-like openings whose outer dimension equaled that length. At the slowest velocities of motion, it was found that subjects' capacity to discriminate direction was comparable for the continuous and split apertures, even though the area of skin stimulated differed as much as 12 fold. In contrast, at all other velocities, P(c) observed with the continuous aperture exceeded that with the split aperture of comparable overall length. For both apertures, P(c) increased with the length of skin traversed. These observations suggested that subjects' discriminative performance may be based solely on positional cues, but only at the slowest velocities of motion (see also Dreyer et al., 1978; Dreyer et al., 1979b). Moreover, increases in the total area of skin contacted were not necessary for the monotonically increasing relationship between P(c) and length of skin traversed.

Ten years later, the author and his colleagues conducted a similar study (Essick et al., 1992 pp. 178-182). Directional d' was assessed for ten velocities from 0.5 to 64 cm/sec at a site overlying the right mental foramen. Brushing stimuli were applied over continuous and split apertures whose overall length equaled 0.75 cm. The total area of skin exposed by the two split apertures was 67% and 34% of that exposed by the 0.75-cm-long continuous aperture. It was found that directional d' observed with the continuous aperture statistically exceeded that with the split apertures over the velocity range 4 to 8 cm/sec. This was interpreted to imply that subjects' capacity to discriminate direction on the face over this velocity range was not due solely to their ability to process information about the overall change in stimulus position.

Essick et al. (1992) concluded that positional cues available at the times of stimulus onset and offset may be unimportant under most conditions of interest to experimenters. Specifically, it was reasoned that if direction discrimination of continuous motion was based solely on information about the overall change in position, the manner in which velocity influenced d' should resemble a "low-pass filter" whose "cut-off" velocity increases with the length of skin traversed (see Essick et al., 1992 pp. 181-182 for detailed discussion). In contrast, for stimuli delivered over lengths of skin for which direction of motion is neither completely ambiguous nor unambiguous, the velocity at which direction is best discriminated remains about the same (cf. section 2.2.1a.). For lengths of skin over which direction is largely unambiguous and that exceed the two-point limen, the prediction based on subjects' use of positional cues is fulfilled. To illustrate, Whitsel et al. (1979; *Fig. 12.* on p. 97) demonstrated that for stimuli delivered across such long chords of skin, P(c) approximated 100% for all velocities less than a relatively high upper limit. Moreover, the limit increased with the length of skin traversed.

The studies described above indicate that use of information about the overall change in position of a natural moving stimulus cannot be excluded for (i) very slow velocities of motion and (ii) for stimuli delivered over long lengths of skin for which direction is largely unambiguous. For many conditions of stimulation, however, subjects' performance cannot be attributed solely to the use of positional cues. This was perhaps shown most convincingly by Gardner and Sklar (1994; section 1.1.2.). Unlike in the studies reviewed above, these investigators employed novel stimuli from which no information about direction could be extracted from the overall change in position. Specifically, an OPTACON-generated edge was delivered to successive positions on the index finger, either toward or away from the fingertip. With only one stimulus pulse applied to each position, directional d' approximated 0.0 when only two rows were successively activated. This was true for apparent velocities over the range 12 to 48 cm/sec. In contrast, directional sensitivity increased with the addition of intervening rows at a rate of 0.81 d' units/row. Variation in the separation of the rows from 1.2 to 2.4 to 4.8 mm had little effect. These findings led Gardner and Sklar to conclude that "direction discrimination is not simply a matter of point localization on the skin, but rather appears to involve the integration of a spatiotemporal sequence, in which additional inputs improve the accuracy of performance" (p. 2426).

2.2. Spatiotemporal parameters

Spatiotemporal parameters include those aspects of stimulation that determine the rate at which mechanoreceptors are activated and at which detailed patterns of discharge are evoked in individual afferents (Edin et al., 1995; Essick and Edin, 1995). Studies to evaluate their impact have concluded that direction discrimination does not result simply from recognition of an absolute change in position: Measures of this sensory capacity do not remain relatively constant up to a critical velocity, beyond which the duration of stimulation becomes the limiting factor. In contrast, studies that have evaluated the impact of stimulus velocity on the threshold length for direction discrimination suggest a much more interesting interplay between change-in-position and time. Specifically, direction appears to be discriminated when the product of the length of skin traversed and duration of stimulation exceeds a criterion constant.

2.2.1. Velocity of stimulus motion

Two different approaches have been employed to evaluate the impact of variations in stimulus velocity on direction discrimination. The simpler approach has been to characterize the manner in which percent correct performance, P(c), or more commonly directional sensitivity, d', varies with stimulus velocity. A second and more complex approach has been to characterize the impact of variation in velocity on the threshold length or distance for direction discrimination. These two approaches will be discussed in turn.

2.2.1a. Impact of velocity on P(c) and d'

Whitsel and colleagues were the first investigators to systematically study the impact of variation in stimulus velocity on discriminative performance. For each subject, test site, and length of skin traversed, data were obtained at multiple velocities of brushing stimuli from 0.75 to 250 cm/sec, a range that extended well beyond that over which direction need be discerned in most daily situations. Although the primary goal was to evaluate the manner in which velocity affected the threshold length for direction discrimination, the impact of velocity on P(c) was also directly examined. Specifically, P(c) was plotted as a function of stimulus velocity with length of skin traversed as a parameter (Dreyer et al., 1978 *thenar and arm sites* on p. 75, 1979b *facial sites on maxillary and mandibular skin* on p. 2054; Whitsel et al., 1979 *volar digit 2* on p. 97). It was found that for relatively short lengths of skin, P(c) approximated 50% (chance performance) for all velocities of motion. For intermediate lengths of skin, plots of the P(c)-versus-velocity relationship were grossly 'bell-shaped', but were skewed toward lower velocities. For long lengths of skin, P(c) approximated 100% up to a relatively high velocity, the limit of which increased with the length of skin traversed (cf. section 2.1.7.).

The distances across the skin which constituted short, intermediate and long lengths differed for the test sites in accord with their innervation densities. For example, the minimum length that met the criterion for 'long' was 1.0 cm on the finger site, but 6.0 cm on the arm site.

In a similar study, the author and B.L. Whitsel evaluated the impact of stimulus velocity on directional sensitivity in d' units (Essick and Whitsel, 1985b pp. 215, 216-217). Brushing stimuli were delivered parallel to the long axis of the thumb of two subjects at five velocities between 1 and 150 cm/sec. All stimuli were delivered across an aperture 1 cm long and 0.5 cm wide positioned at a site on the left mid-thenar eminence. It was found that directional sensitivity was least (0 to 1 d' units) at the lowest and highest velocity employed. The value of d' was highest (3 or more units) at the test velocities 7.4 and 20.1 cm/sec. Importantly, the estimates of d' generated a relatively smooth, inverted U-shaped curve when plotted as a function of the logarithm of velocity.

Ten years ago, the author initiated studies of orofacial sensation (up to that time, all of his work had been conducted on the upper limb). In an early study, directional sensitivity was measured at a site overlying the right mental foramen on the chin of 41 young adults (Essick et al., 1988). A two-interval paradigm was used to obtain estimates of d' at five velocities of motion: 0.5, 2, 6, 12 and 32 cm/sec. Brushing stimuli were applied to a chord of skin 0.75 cm long and 1.0 cm wide, oriented parallel to the inferior border of the mandible. As observed in Essick and Whitsel (1985b), an inverted U-shaped relationship was noted when the d' values were plotted as a function of the logarithm of velocity. This was true for both the averaged data as well as that from individual subjects.

In order to characterize the sensory capacity of individual subjects, the author and D.G. Kelly extensively searched for a quantitative model that described the d'-versus-velocity relationship. Specifically, the degree to which the estimates obtained during a single session fit different families of curves was assessed visually and quantitatively. Although quadratic polynomials, beta functions, logarithmic quadratic polynomials, and Gaussian functions were considered, the fits did not exceed those attained with a subset of the gamma family of curves: The d' measure of sensitivity could be expressed as a function of stimulus velocity, VEL, in the following manner:

$$d' = C \times VEL^{(A-1)} \times e^{(-VEL/B)}$$

where $e^{[z]}$ is the exponential of the argument z, base e. Optimal values of the three positive parameters A, B, and C were estimated by applying standard linear regression techniques, using the linear model obtained by taking logarithms in the equation above:

$$\ln(d') = \ln(C) + (A-1) \times \ln(VEL) + (-1/B) \times VEL$$

where $\ln[z]$ is the natural logarithm of the argument z.

As an example, Figure 3 illustrates the d'-versus-velocity relationship constructed from data obtained during a single testing session. The best-fitting gamma curve and the estimates of the parameters A, B, and C are shown in Figure 3B for the data illustrated in Figure 3A. Four functions of the estimates (hereafter referred to as the 'derived measures') were found to completely describe the d'-versus-velocity relationship and to provide meaningful measures of a subject's capacity to distinguish opposing directions of stimulus motion. Specifically, a subject's peak or maximum directional sensitivity, Y_{mx} (Figure 3C), was predicted by

$$Y_{mx} = (-\beta_1/\beta_2)^{\beta_1} \times e^{\beta_0 - \beta_1}$$

where β_0 is the regression coefficient estimate of $\ln(C)$; β_1, of (A-1); and β_2, of (-1/B). Moreover, the velocity at which the subject was predicted to exhibit maximum directional sensitivity (i.e. his best or modal velocity, V_{md}) was estimated by

$$V_{md} = -\beta_1/\beta_2$$

(Figure 3D). The degree to which stimulus velocity influenced the subject's capacity to distinguish opposing directions of tactile motion was expressed by

$$VT_g^{-1} = \log(\beta_1 + 1/\beta_2^2)$$

where $\log[z]$ is the logarithm of the argument z, base 10 (Figure 3E) and

$$T_1 = 100 \times Y_{mx} \times (\beta_2{}^2/\beta_1)$$

(Figure 3F). The value of $VT_g{}^{-1}$ estimated the global velocity-tuning (this corresponded, in general, to the width of the inverted U-shaped curve relating directional sensitivity to velocity); and VT_1, the local velocity-tuning or 'peakedness' of the tuning curve in the vicinity of the point (V_{md}, Y_{mx}). In addition, the goodness-of-fit of the gamma function model to the data was evaluated by calculating the R^2 value for the linear regression. This value estimated the proportion of variability in the five data points that could be explained by the model.

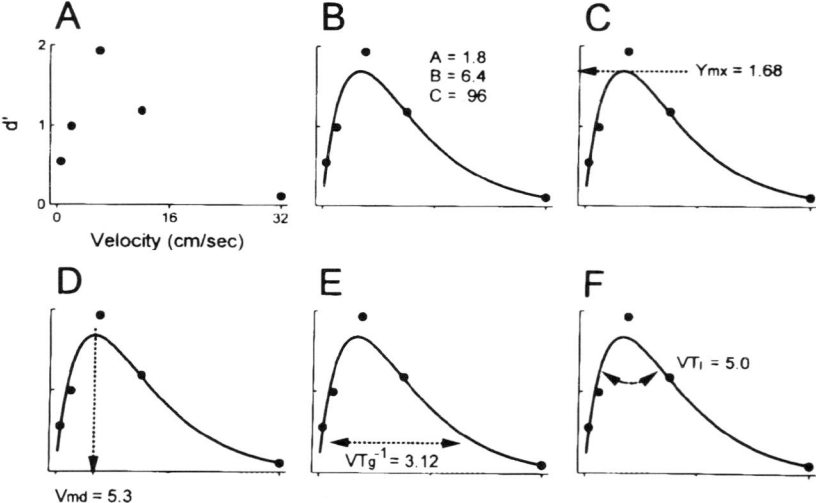

Figure 3. Characterization of the directional d'-versus-velocity relationship from data (**A**) obtained during a single testing session. **B.** The best-fitting gamma curve. **C.** Estimation of the subject's peak directional sensitivity, Y_{mx}. **D.** Estimation of the velocity at which the subject was predicted to exhibit maximum directional sensitivity, V_{md}. **E.** Estimation of $VT_g{}^{-1}$, the global velocity-tuning of the relationship. **F.** Estimation of VT_1, the local velocity-tuning or 'peakedness' of the curve in the vicinity of the point (V_{md}, Y_{mx}).

Application of the gamma curve-fitting procedure to the data obtained from the 41 subjects revealed the following. First, peak sensitivity, Y_{mx}, for the females (mean = 4.05 d' units) statistically exceeded that for the males (mean = 3.26 d' units). Second, the velocity at which the peak occurred, V_{md}, was higher for females (mean = 7.48 cm/sec) than for males (mean = 4.89 cm/sec). Third, the global velocity-tuning of the d'-versus-velocity relationship was less for females than for males (i.e., the curves were broader); however, the curves were equally peaked for both genders. Fourth, for 39 of the 41 subjects, goodness-of-fit to the gamma function model was extremely good (mean R^2 = 0.92, SD 0.08). Last, upper and lower normative limits for each of the four derived measures were calculated and enabled detection and characterization of sensory impairment in patients with damaged mandibular nerves.

Two additional studies were subsequently conducted to evaluate the impact of the parameters of stimulation on the d'-versus-velocity relationship. In the first, the impact of variation in the length of skin traversed was assessed (Essick et al., 1989; section 2.1.4.). Briefly summarized, brushing stimuli were delivered over six lengths of skin ranging from 0.35 to 1.0 cm at the chin site. A protocol similar to that described above was employed for eight subjects. Although subjects' peak sensitivity increased with the length of skin traversed, the velocity at which it occurred, V_{md}, was independent of length and averaged 4.99 cm/sec (SD 1.86). Moreover, the global velocity-tuning of the d'-versus-velocity relationship was less for the longer lengths of skin (i.e., the curves were broader since d' was higher at all velocities of motion); however, the curves were about equally peaked. The mean R^2 goodness-of-fit approximated 0.81, independently of the length of skin traversed.

In the second study, the impact of variation in the test site on the d'-versus-velocity relationship was assessed (Essick et al., 1991; section 2.1.5.). Briefly summarized, four subjects were tested at each of five positions on the right upper limb from the midvolar surface of digit 2 to the skin overlying the midportion of the shoulder. It was found that the velocity of motion at which directional sensitivity was highest, V_{md}, increased with distance proximally along the limb. To illustrate, peak sensitivity occurred at a mean velocity of 5.4 cm/sec on the finger pad (range among the four subject = 1.5 to 9.4 cm/sec). In contrast, peak sensitivity occurred at a mean velocity of 18.6 cm/sec on the proximal forearm (range = 11.5 to 31.2 cm/sec). The velocity on the shoulder was not significantly different from that on the proximal forearm.

2.2.1b. Impact of velocity on threshold length or distance

The importance of assessing the impact of stimulus velocity on direction discrimination was recognized in the earliest experiments. Using identical conditions of stimulus weight and size, Hall and Donaldson (1885, pp. 563-564) measured the distance traversed before one extensively studied subject reported direction of motion at twelve velocities ranging from 0.012 to 1.5 cm/sec. All stimuli were delivered to the volar surface of the right arm. It was found that correct judgments of direction were made for stimuli which moved over 6 to 7 mm of skin, independently of the 120-fold variation in stimulus velocity.

Hall and Donaldson's findings are interpreted by the author to imply that the velocities employed were too low to result in a significant degree of velocity-tuning in the threshold distances. Moreover, the subject (Dr. Donaldson) was likely proficient at using change-of-position information at very low velocities, given his extensive experience with moving tactile stimuli. That threshold distances may be sensitive to velocity over the range employed by Hall and Donaldson was demonstrated by Gould et al. (1979). For a probe glued to the skin, the threshold excursion for direction discrimination on the forearm averaged 0.6 mm for stretch at 1 cm/sec. However, for stretch at 0.1 cm/sec, the threshold excursion was over 2 mm on average.

Whitsel and colleagues first investigated the threshold length for direction discrimination over a wide range of velocities (cf. section 2.2.1a.). Data obtained with natural brushing stimuli were examined by plotting the estimates of the threshold length, referred to as the "critical length" by Whitsel and colleagues, as a function of stimulus velocity with test site as a parameter (see Dreyer et al., 1978 *Fig. 4.* on p. 77; Whitsel et al., 1978 *Figure 8.* on p. 2227; Dreyer et al., 1979b *Fig. 4* on p. 2055; Whitsel et al., 1979 *Fig. 14.* on p. 99). Logarithmic scales were used for both the abscissa and ordinate. The plots were U-shaped and abutted the y-axis (the threshold lengths did not increase at the lowest velocities tested to the extent that they did at the highest velocities tested). By comparing the heights of the plots, the investigators were able to ordinally rank the different test sites according to their relative directional sensitivities (cf. section II.A.5.). In addition, the width of each U-shaped curve, subjectively assessed by visual examination, was interpreted to reflect the degree to which velocity impacted direction discrimination. No quantitative methods were employed to further characterize the relative sensitivities of the sites or velocity-tuning.

The method of data presentation used by Whitsel and co-workers was adopted by the author for threshold lengths determined with the dense-array tactile stimulator. Specifically, estimates of the threshold length were plotted as a function of stimulus velocity using logarithmically scaled axes (e.g., see Essick et al., 1996 *Figure 6*). Similar to those of Whitsel and co-workers, the plots were U-shaped and abutted the y-axis. Closer inspection revealed that over the range 6 to 24 cm/sec, the curve increased linearly with slope equal to 0.5. This suggested that direction was discriminated when the product of the length of skin traversed and the duration of stimulation attained a criterion. This can be appreciated by rearranging the equation of the linear relationship:

$$\log(TL) = 0.5 \times \log(VEL) + \log(\text{constant})$$
$$\Rightarrow \log(TL) = \log[(TL/DUR)^{0.5}] + \log(\text{constant})$$
$$\Rightarrow \log(TL) = \log[(TL/DUR)^{0.5} \times \text{constant}]$$
$$\Rightarrow TL = (TL/DUR)^{0.5} \times \text{constant}$$
$$\Rightarrow TL \times DUR^{0.5}/TL^{0.5} = \text{constant}$$
$$\Rightarrow TL^{0.5} \times DUR^{0.5} = \text{constant}$$
$$\Rightarrow (TL \times DUR)^{0.5} = \text{constant}$$
$$\Rightarrow TL \times DUR = \text{constant}^{2} = \text{criterion product}$$

where DUR is the stimulus duration. Henderson (1971, 1973) earlier reported a similar "space-time tradeoff" for discriminating direction of moving visual stimuli and demonstrated the utility of plotting the threshold distance-versus-velocity relationship on logarithmically scaled axes.

Reciprocity of space and time at threshold does not appear to be limited to movement for which information is provided only by the spatiotemporal sequence of receptor activation. For example, inspection of the data published by Whitsel and co-workers reveals that the critical length increased as a power function of the velocity of natural brushing stimuli with exponent of 0.5. This was true for velocities over a range that always included the highest tested (Dreyer et al., 1978 p. 77; Dreyer et al., 1979b p. 2055; Whitsel et al., 1979 p. 99). The lowest velocity of the range varied among test sites and was lower for sites that exhibited relatively higher directional sensitivities (Whitsel et al., 1979 p. 99).

The findings reviewed above suggest that the general term used by sensory physiologists to describe the process underlying tactile (and visual) motion perception, viz., 'spatiotemporal integration', merits more precise definition and investigation than it has received. To illustrate, the percept of smooth motion from the successive stimulation of discrete positions across the receptor sheet unquestionably reflects the operation of a spatiotemporal integrating or summating process. Attention is shifted from a sequence of precise positions to a continuous unitary percept, for which position at any time is of little interest. In contrast, the spatiotemporal process underlying direction discrimination requires that a criterion *separation* in space and time be exceeded. Direction is discriminated only when a minimum degree of sensory distinctness along spatial and temporal dimensions, or lack of summation, is maintained. Thus, the extent to which the two perceptual processes are subserved by precisely the same central neural mechanisms (and thus merit the same terminology) is presently unclear.

Moreover, it is evident that one psychophysical model, viz., perfect space-time reciprocity at threshold, does not describe subjects' capacity to discriminate direction for all conditions of stimulation. For example, the reciprocity model does not explain the relative insensitivity of direction discrimination to velocity at low stimulus velocities. Nor is it consistent with the insensitivity of directional d' to variations in the temporal parameters (viz., temporal frequency, velocity and duration) or to the total distance traversed by the novel stimuli employed by Gardner and Sklar (1994; section 2.1.4.). Additional experimentation is clearly needed to reconcile the apparent discrepancies in the psychophysical relationships observed with different patterns and modes of stimulation and to establish a model consistent with the diversity of observations.

2.2.2. Relevance of temporal frequency

In previous sections (2.1.5. and 2.2.1a.) it was reported that the velocity of motion at which directional sensitivity was highest increased, whereas relative directional sensitivity decreased, with distance proximally along the upper limb. This suggested that faster stimulus velocities were required on less innervated skin sites to assure delivery of the same amount of information ('bits/sec') to central perceptual processes. To further evaluate this possibility, the relationship between the traverse length required to achieve a criterion level of maximum sensitivity (3 d' units) and the velocity at which the level was attained, V_{md}, was examined (Essick et al., 1991 *Figure 8.* on p. 21). A linear relationship with constant slope (viz., 4.8/sec) was observed. Importantly, this suggested that directional sensitivity was maximal (i.e., Y_{mx} was attained) at an optimal temporal frequency of stimulation.

The psychophysical finding can be interpreted in light of two hypothetical models that explain the motion sensitivity of somatosensory cortical neurons. Both models assume that individual cortical neurons receive input from linear arrays of mechanoreceptive afferents and contribute fundamentally to the percept of motion and direction across the skin. The model proposed by Gardner and colleagues suggests that the robust, motion-sensitive responses of cortical neurons are only evoked when the temporal frequency of mechanoreceptor stimulation exceeds a lower limit (Gardner et al., 1989; 1992). Accordingly, should direction discrimination require the evocation of these responses, V_{md} is predicted to increase with a decrease in innervation density. This is because higher velocities of linear motion would be required to exceed the limit on lesser innervated test sites.

Alternatively, a model developed by Whitsel and colleagues (Whitsel et al., 1991; see also Hollins and Favorov, 1994) suggests that the motion-sensitive and direction-selective responses of somatosensory cortical neurons are attributable to the temporal sequence of lateral interactions in the array of column-shaped cortical aggregates ("segregates") that is activated by a moving tactile stimulus. An optimal temporal frequency of interaction is specified in large part by the properties of the cortical NMDA receptors. Since the skin regions that provide input to adjacent segregates are separated by larger distances for cortex representing proximal body sites than for cortex representing distal body sites, higher linear velocities of motion are required to achieve the same optimal temporal frequency for proximal body sites.

Acknowledgments

The author would like to thank the following individuals for reviewing a previous version of the manuscript: Dr. Ben Edin (Umeå University, Sweden), Dr. Douglas Kelly (University of North Carolina at Chapel Hill), Dr. Håkan Olausson (University of Göteborg, Sweden), Dr. Chuck Vierck (University of Florida at Gainsville), and Dr. Barry Whitsel (University of North Carolina at Chapel Hill). Support for manuscript preparation was provided by National Institutes of Health Grant No. DE07509 and by a gift from Unilever Research.

References

Barlow, S.M. (1987) Mechanical frequency detection thresholds in the human face. Experimental Neurology, 96, 253-261.

Bender, M.B., Stacy, C. and Cohen, J. (1982) Agraphesthesia: A disorder of directional cutaneous kinesthesia or a disorientation in cutaneous space. Journal of the Neurological Sciences, 53, 531-555.

Bliss, J.C., Katcher, M.H., Rogers, C.H. and Shepard, R.P. (1970) Optical-to-tactile image conversion for the blind. IEEE Transactions on Man-Machine Systems, MMS-11, 58-65.

Connor, C.E., Hsiao, S.S., Phillips, J.R. and Johnson, K.O. (1990) Tactile roughness: Neural codes that account for psychophysical magnitude estimates. Journal of Neuroscience, 10, 3823-3836.

Dreyer, D.A, Duncan, G.H. and Wong, C.L. (1979a) Role of position sense in direction detection on the skin. Society for Neuroscience Abstracts, 5, 671.

Dreyer, D.A, Duncan, G.H., Wong, C.L. and Whitsel, B.L. (1979b) Factors influencing capacity to judge direction of tactile stimulus movement on the face. Journal of Dental Research, 58, 2052-2057.

Dreyer, D.A., Hollins, M. and Whitsel, B.L. (1978) Factors influencing cutaneous directional sensitivity. Sensory Processes, 2, 71-79.

Edin, B.B. (1992) Quantitative analysis of static strain sensitivity in human mechanoreceptors from hairy skin. Journal of Neurophysiology, 67, 1105-1113.

Edin, B.B. and Abbs, J.H. (1991) Finger movement responses of cutaneous mechanoreceptors in the dorsal skin of the human hand. Journal of Neurophysiology, 65, 657-670.

Edin, B.B., Essick, G.K., Trulsson, M. and Olsson, K.Å. (1995) Receptor encoding of moving tactile stimuli in humans. I. Temporal pattern of discharge of individual low-threshold mechanoreceptors. Journal of Neuroscience, 15, 830-847.

Essick, G.K. (1983) Tactile Direction Discrimination in Primates. Ph.D. Thesis. Dept. of Physiology, University of North Carolina, Chapel Hill.

Essick, G.K. (1991) Human capacity to process directional information provided by tactile stimuli which move across the skin: Characterization and potential neural mechanisms. In Franzén, O. and Westman, J. (eds) Information Processing in the Somatosensory System, Macmillan Press, London, pp 329-339.

Essick, G.K. (1992) Comprehensive clinical evaluation of perioral sensory function. Oral and Maxillofacial Surgery Clinics of North America, 4, 503-526.

Essick, G.K., Afferica, T., Aldershof, B., Nestor, J., Kelly, D. and Whitsel, B. (1988) Human perioral directional sensitivity. Experimental Neurology, 100, 506-523.

Essick, G.K., Bredehoeft, K.R., McLaughlin, D.F. and Szaniszlo, J.A. (1991) Directional sensitivity along the upper limb in humans. Somatosensory and Motor Research, 8, 13-22.

Essick, G.K., Dolan, P.J., Turvey, T.A., Kelly, D.G. and Whitsel, B.L. (1990) Effects of trauma to the mandibular nerve on human perioral directional sensitivity. Archives of Oral Biology, 35, 785-794.

Essick, G.K. and Edin, B.B. (1995) Receptor encoding of moving tactile stimuli in humans. II. The mean response of individual low-threshold mechanoreceptors to motion across the receptive field. Journal of Neuroscience, 15, 848-864.

Essick, G.K. and McGuire, M.H. (1986) Role of kinetic and static cues in human subjects' evaluation of direction of cutaneous stimulus motion. Society for Neuroscience Abstracts, 12, 14.

Essick, G.K., McGuire, M., Joseph, A. and Franzén, O. (1992) Characterization of the percepts evoked by discontinuous motion over the perioral skin. Somatosensory and Motor Research, 9, 175-184.

Essick, G.K., Rath, E.M., Kelly, D.G., James, A. and Murray, R.A. (1996) A novel approach for studying direction discrimination. In O. Franzén, R. Johansson and L. Terenius, (eds) Somesthesis and the Neurobiology of the Somatosensory Cortex, Birkhäuser Verlag AG, Basel, pp 59-72.

Essick, G.K. and Whitsel, B.L. (1985a) Assessment of the capacity of human subjects and S-I neurons to distinguish opposing directions of stimulus motion across the skin. Brain Research Reviews, 10, 187-212.

Essick, G.K. and Whitsel, B.L. (1985b) Factors influencing cutaneous directional sensitivity: A correlative psychophysical and neurophysiological investigation. Brain Research Reviews, 10, 213-230.

Essick, G.K., Whitsel, B.L., Dolan, P.J. and Kelly, D.G. (1989) Effects of traverse length on human perioral directional sensitivity. Journal of the Neurological Sciences, 93, 175-190.

Gardner, E.P. (1988) Somatosensory cortical mechanisms of feature detection in tactile and kinesthetic discrimination. Canadian Journal of Physiology and Pharmacology, 66, 439-454.

Gardner, E.P., Hämäläinen, H.A., Palmer, C.I. and Warren, S. (1989) Touching the outside world: Representation of motion and direction within primary somatosensory cortex. In Lund, J.S. (ed) Sensory Processing in the Mammalian Brain: Neural Substrates and Experimental Strategies, Oxford University Press, New York, pp 49-66.

Gardner, E.P. and Palmer, C.I. (1989) Simulation of motion on the skin. I. Receptive fields and temporal frequency coding by cutaneous mechanoreceptors of OPTACON pulses delivered to the hand. Journal of Neurophysiology, 62, 1410-1436.

Gardner, E.P., Palmer, C.I., Hämäläinen, H.A. and Warren, S. (1992) Simulation of motion on the skin. V. Effect of stimulus temporal frequency on the representation of moving bar patterns in primary somatosensory cortex of monkeys. Journal of Neurophysiology, 67, 37-63.

Gardner, E.P. and Sklar, B.F. (1986) Factors influencing discrimination of direction of motion on the human hand. Society for Neuroscience Abstracts, 12, 798.

Gardner, E.P. and Sklar, B.F. (1994) Discrimination of the direction of motion on the human hand: A psychophysical study of stimulation parameters. Journal of Neurophysiology, 71, 2414-2429.

Gardner, E.P. and Spencer, W.A. (1972) Sensory funneling. I. Psychophysical observations of human subjects and responses of cutaneous mechanoreceptive afferents in the cat to patterned skin stimuli. Journal of Neurophysiology, 35, 925-953.

Goodwin, A.W., John, K.T., Sathian, K. and Darian-Smith, I. (1989) Spatial and temporal factors determining afferent fiber responses to a grating moving sinusoidally over the monkey's fingerpad. Journal of Neuroscience, 9, 1280-1293.

Goodwin, A.W. and Morley, J.W. (1987) Sinusoidal movement of a grating across the monkey's fingerpad: Effect of contact angle and force of the grating on afferent fiber responses. Journal of Neuroscience, 7, 2192-2202.

Gould, W.R., Vierck ,C.J. and Luck, M.M. (1979) Cues supporting recognition of the orientation or direction of movement of tactile stimuli. In Kenshalo, D.R. (ed) Sensory Function of the Skin of Humans. Proceedings of the Second International Symposium on the Skin Senses. Plenum Press, New York, pp 63-73.

Greenspan, J.D. (1992) Influence of velocity and direction of surface-parallel cutaneous stimuli on responses of mechanoreceptors in feline hairy skin. Journal of Neurophysiology, 68, 876-889.

Hall, G.S. and Donaldson, H.H. (1885) Motor sensations on the skin. Mind, 10, 557-572.

Halpern, L. (1949) Disturbance of dermatokinesthesis in cerebral and spinal diseases. Journal of Nervous and Mental Disease, 109, 1-8.

Henderson, D.C. (1971) The relationships among time, distance, and intensity as determinants of motion discrimination. Perception and Psychophysics, 10, 313-320.

Henderson, D.C. (1973) Visual discrimination of motion: Stimulus relationships at threshold and the question of luminance-time reciprocity. Perception and Psychophysics, 1, 121-130.

Hollins, M. and Favorov, O. (1994) The tactile movement aftereffect. Somatosensory and Motor Research, 11, 153-162.

LaMotte, R.H. and Srinivasan, M.A. (1987a) Tactile discrimination of shape: Responses of slowly adapting mechanoreceptive afferents to a step stroked across the monkey fingerpad. Journal of Neuroscience, 7, 1655-1671.

LaMotte, R.H. and Srinivasan, M.A. (1987b) Tactile discrimination of shape: Responses of rapidly adapting mechanoreceptive afferents to a step stroked across the monkey fingerpad. Journal of Neuroscience, 7, 1672-1681.

LaMotte, R.H. and Srinivasan, M.A. (1991) Surface microgeometry: Tactile perception and neural encoding. In Franzén, O. and Westman, J. (eds) Information Processing in the Somatosensory System, Macmillan Press, London, pp 49-58.

LaMotte, R.H. and Whitehouse, J. (1986) Tactile detection of a dot on a smooth surface: Peripheral neural events. Journal of Neurophysiology, 56, 1109-1128.

LaMotte, R.H., Whitehouse, G.M., Robinson, C.J. and Davis, F. (1983) A tactile stimulator for controlled movements of textured surfaces across the skin. Journal of Electrophysiological Techniques, 10, 1-17.

Loomis, J.M. and Collins, C.C. (1978) Sensitivity to shifts of a point stimulus: An instance of tactile hyperacuity. Perception & Psychophysics, 24, 487-492.

Murray, R.A., Essick, G.K. and Kelly, D.G. (1994) Effect of stimulus force on perioral direction discrimination: Clinical Implications. Journal of Oral and Maxillofacial Surgery, 52, 688-697.

Norrsell, U. and Olausson, H. (1992) Human, tactile, directional sensibility and its peripheral origins. Acta Physiologica Scandinavica, 144, 155-161.

Norrsell, U. and Olausson, H. (1994) Spatial cues serving the tactile directional sensibility of the human forearm. Journal of Physiology (London), 478.3, 553-540.

Olausson, H. (1994) The influence of spatial summation on human tactile directional sensibility. Somatosensory and Motor Research, 11, 305-310.

Olausson, H. and Norrsell, U. (1993) Observations on human tactile directional sensibility. Journal of Physiology (London), 464, 545-559.

Phillips, J.R., Johansson R.S. and Johnson, K.O. (1990) Representation of braille characters in human nerve fibers. Experimental Brain Research, 81, 589-592.

Ray, R.H., Mallach, L.E. and Kruger, L. (1985) The response of single guard and down hair mechanoreceptors to moving air-jet stimulation. Brain Research, 346, 333-347.

Sathian, K., Goodwin, A.W., John, K.T. and Darian-Smith, I. (1989) Perceived roughness of a grating: Correlation with responses of mechanoreceptive afferents innervating the monkey's fingerpad. Journal of Neuroscience, 9, 1273-1279.

Srinivasan, M.A., Whitehouse, J.M. and LaMotte, R.H. (1990) Tactile detection of slip: Surface microgeometry and peripheral neural codes. Journal of Neurophysiology, 63, 1323-1332.

Szaniszlo, J.A., Essick, G.K., Kelly, D.G., Joseph, A.K. and Bredehoeft, K.R. (1996) Evocation and characterization of percepts of apparent motion on the face. Submitted.

Vallbo, Å., Olausson, H., Wessberg, J. and Kakuda, N. (1995) Receptive field characteristics of tactile units with myelinated afferents in hairy skin of human subjects. Journal of Physiology (London) 483.3, 783-795.

Vallbo, Å., Olausson, H., Wessberg, J. and Norrsell, U. (1993) A system of unmyelinated afferents for innocuous mechanoreception in the human skin. Brain Research, 628, 301-304.

Whitsel, B.L., Dreyer D.A. and Hollins, M. (1978) Representation of moving stimuli by somatosensory neurons. Federation Proceedings, 37, 2223-2227.

Whitsel, B.L., Dreyer, D.A., Hollins, M. and Young, M.G. (1979) The coding of direction of tactile stimulus movement: Correlative psychophysical and electrophysiological data. In Kenshalo, D.R. (ed) Sensory Functions of the Skin of Humans. Plenum Press, New York, pp 79-107.

Whitsel, B.L., Favorov, O.V., Kelly, D.G. and Tommerdahl, M. (1991) Mechanisms of dynamic peri- and intra-columnar interactions in somatosensory cortex: Stimulus-specific contrast enhancement by NMDA receptor activation. In Franzén, O. and Westman, J. (eds) Information Processing in the Somatosensory System. Macmillan Press, London, pp 353-369.

Whitsel, B.L., Roppolo, J.R. and Werner, G. (1972) Cortical information processing of stimulus motion on primate skin. Journal of Neurophysiology, 35, 691-717.

Neural Aspects of Tactile Sensation
J.W. Morley (Editor)
1998 Elsevier Science B.V.

Extracting the Shape of an Object from the Responses of Peripheral Nerve Fibers

A.W. Goodwin

When we manipulate objects, our fingers and thumb move in precise patterns that depend on a number of properties of the object being handled. As an illustration, consider the task of screwing a lid on a bottle. The fingers roll over the circular profile of the lid in trajectories that correspond to the shape and size of the lid. Appropriate finger movements would not be possible unless the central nervous system had precise information about the contours of the object. Force is also a critical parameter, and the forces exerted by the fingers must be matched to the nature of the task and to the nature of the object. Large grip forces are appropriate when undoing a tight jar lid, but gentle forces are called for when examining a petal. In addition, the mechanics of grasping requires appropriate relationships between the applied forces and the positions of contact on the digits. Thus, the points of contact on the fingers and the contact forces must be signaled to the brain for successful manipulation. Other properties of the object, such as its compliance and its surface texture, must also be taken into account by the brain.

During active manipulation, a number of different types of mechanoreceptors will relay information to the central nervous system. The positions and velocities of the fingers can be determined from the angles and angular velocities of the joints involved in the movement. There are at least three groups of receptors with responses that depend on joint angle and velocity. Receptors in the joints themselves respond, but their role in fine finger movements is still not clear (Burke et al., 1988; Edin, 1990). Spindles in the extrinsic and intrinsic hand muscles respond to the lengths and velocities of these muscles which, in turn, determine the angles and angular velocities of the joints (Gandevia et al., 1992). More recently it has been shown that cutaneous receptors in the regions of the joints and in the hairy skin of the hand are sensitive to the positions and velocities of the fingers (Hulliger et al., 1979; Edin and Abbs, 1991).

Golgi tendon organs, which are sensitive to muscle tension, could relay information about forces exerted by the fingers (Edin and Vallbo, 1990). All these signals from the peripheral receptors discussed above are available to the sensorimotor system controlling the hand movements. Efferent signals from centrally generated motor commands are also likely to play a role (McCloskey, 1991; Gandevia et al., 1992). In addition, there is a wealth of precise information about the stimulus parameters provided by cutaneous mechanoreceptors in the fingerpads that, until recently, has been given little attention. The information signaled by afferents from these cutaneous receptors is the subject of this article.

1. Definition of shape and size

Of all the parameters of a manipulated object, shape and size are perhaps the most obvious. The finger movements needed to manipulate a sphere are different from those needed to manipulate a cube or a tetrahedron. Handling a small sphere will require a different pattern of movements from those used to handle a large sphere. There are many ways of defining the shape of an object (Lord and Wilson, 1984). The most direct definition is in terms of the 3-dimensional Cartesian coordinates of each point on the surface. This is certainly a complete definition of the surface but it is not economical, requiring a large number of points even for something as simple as a sphere. The computational inefficiency of such a description makes it an unlikely choice for the brain to use. An alternative description, which has many advantages for use in the tactile system, defines shape in terms of the local curvature of each point on the surface. Computations are simplified since only changes, or significant changes, in local curvature need be processed (Richards et al., 1986). For example, over the entire face of a cube the curvature is constant and only changes at an edge or at a vertex. For a sphere the curvature is invariant.

Shape and size are usually considered as two independent attributes of an object (Koenderink and van Doorn, 1992). A large cube and a small cube have the same shape, consisting of alternating flat faces and rectangular edges, and differ only by a scaling factor which is the ratio of the lengths of the edges. In the case of two spheres of different size, the shape is the same for both and consists of a surface of constant curvature. The scaling factor for the size difference is the ratio of the radii or, equivalently, the ratio of the curvatures. Curvature defines a sphere completely; the shape is defined by the constancy of the curvature and the size by its magnitude. The mechanism for extracting shape and size differs for different computational schemes (Marshall, 1989). Whatever scheme is used, it must be possible to determine both of these parameters.

The idea of describing an object in terms of its local curvatures dates back at least to Gauss (Lord and Wilson, 1984; Hoffman and Richards, 1984). Such a description is attractive to computational scientists as it has many advantages. For most objects the definition is compact and leads to efficient schemes for computing and identifying shapes (Marr and Ninshihara, 1978; Asada and Brady, 1986). In addition, the description does not change with rotation or translation of the object. This is an important consideration because manipulated objects often roll along and move tangentially across the fingerpads. If the neural description of the object was not invariant to translation and rotation, the computational task for the brain would be complicated.

2. Active touch

Manipulation of objects by humans involves a temporal sequence of changing conditions of grasp and changing sensory feedback. It is only by integrating information acquired during such an exploration that the brain can use sensory signals to build up a complete description of the object. Thus as we move our fingers over an object, the relationships of the fingers and thumb to each other and their position in space change in accordance with the shape of the object. This form of tactile exploration, such as we may use to explore a sculpture, was addressed by early psychologists and is termed active touch (Gibson, 1962; Katz, 1989; Roland and Mortensen, 1987).

It is commonly recognized that during active touch, characteristics of the object's shape may be conveyed to the brain by receptors responding to the changing angles of the joints in the hand. Receptors in the joints themselves and in the muscles moving the fingers are usually assigned the major role. In addition to the sequence of joint angles, there is another potential source of information for the definition of shape. If the cutaneous receptors in the fingerpads were able to signal the local curvature of the object in contact with the skin, then the sequence of local curvatures relayed during active touch would also provide a description of the object's shape. In general, the brain could combine information from all the various types of peripheral receptors and from efferent motor commands to obtain the final perception of the explored object.

3. Local shape

The argument for the importance of cutaneous receptors becomes more compelling when we consider manipulation in the broader sense. Sometimes we explore objects simply to determine their shape, but usually tactile manipulation is part of a more complex sensorimotor act. Consider an everyday task such as doing up a button; the aim is not to come up with some final perception of the shape of the button but rather to execute the sensorimotor task satisfactorily. To accomplish this, the brain must be aware of the local shape of the button throughout the manipulation so that future finger movements can be planned appropriately. If the curvature were underestimated then the fingers would move off the button, and if it were overestimated then the planned trajectories would be too curved leading to inappropriate forces and inappropriate relative motion of the fingers and the button. Thus the information must be available to the brain throughout the manipulation A global picture obtained by integration at the end of the exploration cannot be used retrospectively to control the finger movements.

The crucial role that the cutaneous receptors in the fingerpads play in measuring local curvature is stressed in Figure 1. At the instant of time illustrated, all signals related to joint angles would be the same for the sphere and the cube. These include responses of muscle spindles and activity in cutaneous receptors around the joints and in the hairy skin. However, the patterns of finger movements required to manipulate the two objects are quite different. Only the skin in contact with the objects can differentiate the local curvatures at the contact points.

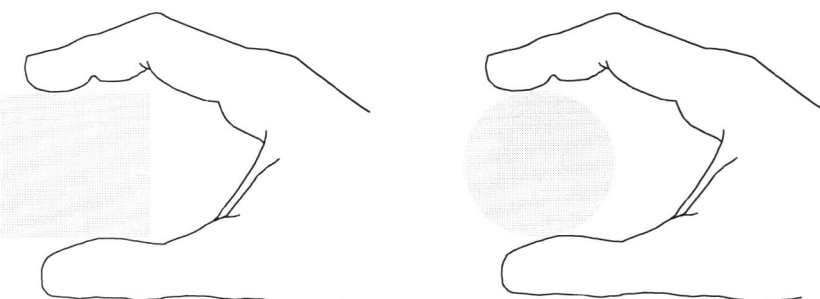

Figure 1. The positions of the finger and thumb (angles of the joints and lengths of the muscles) are the same when grasping the cube as when grasping the sphere. Differences in the shapes of the two objects can be determined from the responses of the mechanoreceptors in the fingerpads.

Common objects may have complex shapes with local variations superimposed on the global geometry. For example, a pebble with an ellipsoidal global shape may have protuberances, indentations or flat spots at various points on the surface changing the local shape at those points. To handle such objects successfully, these local variations must be taken into account by the sensorimotor control system. The cutaneous receptors are ideal candidates for signaling this information.

4. Receptors in glabrous skin

Glabrous skin covering the fingerpads is highly specialized (Halata, 1975). As shown in Figure 2, the superficial epidermis is arranged in a pattern of alternating grooves and ridges which is reflected in the outermost thick cornified layer, resulting in a pattern of fingerprints which is unique to the individual. The ridges continue into the lower epidermis as deep wide glandular ridges which effectively protrude into the dermis. Beneath the grooves of the superficial epidermis there are adhesive ridges protruding into the dermis. These are parallel to the glandular ridges but are not as wide or as deep. A system of cross ridges runs perpendicular to the glandular and adhesive ridges. Crypts or dermal papillae are situated in the dermis between the glandular and adhesive ridges. Thus the dermis is also arranged in a pattern of grooves and ridges but it is more complex than the pattern in the epidermis.

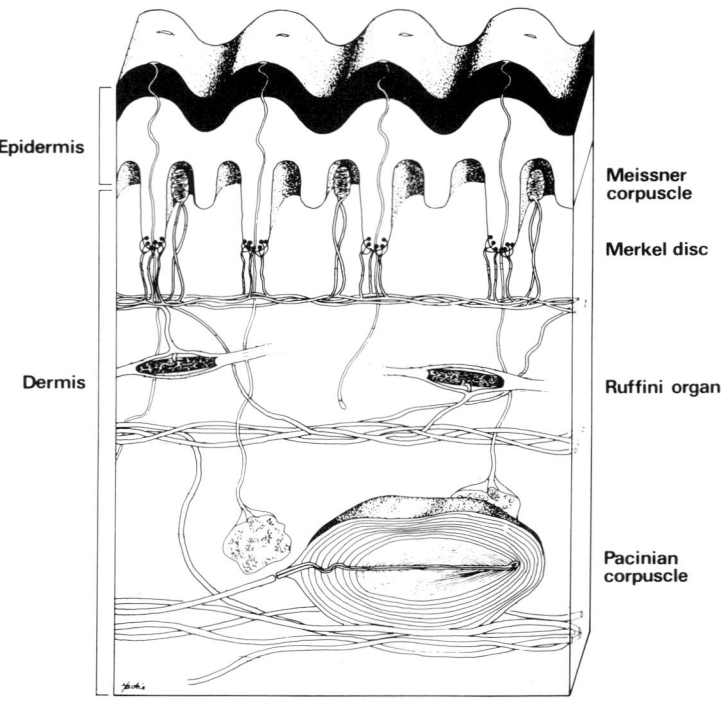

Figure 2. Glabrous skin is arranged in a pattern of alternating grooves and ridges. Four types of low-threshold mechanoreceptors are located at characteristic locations in the skin. From Darian-Smith (1984).

It is apparent that the mechanics of the skin are complicated even without considering the details of the attachments of the receptors. Skin is not a uniform isotropic structure. Moreover, it is not flat and its curvature varies over the fingerpad. Nevertheless, important advances in our understanding of skin mechanics have resulted from models based on assumptions of flat uniform skin (Phillips and Johnson, 1981; Srinivasan, 1989).

There are four types of receptors that are activated by non-noxious mechanical stimuli. Merkel cells and associated disk-shaped nerve terminals are found in the epidermis at the base of the glandular ridges where groups of up to 10 Merkel cells are found (Halata, 1975). Meissner corpuscles are found in the crypts of the dermis lying between the glandular and adhesive ridges. They are oval in shape (about

150x50 micrometers) with the long axis parallel to the glandular ridges (Chouchkov, 1973). Although the Merkel complexes are in the epidermis and the Meissner corpuscles are in the dermis, the arrangement of the ridges results in the skin surface being closer to the Meissner corpuscle than to the Merkel complex. In models of skin mechanics that assume uniform properties of the skin, the depth is greater for the Merkel complexes than for the Meissner corpuscles. In the future, more realistic models will have to take into account the obvious anisotropic structure of skin and the complex effects of the various ridges.

In the subpapillary dermis, a deeper layer of the dermis, an elongated encapsulated receptor is found. This is the Ruffini corpuscle and its long axis is usually oriented parallel to the skin surface (Miller et al., 1958). The fourth type of receptor is the Pacinian corpuscle. This is an encapsulated receptor with a central nerve fiber surrounded by numerous concentric layers of lamellar cells (Bell et al., 1994). These large ovoid corpuscles, up to a millimeter or more in length, are found deep in the dermis and in the tissues beneath the skin.

5. Primary afferent fibers

The four groups of low-threshold mechanoreceptors in the glabrous skin of the hand are innervated by large myelinated axons that travel in the median nerve or the ulnar nerve. These fibers have conduction velocities ranging from about 20 to about 80 ms^{-1} (Vallbo and Johansson, 1984; Mackel, 1988). There is considerable convergence and divergence between the receptors and the primary afferent fibers. A number of Merkel disks are innervated by a single branching axon (Chambers et al., 1972). Each Meissner corpuscle may be innervated by several axons, and a single axon may innervate several Meissner corpuscles (Cauna, 1956). A single axon innervates only one Ruffini ending (Chambers et al., 1972). Pacinian corpuscles are usually supplied by a single axon with each axon supplying a single corpuscle (Cauna and Mannan, 1958).

The primary afferent fibers form four distinct functional classes. They can be classified using relatively simple stimuli (Talbot et al., 1968; Darian-Smith, 1984; Vallbo and Johansson, 1984). For two of the classes, a punctate probe indenting the skin elicits a response during the dynamic phases when the probe is moving into the skin or out of the skin but elicits no response during the static phase when the probe remains indented in the skin. Such fibers, termed rapidly (or fast) adapting afferents, respond vigorously to vibrating stimuli. One class responds optimally to vibration at frequencies around 40 Hz. These terminate in Meissner corpuscles and are termed fast adapting type I afferents (FAI) by some investigators and simply rapidly adapting afferents (RA) by others. The second class of fibers responds optimally to vibration at frequencies in excess of 200 Hz. These terminate in Pacinian corpuscles and are termed either fast adapting type II afferents (FAII) or simply Pacinian afferents. The above 2 classes can also be distinguished on the basis of the spatial characteristics of their receptive fields. For a punctate probe, an FAI has a relatively small receptive field with a rapidly rising threshold as the probe moves away from the center of the receptive field. For an FAII the receptive field is larger and the threshold rises more gradually as the probe moves away from the receptive field center.

The remaining two classes of fibers respond throughout the static phase of indentation and are referred to as slowly adapting afferents. The two classes can be distinguished on the basis of the spatial characteristics of their receptive fields. For the slowly adapting type I afferents (SAI), the receptive fields are small and thresholds rise rapidly as the probe moves away from the receptive field center. For SAIIs, the receptive fields are larger and thresholds rise more gradually away from the receptive field center. In addition, SAIIs respond well to lateral stretch of the skin and have a more regular discharge than do SAIs. The SAIs and SAIIs are supplied by Merkel complexes and Ruffini corpuscles respectively.

Since each receptor does not have its own axon to convey information to the spinal cord, it is the innervation density of the primary afferent fibers that is important in determining resolution rather than the density of the receptors. Two of the afferent types, SAI and FAI, have a high innervation density on the fingertips estimated in the human at 0.70 and 1.40 afferents mm^{-2} respectively (Johansson and Vallbo, 1979). The density of these afferents decreases for progressively more proximal parts of the hand. The density of the

SAIIs and FAIIs is much lower, estimated at 0.09 (excluding those around the nail) and 0.21 mm^{-2} respectively (Johansson and Vallbo, 1979). Similar estimates have been obtained in the monkey (Darian-Smith and Kenins, 1980).

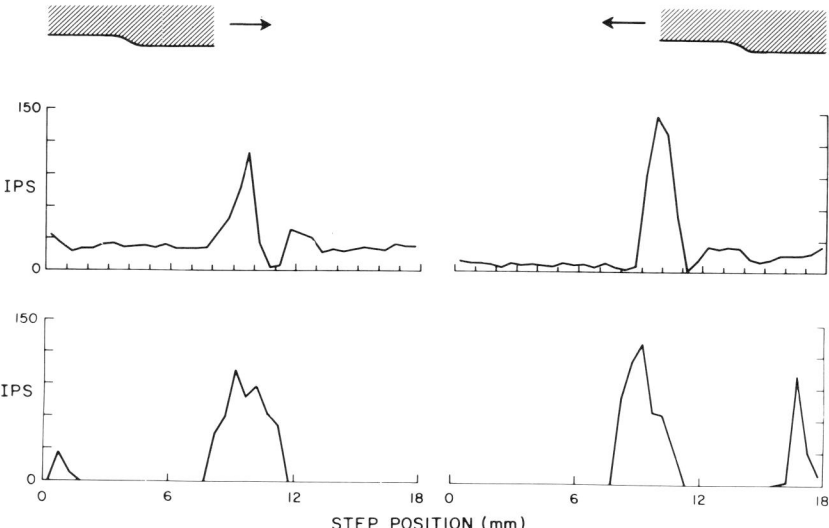

Figure 3. The step (top) was stroked across the receptive field of an SAI (middle) and an FAI (bottom). Histograms show the mean discharge rate (impulses per second or IPS) in 0.5 mm bins. In the left panel the step was stroked from the high side to the low side (left to right) and in the right panel from the low to the high side (right to left). Modified from LaMotte and Srinivasan (1987a,b).

6. Responses to sinusoidally shaped steps

When humans manipulate objects, all four mechanoreceptor types are activated. It is not easy to predict, from the early data gathered with indenting and vibrating probes, how the parameters of the manipulated object would affect the responses of the afferents. The first direct evidence that responses depend on the shape of the stimulus was the demonstration that SAIs showed an enhanced response to edges (Vierck, 1979; Phillips and Johnson, 1981; Johansson et al., 1982). A smaller edge enhancement was reported for FAIs (Johansson et al., 1982).

An edge is a region of high curvature. LaMotte and Srinivasan (1987a,b) extended this notion by manufacturing steps with profiles that were sinusoidal in shape. The curvature across such a step varies. By keeping the height of the step constant and changing the width, they were able to generate steps containing a range of curvatures. These surfaces were stroked across the receptive fields of SAIs and FAIs innervating monkey's fingerpads.

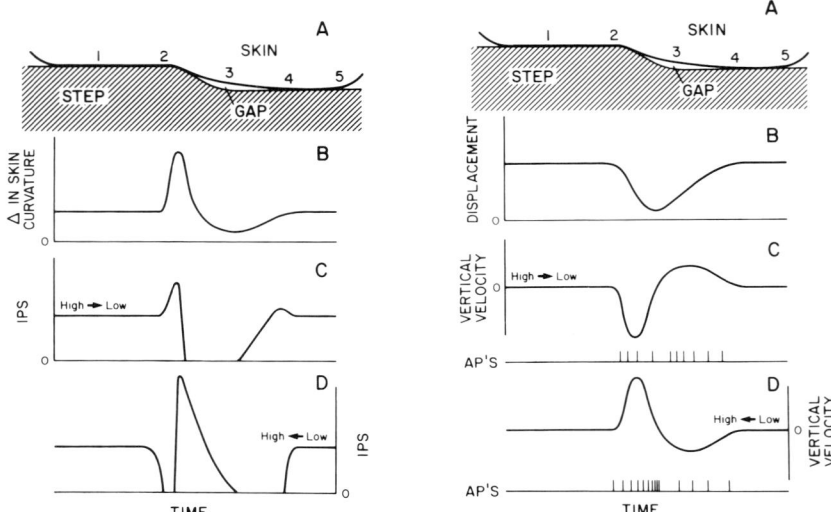

Figure 4. Profiles of the skin in contact with a step are shown at the top (A). Left panel. For an SAI, the pertinent parameters are the amount of curvature (B) and its rate of change; together they predict the responses for a movement from high to low (C) and from low to high (D). Right panel. For an FAI, the skin displacement (B) leads to the predominant parameter, vertical velocity, shown for a high to low movement (C) and a low to high movement (D). Tick marks show the occurrence of hypothetical action potentials for the FAI in C and D. Modified from LaMotte and Srinivasan (1987a,b).

As the steps were stroked over the receptive field, the afferents responded in an apparently complex manner that depended on the afferent type, the step width and the stroke velocity. As can be seen from the responses of a typical SAI and a typical FAI in Figure 3, the responses when the step was stroked from left to right (left panel) were different from those when the step was stroked from right to left (right panel). By relating these responses to profiles of the skin during the stroke, LaMotte and Srinivasan were able to explain their results in terms of a few key parameters as shown in Figure 4. The major features of the SAI responses were predicted by the amount and rate of change of skin curvature at the center of the receptive field. For the FAIs, the major features of the responses could be explained by a predominant sensitivity to vertical velocity at the receptive field center. They also measured the capacity of human subjects to discriminate steps with different widths. Comparison of this capacity with features of the afferent responses (like burst width, pause width and burst rate) led to the conclusion that both the SAIs and the FAIs may have contributed to the human ability.

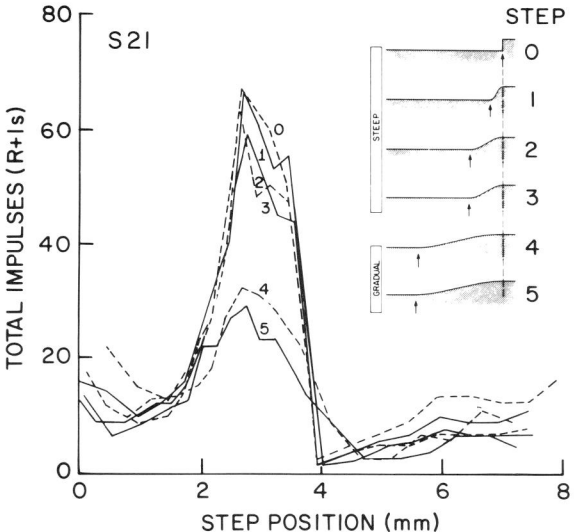

Figure 5. Profiles across the receptive field of an SAI responding to steps with sinusoidal shapes and different widths. Steps are shown in the inset - arrows and vertical line mark the beginning and end respectively of the sinusoids. The response is the number of impulses accumulated after 1 second following the ramp of indentation. From Srinivasan and LaMotte (1987).

In a second series of experiments, Srinivasan and LaMotte (1987) used the same sinusoidal steps but indented them vertically into the skin. The absence of lateral scanning motion simplified the interpretation of these experiments. Profiles of responses across the receptive field of a typical SAI (Figure 5) emphasize the importance of curvature. The profiles were sharper and higher for the steps with the smaller widths (greater curvature at corresponding points). In contrast, FAI responses were small and poorly modulated. Human performance was measured in matching psychophysical experiments and could be accounted for by spatial and intensive features of the SAI responses. For a sinusoidal step, the curvature is complex and is not constant over the contact area with the skin; this complicates the interpretation of the above data. Cylinders have an advantage over sinusoidal steps in that the radius of a cylinder is the same all over the contact area. The dependence of SAI responses on curvature was reinforced by observations with cylinders indenting the fingerpad (Srinivasan and LaMotte, 1991). Responses of SAIs increased as the curvature of the cylinder increased, that is as the radius decreased.

A number of questions arise from the above studies. How precisely can humans discriminate the shapes of objects and, in particular, differences in curvature? How do the cutaneous afferents in the fingerpads relay sufficiently precise information to the brain? How do the afferent fibers signal unambiguous changes in curvature when other parameters of the stimulus change as well? How are these other parameters relayed to the brain by the cutaneous afferents?

To answer these questions, we performed a series of experiments in which quantitative psychophysical measurements in humans were compared with quantitative measures of neural activity recorded from single fibers in monkeys' peripheral nerves.

7. Psychophysical experiments with spheres

A stimulator based on a balanced beam (Figure 6) was used to present stimuli passively to an immobilized fingerpad in human psychophysical experiments. A curtain obscured the subject's view of the stimulator and care was taken to eliminate auditory and other extraneous cues. This arrangement guaranteed that the only information relayed to the brain originated from mechanoreceptors in the skin of the fingerpad. The stimulator allowed control of 3 critical parameters. Contact force was changed by changing the weights on the beam; the position of contact was adjusted via micrometers and dial indicators; and the shape was determined by selecting different objects. We chose to use spheres as our stimuli since they have a number of advantages. First, curvature is constant over the entire surface so that when a sphere contacts the skin, the curvature of the skin over the whole contact area is the same. Second, the shape (and size) of the sphere is specified by a single parameter; either the radius or, equivalently, the curvature which is simply the reciprocal of the radius (Figure 7). Third, it is relatively simple to manufacture a sphere of specified radius.

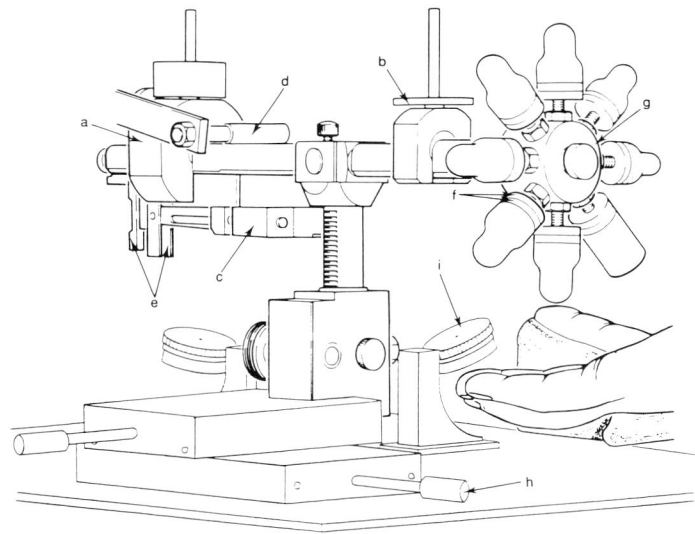

Figure 6. Balanced beam stimulator allowing control of contact force via weights *b*, and of position via micrometers *h* and dial indicators *i*. The hub *g* held spheres of different radii. Smooth contact was achieved by a damper *d*.

The ability of humans to discriminate changes in curvature was measured using a forced-choice paradigm (Goodwin et al., 1991; Goodwin and Wheat, 1992a). Either the same stimulus (same sphere) was presented twice (S_s) or two different spheres were presented (S_d). Subjects were required to respond that the stimuli were the same (R_s) or different (R_d). Two conditional probabilities were calculated; the probability of the subject responding that the stimuli were different when they were different, $p(R_d/S_d)$, and of responding that they were different when in fact they were the same, $p(R_d/S_s)$.

The conditional probabilities for a subject discriminating comparison spheres of increasing curvature from a standard sphere of curvature 287 m^{-1} (radius 3.48 mm) are shown in Figure 8 (left). Standard decision theory was used to obtain a bias-free measure of discrimination ability (d') shown on the right. A d' value of 1.35 corresponds to a correct response 75% of the time in the absence of bias. This general technique was used in all our assessments of discrimination and Figure 8 is typical of all response curves. As seen in

Curvature (m^{-1})

0	81	172	256	340	521	694

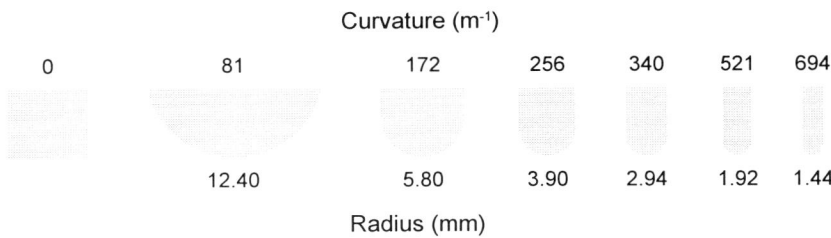

	12.40	5.80	3.90	2.94	1.92	1.44

Radius (mm)

Figure 7. The shape (and size) of a sphere is completely described by its radius or, equivalently, by its curvature which is the reciprocal of the radius.

Table 1, humans were able to discriminate flat surfaces from gently curved convex surfaces and from gently curved concave surfaces. Two spheres could be discriminated if their curvatures differed by about 10%.

8. Responses of single nerve fibers

In the human psychophysical experiments, all judgments had to be made solely on the basis of information conveyed by the cutaneous mechanoreceptive afferents. To characterize these afferents, responses were recorded from primary afferent fibers in anaesthetized monkeys (Goodwin et al., 1995). Single fibers innervating mechanoreceptors in the fingerpad were isolated by micro-dissection of the median nerve. Each fiber was classified according to the criteria described previously and was only studied if its receptive field was on the central portion of a fingerpad. In these experiments the stimulator was the same as that used in the human studies. The spheres, shown in Figure 7, ranged in curvature from 0 m^{-1} (flat) to 694 m^{-1} (radius 1.44 mm).

	Curvature (m^{-1})			
	Flat-Convex	Flat-Concave	Convex-Convex	Convex-Convex
Standard	0.0	0.0	144	287
Comparison (75%)	4.89	-5.40	158	319
Difference Limen	4.89	-5.40	14	32
Weber Fraction			0.097	0.11

Table 1. Human ability to discriminate curved surfaces. Difference limen is the difference between the curvatures of standard and comparison stimuli that could be discriminated with a probability of 75% in the absence of bias (d'=1.35). Weber fraction is the difference limen divided by the standard.

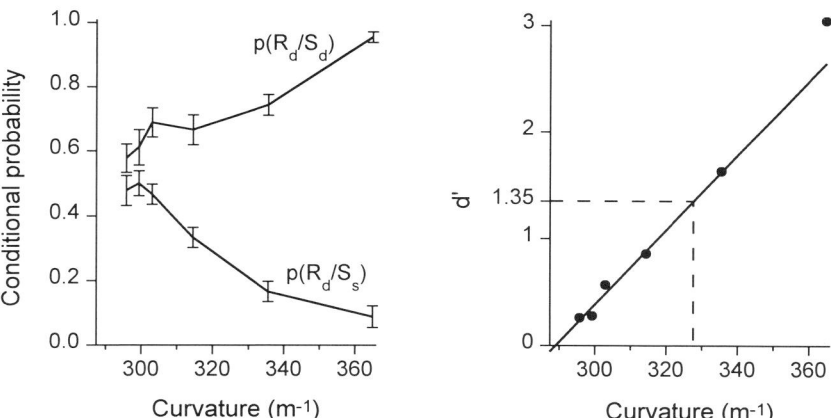

Figure 8. A typical human subject discriminating spheres of increasing curvature from a standard sphere of curvature 287 m^{-1}. The d' function extracted from the conditional probabilities is approximately linear.

The stimulator arm was attached to a damper (Figure 6) which ensured that contact with the stimulus was smooth; contact velocity was about 20 mms^{-1}. For such stimuli, SAIs responded vigorously and it is these afferents that will be analyzed here. About 50% of the FAIs did not respond at all; the contribution of the remaining 50% will be discussed later. FAIIs did not respond, and monkeys do not have SAIIs (see later).

The responses of a typical SAI when spheres with different curvatures were applied to the center of the receptive field at constant contact force are shown in Figure 9. In this figure, and in subsequent figures, the response is taken as the number of action potentials following contact and occurring during the first second of response. This corresponds to the one second contact time used in the psychophysical experiments. When contact force remained constant, an increase in the curvature of the sphere (or equivalently, a decrease in the radius of the sphere) resulted in a monotonic increase in response. Some saturation was evident at the higher curvatures. This is in keeping with the results of Srinivasan and LaMotte (1987,1991). The small standard errors indicate the consistency of responses to repeated stimuli.

As seen in Figure 9, the response of this afferent also changed when the contact force changed. These two parameters, curvature and force, are not signaled uniquely by the responses of this afferent as illustrated by the following example. The response of 56 impulses to a sphere with curvature 200 m^{-1} applied with a contact force of 15 g wt would have increased to a response of 73, either when the contact force increased to 20 g wt (with no change in curvature) or when the curvature increased to 296m^{-1} (with no change in contact force).

It is obvious that a single afferent fiber does not provide unambiguous information about either the shape of the object in contact with the receptive field or the contact force. This is in marked contrast to human behavior. We know from everyday experience that we are able to extract both the shape of an object and the force with which we are contacting it. This property was quantified in scaling experiments in which human subjects were presented with spheres that varied both in curvature and in contact force (Goodwin et al., 1991b; Goodwin and Wheat, 1992b). In one series of experiments, subjects were asked to estimate stimulus curvature. Their perceived curvature varied monotonically with the actual curvature of the stimulus (Figure 10, left), and the variations in contact force had little effect. In a second

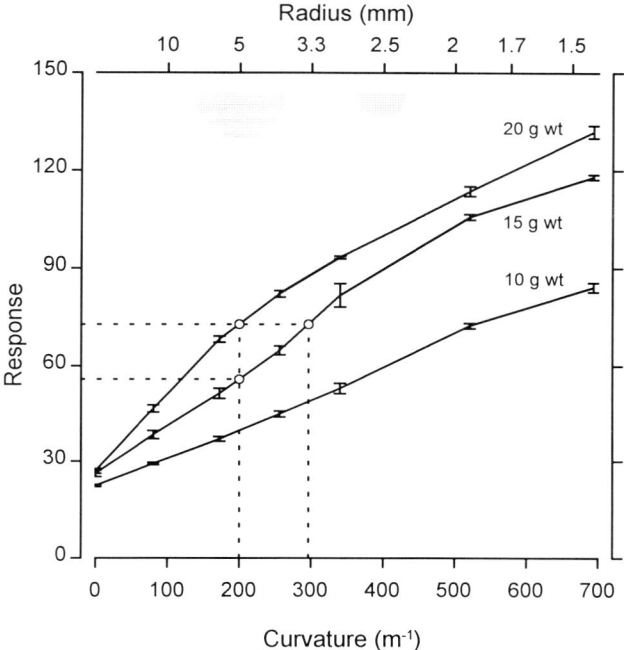

Figure 9. Responses (mean±SEM, n=8) of a single SAI to spherical stimuli comprising 7 curvatures and 3 contact forces. Response is the number of action potentials in the first second following contact.

series, the subjects were asked to scale for contact force. Perceived contact force varied monotonically with applied contact force (Figure 10, right), and varying the curvature of the stimulus had little effect. Even when curvature and force were randomized, humans could somehow extract both of the parameters from the responses of the cutaneous afferents.

9. Responses of populations of fibers

Some SAIs are more sensitive than others and respond more vigorously to the same stimulus. In our sample, the response of the most sensitive fiber was more than 4 times that of the least sensitive fiber. This sensitivity can be accounted for by a scaling factor which is different for each fiber. Apart from the variation in scaling factors, all SAIs behaved in the same way when the curvature or the force of the

stimulus was varied; the functional relationships were the same as for
the fiber illustrated in Figure 9.Thus the effect of curvature and contact

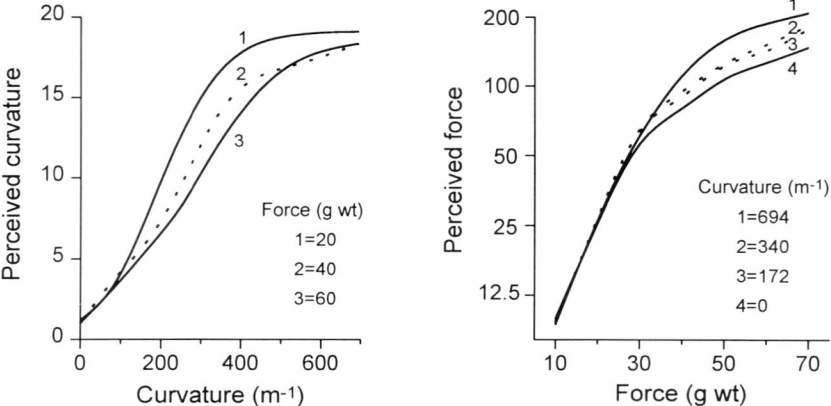

Figure 10.   When presented with stimuli consisting of a mixture of
curvatures and contact forces, human subjects were able to scale either for
curvature (left panel) or for contact force (right panel).

force can be specified quantitatively for all SAIs by the single equation:

$$response=s\phi[1.91-1.62\exp(-0.00243\kappa)]$$

where s is the sensitivity of the fiber, κ is the curvature of the stimulus,
and the constant ϕ has the value 0.792, 1.01 and 1.21 for contact forces
of 10, 15 and 20 g wt respectively.

    The third parameter of the stimulus is its position in the receptive
field. For a single SAI afferent, a change in the position of a sphere in
the receptive field resulted in a change in response. As shown in Figure
11A, the response to the most curved sphere (curvature 694 m^{-1}, radius
1.44 mm) was greatest when the sphere was located at the center of the
receptive field and decreased as the sphere moved further from the
center. When the curvature of the sphere decreased, the receptive field
profile changed in two ways as illustrated for a curvature of 172 m^{-1}
(radius 5.81 mm).   First, the response evoked at the center of the
receptive field decreased; this is a reflection of the decrease in response
seen in Figure 9 as the curvature of the stimulus decreased from 694 m^{-1}
to 172 m^{-1}. Second, for the less curved sphere, the decline in response

with increasing distance from the center of the receptive field was more gradual than for the more curved sphere.

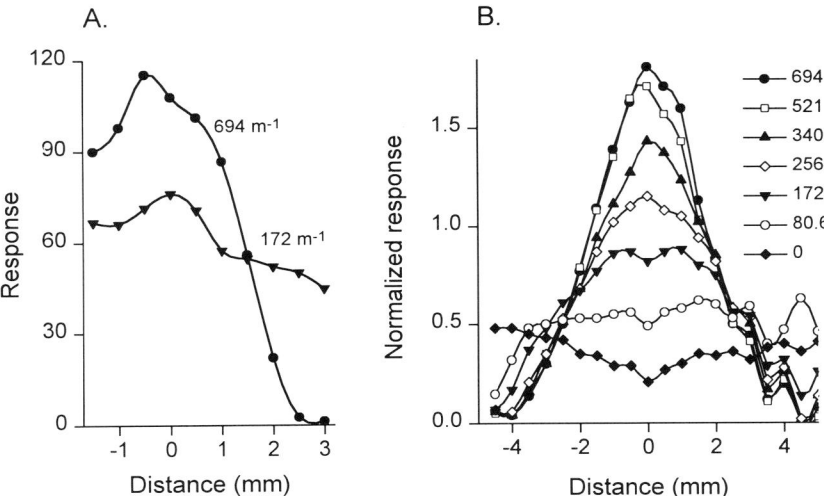

Figure 11. Receptive field profiles shown for a single SAI in A and averaged over the sample of SAIs (after normalizing to eliminate fiber sensitivity) in B. Distance is measured from the center of the receptive field in a direction parallel to the long axis of the finger (y direction). Legend in B shows curvatures in units m^{-1}.

All SAIs exhibited the same characteristic position-response profiles as the fiber shown in Figure 11A except for a scaling factor determined by the sensitivity of the afferent. These profiles are shown in Figure 11B for all 7 curvatures; the profiles were first normalized to eliminate the sensitivity of each individual afferent and then averaged across the sample of SAIs. A decrease in curvature resulted in a decrease in the height of the profile and also resulted in a broader profile. The profiles in Figure 11B are illustrated for the case where the sphere moved away from the receptive field center along a line parallel to the long axis of the finger (y direction). A similar decrease in response resulted when the sphere moved away from the center of the receptive field in the orthogonal direction, that is in a direction perpendicular to the long axis of the finger (x direction).

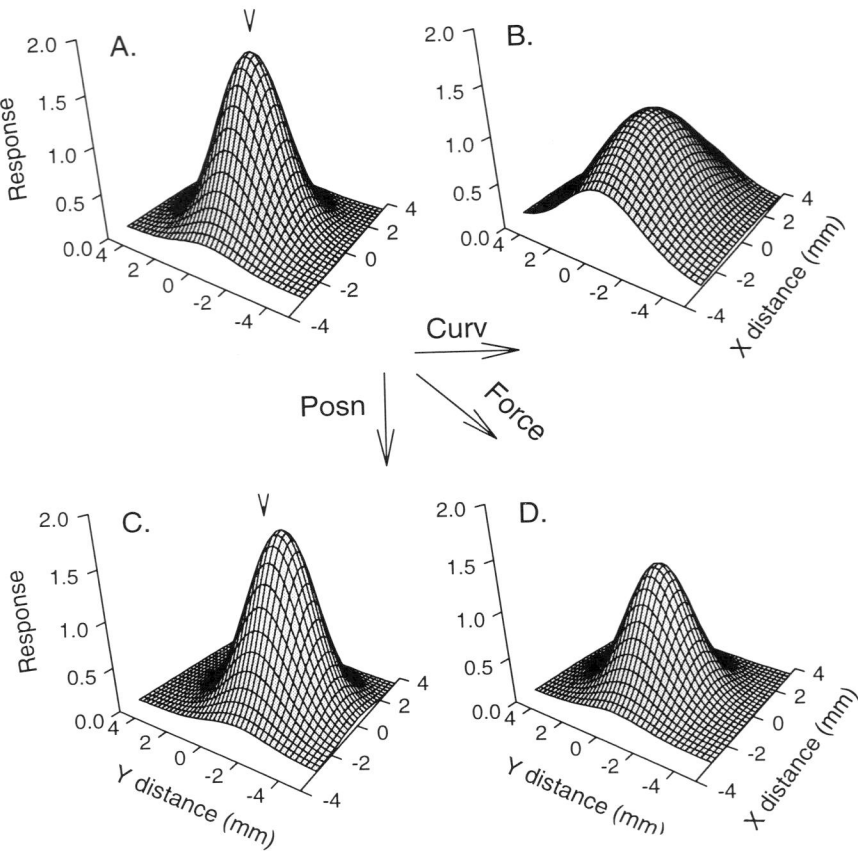

Figure 12.   Changes in the SAI population response when different
parameters of the stimulus change.   From A to B the curvature of the
stimulus changes from 694 m^{-1} to 256 m^{-1}. From A to C the position of the
sphere changes by 1 mm in the y direction (arrowhead shows y=0). From A
to D the contact force decreases. Distances x and y are measured orthogonal
to and parallel to the long axis of the finger respectively.

    The complete receptive field profile of an afferent is defined in two
spatial dimensions. The distances x and y are measured in directions
orthogonal to and parallel to the long axis of the finger respectively; the
origin is at the receptive field center. Since the receptive field profiles
were the same for all SAIs, except for the scaling factor determined by
that afferent's sensitivity, the receptive field profile for any SAI can be
specified quantitatively by the equation:

$$z=a\exp(-bx^2-cy^2)$$

This equation defines the normalized response z at any point in the receptive field with coordinates x and y; the constants a, b, and c depend on the curvature of the sphere.

Receptive field profiles may be interpreted in a more general way. In Figure 11 the responses of a single fiber are shown for various positions of the sphere in the receptive field. The reciprocal interpretation is that the sphere is always located at zero, and that the profiles show responses of a succession of afferents with receptive field centers located at increasing distances from zero. This is possible because all SAIs have the same profiles of normalized responses. Hence these profiles can be interpreted as the profiles of response across the population of active SAIs. The fact that individual SAI responses will also depend on the sensitivity of the fiber is easily dealt with (see later).

The skin is a two-dimensional sheet. When a sphere contacts the skin, a population of SAI afferents spanning the two-dimensional sheet will be activated. This population response is shown for a sphere with a curvature of 694 m^{-1} in Figure 12A, and for a sphere with a curvature of 256 m^{-1} in Figure 12B. Profiles in Figure 11B (for curvatures 694 and 256 m^{-1}) are cross-sections of Figure 12 (A and B respectively) along the line x=0. When the profiles are viewed for both dimensions on the skin it is even more evident that as the curvature or shape of the sphere changes, the shape of the population response changes in correspondence. A more curved sphere (Figure 12A compared to 12B) results in a "more curved" population response (an increase in height and a decrease in width).

A change in contact force also changed the population response but the effect was different from that produced by a change in curvature. The difference is easiest to see by first inspecting profiles in one direction on the finger for a single SAI. In Figure 13 profiles for a typical SAI are shown along a line through the center of the receptive field and parallel to the long axis of the finger. At each point in the receptive field, an increase in contact force increased the response so

A.W. Goodwin

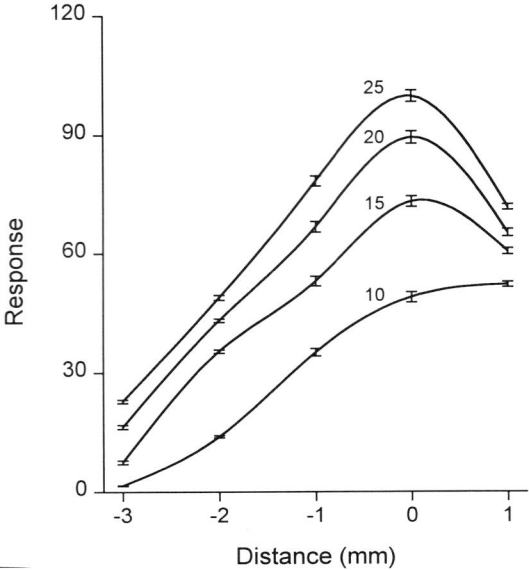

Figure 13. Profiles for a single SAI along a line through the center of the receptive field and parallel to the long axis of the finger.   Responses (mean±SEM, n=8) are shown at 4 contact forces (in units g wt).

that the whole profile was scaled upwards with little change in the shape of the profile.   This was true for all the SAIs, and the predominant effect of force on the population response was a scaling of the profile without a change in shape.   Profiles across both dimensions of the skin illustrated in Figure 12 show this clearly; a decrease in force scales the profile shown in A to that shown in D.

Inspecting Figure 12A, B, and D it is obvious that, although single afferents convey ambiguous information about the shape and force of a contacting sphere, the whole population provides the brain with unambiguous information.   The shape of the sphere is represented by the shape of the population response and the contact force is represented by the overall magnitude of the response.

10. Position of an object on the skin

When we handle an object, our fingers roll over it and move tangentially over it so that the position of the object on the skin changes. Successful manipulation depends on the sensorimotor control system having an accurate knowledge of the object's position. A critical factor that limits the precision of position information is the afferent innervation density. The estimated innervation density of the SAIs is about 0.70 afferents mm^{-2} (Johansson and Vallbo, 1979). Assuming a square mosaic of innervation, the resulting separation between adjacent receptive field centers is 1.2 mm and the sampling theorem predicts an upper limit of resolution of 2.4 mm. A simplistic interpretation would place the resolution of position on the skin at 2.4 mm. However, experiments by Loomis (1979) have shown an ability to localize stimuli with far greater precision. The term hyperacuity has been used for corresponding phenomena in the visual system (Westheimer, 1981).

We measured the capacity of humans to discriminate the positions of spheres contacting the fingerpad. In forced-choice psychophysical experiments, the stimulator shown in Figure 6 was used to present the stimuli passively to the fingerpad (Wheat et al., 1995). Information transmitted to the brain was thus limited to that originating in the cutaneous mechanoreceptors, relayed principally by the SAIs. Two spheres were used. For the less curved sphere (curvature 172 m^{-1}, radius 5.80 mm) the discrimination threshold averaged over 7 subjects was 0.55 mm. For the more curved sphere (curvature 521 m^{-1}, radius 1.92 mm) performance was more acute with a discrimination threshold of 0.38 mm. These thresholds are significantly lower than thresholds predicted from simple considerations such as the sampling theorem. To elucidate the neural mechanisms underlying this behavior, we performed corresponding experiments in the monkey, recording the responses of single SAIs (Wheat et al., 1995).

In Figure 14A, the solid line shows the responses of a single SAI to a sphere (the standard) placed at different positions in the receptive field along a line through the receptive field center and parallel to the long axis of the finger. At each position, a second sphere (the comparison) was located 0.5 mm proximal to the first sphere and elicited the responses shown by the broken line. Using the argument of reciprocity discussed above, these responses may also be viewed as the profiles across a population of SAIs. The shift in position of the second sphere (the comparison) resulted in a corresponding shift in the response profile. For the less curved sphere, illustrated in Figure 14B, the profiles were lower and broader as seen above. Nevertheless, a similar profile shift occurred for the less curved sphere and is most obvious around the skirts.

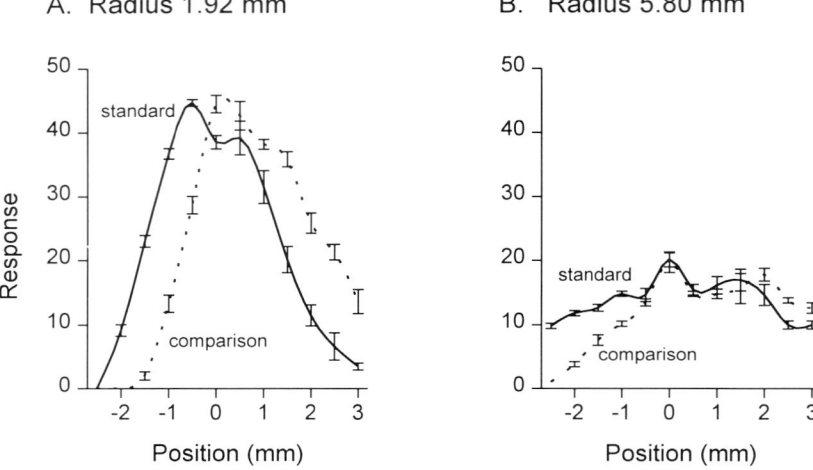

Figure 14. The comparison sphere was positioned 0.5 mm proximal to the standard sphere. Profiles show responses (mean±SD, n=6) along a line through the receptive field center and parallel to the long axis of the finger. Two spheres were used with curvature 521 m⁻¹ (radius 1.92 mm) and 172 m⁻¹ (radius 5.80 mm) respectively.

Stimuli were presented 6 times at each point in Figure 14. Standard deviations shown by the error bars are small, indicating a low variability or high degree of precision in the responses of peripheral nerve fibers. As the difference between the positions of the two spheres (standard and comparison) decreases, the shift between the two profiles will decrease. A point is reached where the shift is no longer discernible within the noise (indicated here by the standard deviation of the responses). Even from this simple figure three facts are evident at a qualitative level. First, the profile shift could provide a basis for the brain to determine a change in the position of the sphere with precision. Second, the response to the more curved sphere is greater than the response to the less curved sphere. This may result in higher signal to noise ratios and explain why the discrimination threshold measured in the human psychophysical experiments was lower for the more curved sphere. Third, the apparent hyperacute performance is probably a result of the relatively broad response profiles. Judgments of whether or not a profile has shifted can utilize information from a number of afferents defining the profile. The shift in the whole profile is far more sensitive than would be predicted from simple considerations of afferent spacing.

When an object contacts the skin, a population of afferents is activated over the two dimensions of the skin. A change in the position of the object is reflected in a shift of the profile of activity across the two dimensions. Figure 12 (A and C) illustrates the change that occurs when a sphere with curvature 694 m^{-1} is shifted by 1 mm along a line parallel to the long axis of the finger (y direction). The arrowheads indicate the position of the origin (x=0, y=0). As emphasized previously, there is ambiguity in the responses of a single afferent since a change in a single fiber's response could be due to a change in curvature, contact force, or position on the skin. However, it is clear from Figure 12 that the position of a sphere on the skin is represented unambiguously in the population response.

11. Extraction of stimulus parameters from population responses

The population response profiles illustrated in Figure 12 provide the basis for the brain to extract the parameters of the stimulus. Any change in the stimulus results in a change in the profile which can be quantified by non-specific measures such as the volume between the two profiles. In human psychophysical experiments where the only difference between two stimuli is a change in a single parameter, for example a change in position, such a non-specific measure could be used. However, in normal manipulation such a measure would provide ambiguous information. For a sphere, the volume between two population profiles could be indicative of a change in position or of a change in curvature or of a change in contact force. As indicated previously there are specific features of the population response that can signal these parameters independently (or largely independently) of the other parameters. Our experiments have enabled us to quantify the profiles, and the next step is an examination of the sorts of computations that the brain could perform on the profiles to extract specific information about the individual stimulus parameters. A few simple illustrations are given below.

The position of a sphere on the skin is evident from the position of the profile in the population. Thus the brain might compute two values from the profile, one giving the x coordinate of the contact point and the other giving the y coordinate. The x and y coordinates of the centroid of the profile are simple to calculate and are obvious candidates for the x and y position of the stimulus. For Figure 15A, profiles were generated for spheres positioned over a 1 mm range along a line parallel to the long axis of the finger. The y component of the centroid was computed and is plotted against the position of the sphere. Although this is a simple calculation, it determines the position of the object on the skin accurately. Moreover, if the calculation is repeated on profiles for a range of curvatures and a range of contact forces, it is found that the centroid is not affected by changes in the curvature of the sphere or by changes in the contact force; the centroid always determines the position of the stimulus.

A measure of the population response that corresponds to its shape will indicate the curvature of the sphere. One possibility for this measure is the second moment of the profile (analogous to the moment of inertia) which is also simple to calculate. Figure 15B, which shows the second moment for the range of curvatures used in our experiments, is strikingly similar to the human psychophysical function shown in Figure 10. This too is a unique measure in that the second moment is an indicator of curvature regardless of the position of the sphere and regardless of its contact force. Finally, measures of the overall activity of the profiles correspond to the contact force of the stimulus and are largely independent of the position or the curvature of the sphere.

The computations explored above demonstrate three relatively simple measures of the population response, each of which varies systematically with changes in one of the three stimulus parameters and is insensitive to changes in the remaining two parameters. This analysis facilitates a quantitative understanding of how the afferent fiber population response provides the unambiguous information evident from the psychophysical experiments. Since we have a quantitative description of the afferent responses, we have been able to

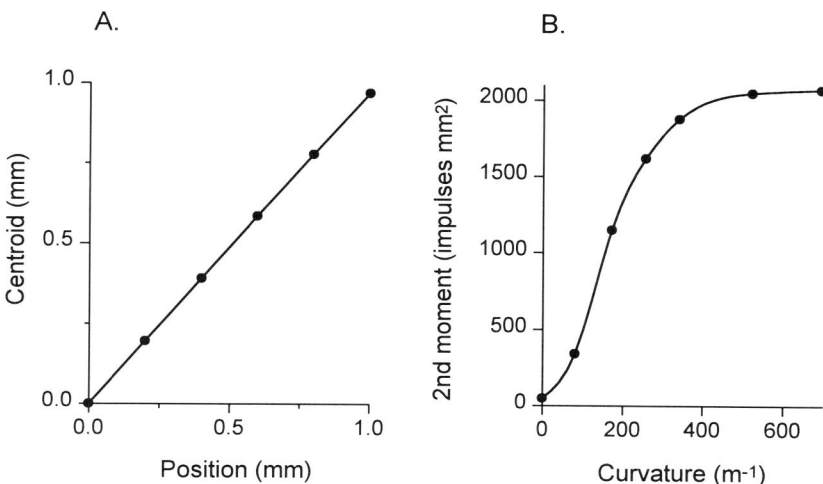

Figure 15. A. The y coordinate of the centroid of the SAI population response is an effective measure for the y coordinate of the position of a sphere on the skin. B. The curvature of the sphere can be determined from the 2nd moment of the population response.

develop a computer model to investigate such measures. Another benefit of the model is that it allows us to examine the effects on resolution of factors like innervation density and noise at various levels of the nervous system. The fact that different SAIs have different sensitivities can be accounted for easily by allowing the sensitivity of individual afferents to vary randomly with the same statistical distribution as that measured in our experiments.

12. Other afferent types

SAIs respond vigorously to our stimuli and have a high innervation density in the fingerpads; they are the most likely source of the majority of high resolution information about the stimuli. Although the FAIs also have a high innervation density, 50% of them did not respond to the stimuli, and those that did respond showed inconsistent force and curvature effects. They may provide information about the position of the object on the skin (Wheat et al., 1995). None of the FAIIs responded. The SAIIs have not been studied since they are not found in the glabrous skin of the monkey. In humans, the known receptive field characteristics of SAIIs, and their low innervation density, make them an unlikely source of information about position or curvature (Vallbo and Johansson, 1984b; Johansson and Vallbo, 1979b). It is possible that they provide some information about contact force.

13. Scanned stimuli

Normally, active touch is used to explore objects and to manipulate them. This results in some motion of the object across the skin, the degree of movement varying with the nature of the particular task. Some of this motion will be of a rolling nature and some will be of a translational or scanning nature. Here the population response profiles will be moving within the population of afferents. The movement itself, and factors like the asymmetry of skin mechanics, will introduce additional complications. LaMotte et al. (1994) investigated the responses of SAIs and RAIs when cylinders, spheres and ellipsoids were scanned over the fingerpads of monkeys. Their results indicate that the shapes of these objects can be clearly identified in the population responses.

14. Conclusion

If we execute a simple task like grasping a button with our eyes closed, we have no difficulty determining the position of the button on the fingertips. We are also aware of the force we are using to grasp the button and can determine its local curvature. Information of this nature is available to us during the countless manipulations we make in our daily lives. Even if we do not use these signals consciously, the sensorimotor control system depends on them for precise movements of the fingers and thumb. Detailed information about the position, shape and contact force of handled objects is relayed by the cutaneous mechanoreceptive afferents innervating the fingers and thumb. Although the signals from single fibers are ambiguous, each of the three parameters is represented unambiguously in the responses of populations of fibers. It is relatively simple to extract precise details about the stimulus from these population responses.

Acknowledgments

This research was supported by a grant from the National Health and Medical Research Council of Australia.

References

Asada, H. and Brady, M. (1986) The curvature primal sketch. IEEE Transactions on Pattern Analysis and Machine Intelligence, 8, 2-14.

Bell, J., Bolanowski, S. and Holmes, M.H. (1994) The structure and function of Pacinian corpuscle: a review. Progress in Neurobiology, 42, 79-128.

Burke, D., Gandevia, S.C. and Macefield, G. (1988) Responses to passive movement of receptors in joint, skin and muscle of the human hand. Journal of Physiology, 402, 347-361.

Cauna, N. (1956) Nerve supply and nerve endings in Meissner's corpuscles. American Journal of Anatomy, 99, 315-350.

Cauna, N. and Mannan, G. (1958) The structure of human digital Pacinian corpuscles (*Corpuscula lamellosa*) and its functional significance. Anatomy, 92, 1-20.

Chambers, M.R., Andres, K.H., Duering, M.V. and Iggo, A. (1972) The structure and function of the slowly adapting type II mechanoreceptor in hairy skin. Quarterly Journal of Experimental Physiology, 57, 417-445.

Chouchkov, C.N. (1973) The fine structure of small encapsulated receptors in human digital glabrous skin. Journal of Anatomy, 114, 25-33.

Darian-Smith, I. (1984) The sense of touch: performance and peripheral neural processes. In Brookhart, J.M., Mountcastle, V.B., Darian-Smith, I and Geiger, S.R. (eds.) Handbook of Physiology - The Nervous System III, American Physiological Society, Bethesda, pp739-788.

Darian-Smith, I. and Kenins, P. (1980) Innervation density of mechanoreceptive fibres supplying glabrous skin of the monkey's index finger. Journal of Physiology, 309, 147-155.

Edin, B.B. (1990) Finger joint movement sensitivity of non-cutaneous mechanoreceptor afferents in the human radial nerve. Experimental Brain Research, 82, 417-422.

Edin, B.B. and Vallbo, Å.B. (1990) Muscle afferent responses to isometric contractions and relaxations in humans. Journal Neurophysiology, 63, 1307-1313.

Edin, B.B. and Abbs, J.H. (1991) Finger movement responses of cutaneous mechanoreceptors in the dorsal skin of the human hand. Journal of Neurophysiology, 65, 657-670.

Gandevia, S.C., McCloskey, D.I. and Burke, D. (1992) Kinaesthetic signals and muscle contraction. Trends in Neuroscience, 15, 62-65.

Gibson, J.J. (1962) Observations on active touch. Psychological Review, 69, 477-491.

Goodwin, A.W., John, K.T. and Marceglia, A.H. (1991) Tactile discrimination of curvature by humans using only cutaneous information from the fingerpads. Experimental Brain Research, 86, 663-672.

Goodwin, A.W. and Wheat, H.E. (1992a) Human tactile discrimination of curvature when contact area with the skin remains constant. Experimental Brain Research, 88, 447-450.

Goodwin, A.W. and Wheat, H.E. (1992b) Magnitude estimation of force when objects with different shapes are applied passively to the fingerpad. Somatosensory and Motor Research, 9, 339-344.

Goodwin, A.W., Browning, A.S. and Wheat, H.E. (1995) Representation of curved surfaces in responses of mechanoreceptive afferent fibers innervating the monkey's fingerpad. Journal of Neuroscience, 15, 798-810.

Halata, Z. (1975) The mechanoreceptors of the mammalian skin. Ultrastructure and morphological classification. Advances in Anatomy, Embryology and Cell Biology, 50, 1-77.

Hoffman, D.D. and Richards, W.A. (1984) Parts of recognition. Cognition, 18, 65-96.

Hulliger, M., Nordh, E., Thelin, A.-E. and Vallbo, A.B. (1979) The responses of afferent fibres from the glabrous skin of the human hand during voluntary finger movements in man. Journal of Physiology, 291, 233-249.

Johansson, R.S. and Vallbo, A.B. (1979) Tactile sensibility in the human hand: relative and absolute densities of four types of mechanoreceptive units in glabrous skin. Journal of Physiology 286, 283-300.

Johansson, R.S., Landstrom, U. and Lundstrom, R. (1982) Sensitivity to edges of mechanoreceptive afferent units innervating the glabrous skin of the human hand. Brain Research, 244, 27-32.

Kappers, A.M.L., Koenderink, J.J. and Lichtenegger, I. (1994) Haptic identification of curved surfaces. Perception and Psychophysics, 56, 53-61.

Katz, D. (1989) The World of Touch. Translated by L.E. Krueger, Lawrence Erlbaum, Hillsdale.

Koenderink, J.J. and van Doorn, A.J. (1992) Surface shape and curvature scales. Image and Vision Computing, 10, 557-564.

LaMotte, R.H. and Srinivasan, M.A. (1987a) Tactile discrimination of shape: responses of slowly adapting mechanoreceptive afferents to a step stroked across the monkey fingerpad. Journal of Neuroscience, 7, 1655-1671.

LaMotte, R.H. and Srinivasan, M.A. (1987b) Tactile discrimination of shape: Responses of rapidly adapting mechanoreceptive afferents to a step stroked across the monkey fingerpad. Journal of Neuroscience, 7, 1672-1681.

LaMotte, R.H., Srinivasan, M.A., Lu, C. and Klusch-Petersen, A. (1994) Cutaneous neural codes for shape. Canadian Journal of Physiology and Pharmacology, 72, 498-505.

Loomis, J.M. (1979) An investigation of tactile hyperacuity. Sensory Processes, 3, 289-302.

Lord, E.A. and Wilson, C.B. (1984) The mathematical description of shape and form. Ellis Horwood Limited, Chichester.

Mackel, R. (1988) Conduction of neural impulses in human mechanoreceptive cutaneous afferents. Journal of Physiology, 401, 597-615.

Marr, D. and Ninshihara, H.K. (1978) Representation and recognition of the spatial organization of three-dimensional shapes. Proceedings of the Royal Society London, B200, 269-294.

Marshall, S. (1989) Review of shape coding techniques. Image and Vision Computing, 7, 281-294.

McCloskey, D.I. (1991) Corollary discharges: motor commands and perception. In Brookhart, J.M., Mountcastle, V.B., Brooks, V.B. and Geiger, S.R. (eds.) Handbook of Physiology - The Nervous System II, American Physiological Society, Bethesda, pp 1415-1447.

Miller, M.R., Ralston, H.J. and Kasahara, M. (1958) The pattern of cutaneous innervation of the human hand. American Journal of Anatomy, 102, 183-197.

Phillips, J.R. and Johnson, K.O. (1981a) Tactile spatial resolution. II. Neural representation of bars, edges, and gratings in monkey primary afferents. Journal of Neurophysiology, 46, 1192-1203.

Phillips, J.R. and Johnson, K.O. (1981b) Tactile spatial resolution. III. A continuum mechanics model of skin predicting mechanoreceptor responses to bars, edges, and gratings. Journal of Neurophysiology, 46, 1204-1225.

Richards, W., Dawson, B. and Whittington, D. (1986) Encoding contour shape by curvature extrema. Journal of the Optical Society of America, 3, 1483-1491.

Roland, P.E. and Mortensen, E. (1987) Somatosensory detection of microgeometry, macrogeometry and kinesthesia in man. Brain Research Reviews, 12, 1-42.

Srinivasan, M.A. (1989) Surface deflection of primate fingertip under line load. Journal of Biomechanics, 22, 343-349.

Srinivasan, M.A. and LaMotte, R.H. (1987) Tactile discrimination of shape: responses of slowly and rapidly adapting mechanoreceptive afferents to a step indented into the monkey fingerpad. Journal of Neuroscience, 7, 1682-1697.

Srinivasan, M.A. and LaMotte, R.H. (1991) Encoding of shape in the responses of cutaneous mechanoreceptors. In Franzen O. and Westman J. (eds) Information Processing in the Somatosensory System, MacMillan Press, London, pp 59-69.

Talbot, W.H., Darian-Smith, I., Kornhuber, H.H. and Mountcastle, V.B. (1968) The sense of flutter-vibration: comparison of the human capacity with response patterns of mechanoreceptive afferents from the monkey hand. Journal of Neurophysiology, 31, 301-334.

Vallbo, A.B. and Johansson, R.S. (1984) Properties of cutaneous mechanoreceptors in the human hand related to touch sensation. Human Neurobiology, 3, 3-14.

Vierck, C.J. (1979) Comparison of punctate, edge and surface stimulation of peripheral, slowly adapting, cutaneous afferent units of cat. Brain Research, 175, 155-159.

Westheimer, G. (1981) Visual Hyperacuity. Progress in Sensory Physiology, 1, 1-30.

Wheat, H.E., Goodwin, A.W. and Browning, A.S. (1995) Tactile resolution: peripheral neural mechanisms underlying the human capacity to determine positions of objects contacting the fingerpad. Journal of Neuroscience, 15, 5582-5595.

Neural Aspects of Tactile Sensation
J.W. Morley (Editor)
© 1998 Elsevier Science B.V. All rights reserved

The Signalling of Touch, Finger Movements and Manipulation Forces by Mechanoreceptors in Human Skin

V. G. Macefield

Thirty years ago, research into tactile sensibility in human subjects was limited to psychophysical studies. Such an approach continues to be immensely useful, but in 1968, Karl-Erik Hagbarth and Åke Vallbo, at the University Hospital in Uppsala, Sweden, published a series of papers describing their multi-unit recordings from cutaneous and muscle afferents in awake human subjects, using tungsten microelectrodes inserted through the skin into cutaneous or motor fascicles of peripheral nerves (Hagbarth and Vallbo, 1968; Vallbo and Hagbarth, 1968). In the next two years these researchers published their first reports on the firing properties of single primary afferent axons supplying mechanoreceptors in the skin and muscles (Hagbarth and Vallbo, 1969; Knibestol and Vallbo, 1970; Vallbo, 1970a; Vallbo, 1970b). The technique of "microneurography" was born, and has since been applied to detailed analyses of the discharge behaviour of single mechanosensitive endings supplying the face and teeth (Johansson et al., 1988a; Johansson et al., 1988b; Trulsson et al., 1992), and the digital joints (Burke et al., 1988; Macefield et al., 1990; Edin, 1990), as well as single motor axons supplying skeletal muscles (Gandevia et al., 1990; Macefield et al., 1993). Although tungsten microelectrodes record preferentially from large-diameter axons, unitary recordings have also been made from non-myelinated axons (C-fibres) in human peripheral and cranial nerves, both afferent (Ochoa and Torebjork, 1989; Torebjörk and Ochoa, 1990; Nordin, 1990; Yarnitsky et al., 1992; Vallbo et al., 1993; Simone et al., 1994), and efferent (Hallin and Torebjörk, 1974; Macefield et al., 1994; Macefield and Wallin, 1996).

The tactile system subserves not only the sense of "touch" in the broadest interpretation of the word, which implies that a stimulus generates a perceptive quality, but is also important in motor control, which - depending on the task - may or may not require conscious attention. The technique of microneurography has allowed us to examine how primary afferent fibres convey sensory information to the central nervous system, and to interpret changes in motor behaviour as a consequence of the pattern of afferent activity. In this chapter I shall review what we have learnt from microneurographic recordings from single tactile afferents in conscious human subjects. Most of the emphasis will be on afferents supplying the glabrous skin, as the volar aspect of the hand is the site most frequently studied. The material will be presented in five sections: the first dealing with the capacities of cutaneous afferents in glabrous skin to signal the various mechanical parameters encountered during "touch", the second with tactile afferents in non-glabrous skin, the third with signalling movements of the fingers, and the fourth with the higher forces associated with manipulative actions. The final section will address the issue of sensory specificity, as elaborated by studies examining the perceptual responses of subjects to selective stimulation of single sensory axons.

1. Properties of tactile afferents in human glabrous skin

1.1. Receptor types

Histologically, four types of specialized mechanoreceptor terminal can be identified in human glabrous skin, two located superficially and two deeper (Miller et al., 1958). In the upper layers of skin there are groups of expanded disc-like endings that arise from branched axons and which are closely associated with specialized cells in the basal layer of the epidermis (Merkel cell-neurite complexes, Fig. 1); within the intradermal papillae lay elipsoidal encapsulated endings (Meissner's corpuscles), the long axis of which is oriented normal to the skin, which are supplied by 2-6 myelinated axons (Darian-Smith, 1984).

In the subpapillary dermis one finds the encapsulated Ruffini and Pacinian corpuscles, both endings originating from a single axon. The Ruffini corpuscle, which is morphologically similar to the Golgi tendon organ, is oriented with its long axis in the plane of the skin and forms mechanical linkages with the longitudinally arranged collagen fibres that course through the dermis (Darian-Smith, 1984). The Pacinian corpuscle, which is composed of concentric lamellae around a central core, is situated in deeper layers of the dermis and subcutaneous tissues. In addition to these specialized endings, free nerve endings are found within the epidermis and dermis.

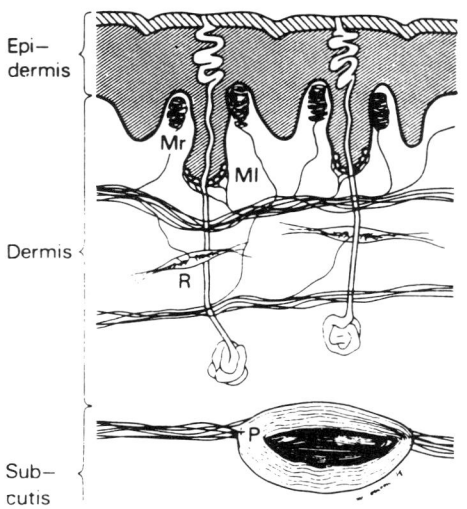

Figure 1. Histological representation of the types of mechanoreceptor seen in human glabrous skin. In the upper layers are the Meissner corpuscles (Mr) and the Merkel cell neurite complexes (Ml); the Ruffini endings (R) and Pacinian corpuscles (P) are located deeper. Reproduced with permission from Johansson and Vallbo (1983).

In support of the histological findings, microelectrode recordings from the median and ulnar nerves have also revealed the existence of four classes of low-threshold mechanosensitive afferent in the glabrous skin of the human hand: two classes of afferent that adapt rapidly to a sustained indentation of the skin ("fast-adapting") - types FA I and FA II - and two classes of slowly adapting afferent - SA I and SA II. These have been classified according to their responses to sustained indentation and to the sizes of their receptive fields: type I afferents possess small, well-defined receptive fields, whereas the receptive fields of the type II afferents are large with borders that are difficult to delineate.

The most common class encountered in recordings from the median nerve, which supplies most of the glabrous skin of the hand, is the FA I (43%), followed by the SA I (25%), SA II (19%) and FA II (13%) classes (Johansson and Vallbo, 1979b). It is believed that the FA I afferent (also referred to as "RA" or "QA") supplies the Meissner corpuscle, and the SA I afferent the Merkel cell-neurite complex

(Johansson, 1978; Johansson and Vallbo, 1983), based on their behavioural similarities with afferents recorded in the cat and monkey (Darian-Smith, 1984). The receptors belonging to the FA II ("PC") and SA II afferents are believed to be the Pacinian corpuscle and Ruffini ending, respectively (Johansson, 1978; Johansson and Vallbo, 1983).

In addition to these myelinated afferents, the glabrous skin areas of the hand and face (forehead) are supplied by non-myelinated afferents, as judged by conduction velocities within the C-fibre range. However, whereas those of the hand respond only to noxious mechanical and thermal stimuli (Ochoa and Torebjork, 1989; Torebjörk and Ochoa, 1990; Yarnitsky et al., 1992), non-myelinated afferents travelling in the supraorbital nerve (which supplies the forehead) can be divided into two types: one that responds only to noxious stimuli, and another that responds to weak mechanical stimuli (Nordin, 1990).

1.2. General features of microelectrode recordings from human nerves

The tungsten microelectrodes used for percutaneous recordings from human nerves typically have a shaft diameter of 200 mm and a tapered point that is mechanically or electrolytically sharpened to a diameter of a few microns. The majority of recordings from large-diameter axons have positive-going spikes, which reflects impalement of the myelin sheath by the electrode tip in the internode region (Vallbo, 1976; Brink and Mackel, 1993; Inglis et al., 1996). Negative-going spikes, indicating proximity of the electrode tip to a node of Ranvier, occur in only a few percent of unitary recordings (Vallbo, 1976; Brink and Mackel, 1993; Inglis et al., 1996).

Close examination of spike morphology reveals that 75% of all single-unit recordings from myelinated axons exhibit a secondary peak either on the rising or falling phase of the spike (Vallbo, 1976; Inglis et al., 1996). Whereas the major component of the spike represents the inward currents occuring at the nearby node of Ranvier, the secondary peak reflects action currents at the next node. This is illustrated in Fig. 2B. The significance of this observation is that the majority of unitary spikes recorded from microelectrodes inserted into human peripheral nerves conduct past the electrode tip, i.e. the recorded sensory signals progress towards the central nervous system unimpeded by the presence of the microelectrode (Vallbo, 1976; Inglis et al., 1996).

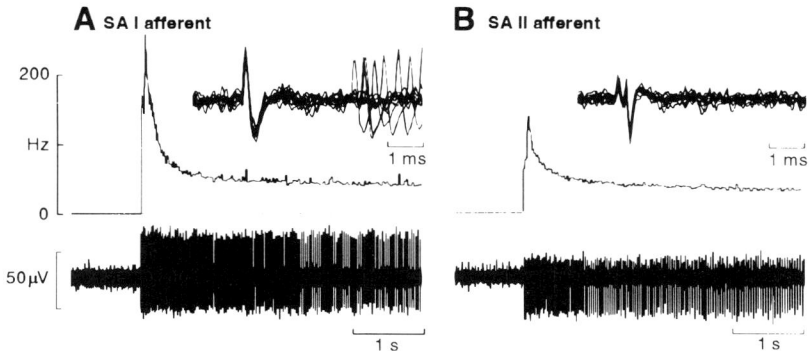

Figure 2. Nerve and instantaneous frequency responses of two slowly-adapting afferents in the tip of the index finger to a constant force stimulus. Spikes generated during the initial phase of the response are superimposed in the insets.

By contrast with spikes recorded from myelinated axons, the initial phase of spikes recorded from single C-fibres in human nerves is always negative (Torebjörk and Ochoa, 1990; Nordin, 1990; Yarnitsky et al., 1992; Vallbo et al., 1993; Simone et al., 1994; Macefield et al., 1994; Macefield and Wallin, 1996).

1.3. Firing properties of tactile afferents in human glabrous skin

The type I class of rapidly-adapting afferent (FA I) can be activated by discrete punctate stimuli in a small, well-defined area of glabrous skin. These afferents are particularly sensitive to light stroking across the skin, responding to local shear forces and incipient or overt slips within the receptive field. This is illustrated in Fig. 3, in which a hand-held force transducer was moved slowly across the skin.

The type II rapidly-adapting afferents (FA II), like their PC counterparts in experimental animals, are exquisitely sensitive to brisk mechanical transients. Characteristically, FA II afferents can be stimulated by tapping over areas remote from the site of maximal mechanosensitivity (Fig. 4B). Unlike the FA I afferents, FA II afferents respond vigorously to blowing over the receptive field (Fig 4A), responding to the fricative quality of the airflow generated by the experimenter blowing through pursed lips onto the receptive field area (they do not respond when blowing through a straw, for example). Instantaneous firing rates are typically higher for the FA II afferents than for the FA I afferents.

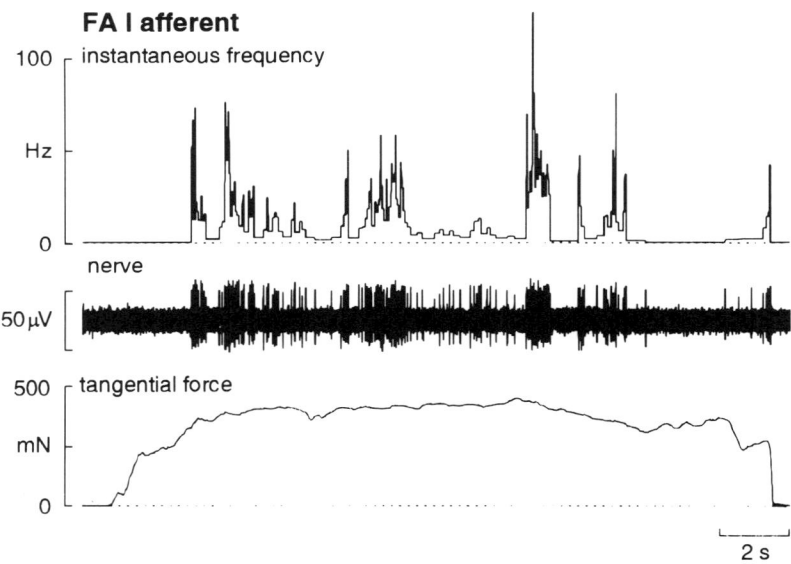

Figure 3. Responses of an FA I afferent in the distal phalanx of the ring finger to stroking across the receptive field.

Slowly-adapting type I afferents (SA I) characteristically have a high dynamic sensitivity to indentation stimuli applied to a discrete area, and often respond with an off-discharge during release. Fig. 5 shows the response of an SA I afferent located over the MCP joint of the index finger to weak compressive forces applied over the receptive field. Superimposed on the instantaneous frequency record is a linear regression calculated between force, its first time-derivative, and the firing rate. Both static and dynamic components of the stimulus are represented in the response.

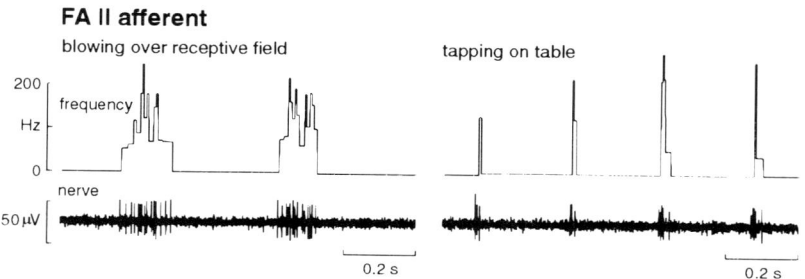

Figure 4. Responses of an FA II afferent in the proximal phalanx of the index finger to blowing over the receptive field (A) and tapping on the table supporting the hand (B).

While the slowly-adapting type II afferents do respond to forces applied normal to the skin, a unique feature of the SA II afferents is their capacity to respond also to lateral skin stretch. Many possess directional sensitivity, the discharge of some afferents increasing with stimuli applied in certain directions, but decreasing in others (Knibestol and Vallbo, 1970; Knibestol, 1975; Johansson, 1978). Figure 6 shows the responses of a SA II afferent located in the tip of the thumb to skin stretch stimuli applied at three different rates over the metacarpophalangeal joint of the thumb - some 60 mm from the centre of the afferent's receptive field. It is apparent that SA II afferents do possess dynamic sensitivity, although little of the first time-derivative of the force signal is reflected in the linear regression shown superimposed on the instantaneous frequency record. Because of this lower dynamic sensitivity, peak firing rates are typically lower for the SA II afferents than for the SA I afferents (*see* Fig. 2).

The representation of SA II afferents in a recording sample is very much dependent on the methods used to search for tactile afferents in different investigations: 13% (Phillips et al., 1992), 18% (Burke et al., 1988), 19% (Johansson and Vallbo, 1979b), 32% (Hulliger et al., 1979b), 51% (Macefield et al., 1990). The high representation of this class in the latter study can be attributed to the routine use of skin stretch

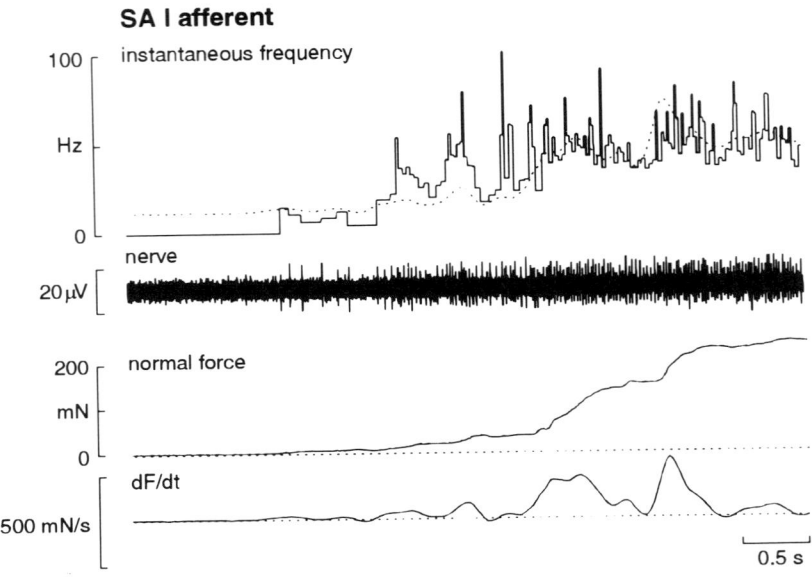

Figure 5. Responses of an SA I afferent located over the MCP joint of the index finger to an increasing force applied normal to the skin. The first time derivative (dF/dt) is shown in the bottom trace.

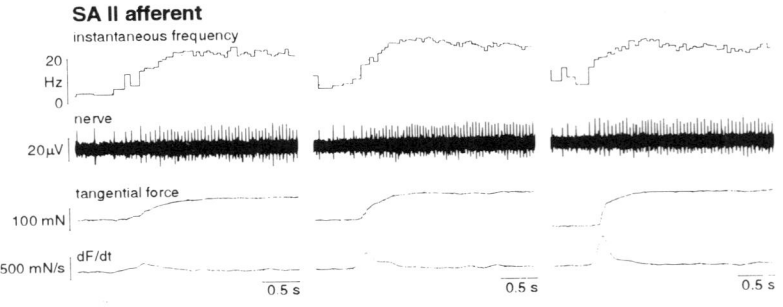

Figure 6. Responses of an SA II afferent located in the tip of the thumb to remote skin stretch applied at the base of the thumb. Skin stretch was applied at three rates, indicated by the first time derivative (dF/dt).

and joint rotation in the searching procedure. A small proportion of SA II afferents are spontaneously active at rest, presenting a characteristically regular discharge. Indeed, regularity of discharge is one of the features that is used to classify SA II endings in cat hairy skin (Chambers et al., 1972), and is used to differentiate between SA I and SA II afferents in human skin. Interestingly, using this measure, SA II afferents are not found in the glabrous skin of the cat footpad, although they are present in the hairy skin (Ferrington, 1985). Likewise, afferents with properties of the SA II afferents have not been encountered in recordings from the glabrous skin of the monkey hand (Darian-Smith, 1984).

As noted above, mechanosensitive non-myelinated afferents of the face can be classified into two groups: those responding only to noxious mechanical stimuli and those responding to gentle tactile stimulation. The latter group respond in a slowly-adapting fashion to forces applied normal to the skin, but unlike the slowly-adapting myelinated afferents, typically generate a sustained 'off-discharge' following a brisk percussive stimulus (Fig. 7). Stroking the skin is a particularly effective stimulus: peak firing rates of 60-100 Hz can be achieved by light stroking of the forehead, the slower the stimulus the greater the response (Nordin, 1990).

1.4. Receptive fields

The type I tactile afferents have small circular or ovoid receptive fields with distinct borders, and have a higher innervation density in the tips of the digits than more proximally within the hand (Johansson and Vallbo, 1979b); this same proximo-distal increase in density of the type I afferents has also been observed in the monkey hand (Darian-Smith and Kenins, 1980). In addition to this density gradient, receptive fields are smaller on the distal phalanx of the fingers than on the palm: for the FA I afferents mean field areas on the distal phalanx are 39.3 ± 7.5 (SE) mm2 and 75.3 ± 24.1 mm2 on the palm (Knibestol, 1973); for the SA I afferents the corresponding values are 20.1 ± 3.1 mm2 and 87.8 ± 28.4 mm2 for the SA I afferents (Knibestol, 1975).

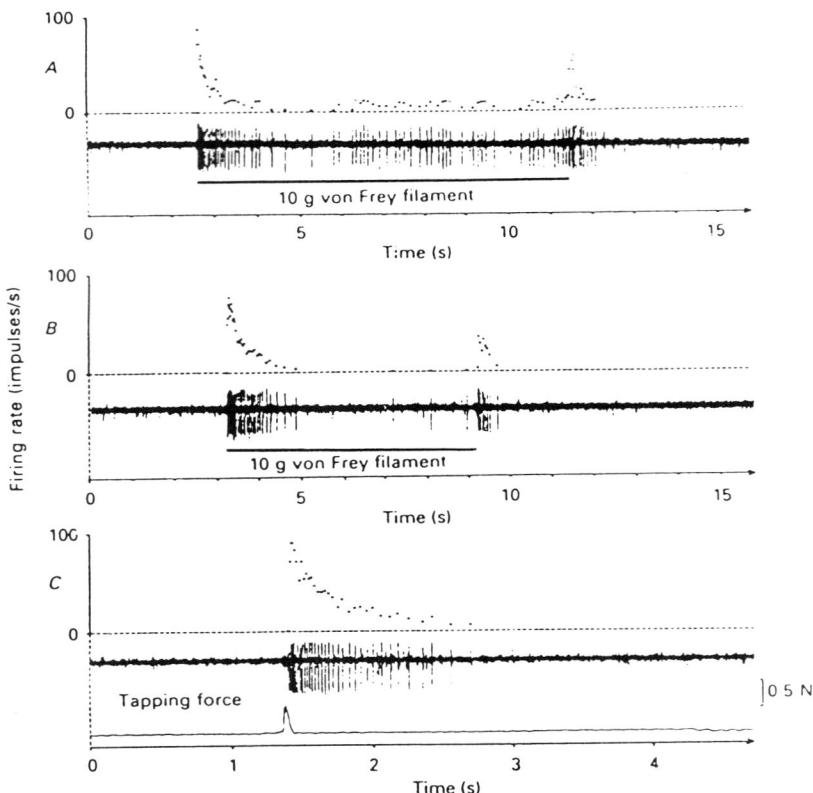

Figure 7. Responses of a low-threshold non-myelinated cutaneous afferent in the forehead to mechanical stimulation. Reproduced with permission from Nordin (1990).

When receptive fields are mapped with a servo-controlled stimulator that moves across the skin, the iso-sensitivity profiles reveal several small zones of maximal sensitivity within the receptive field area: 12-17 such zones for the FA I afferents but only 4-7 for the SA I afferents (Johansson, 1978). An example is shown in Fig. 8. Scanning the field area with an embossed dot array also revealed several zones of maximal sensitivity within the field, although the size of the dots limited the spatial resolution obtained by this method: 5-8 zones were found within the FAI fields and 3-5 zones were found within the SAI

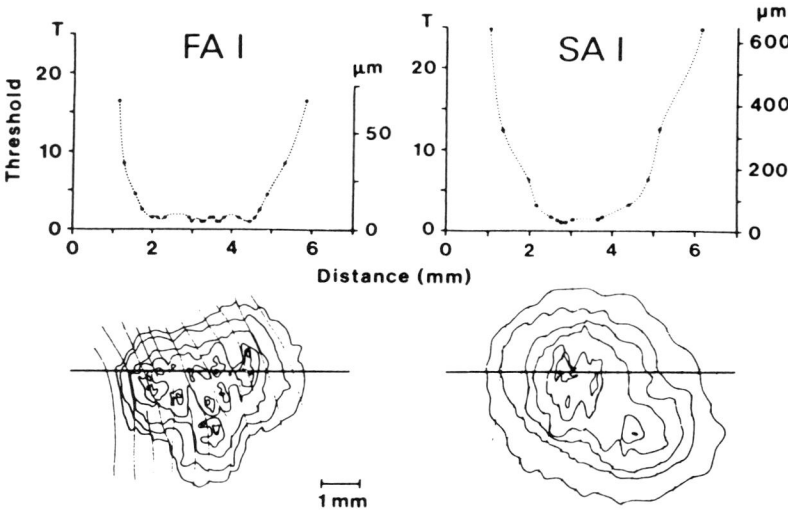

Figure 8. Receptive field maps of an FA I and SA I afferent measured by controlled indentations with a hemispherical probe (diameter 0.4 mm). Zones in which the afferents responded to a particular indentation are connected by isosensitivity lines. In the upper panels are shown the variation in threshold and indentation amplitude across the middle of the receptive field. Reproduced with permission from Johansson and Vallbo (1983).

fields (Phillips et al., 1992). It is believed that these "hot-spots" represent the individual Meissner corpuscles supplied by a single FA I afferent, and the Merkel cell neurite complexes innervated by each SA I axon (Johansson, 1978; Phillips et al., 1992).

By contrast to the type I afferents, the type II afferents have large, poorly-defined receptive fields with obscure borders and, usually, a single zone of maximal sensitivity (Fig. 9). In addition, their density is fairly uniform throughout the hand, with the exception of a specific representation of SA II afferents associated with the nail beds (Knibestol, 1975; Johansson and Vallbo, 1979b). As noted above, both classes of type II afferent can respond to stimuli applied outside their receptive field: FA II afferents respond to brisk mechanical stimuli and SA II afferents respond to lateral skin stretch.

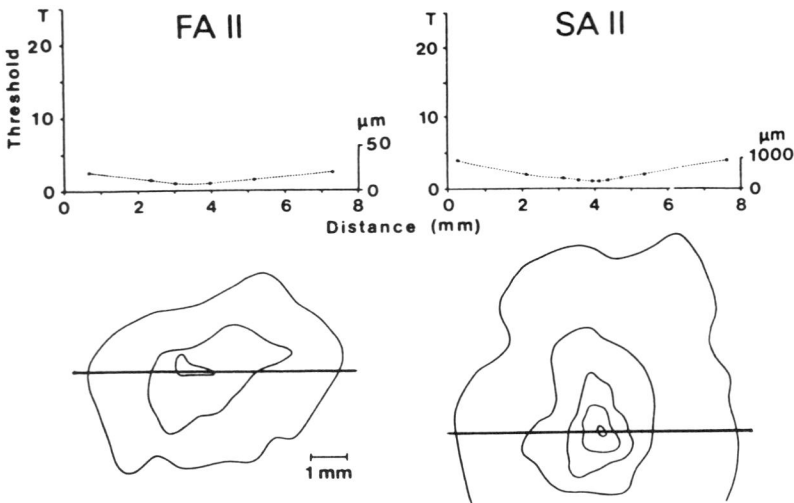

Figure 9. Receptive field maps of an FA II and SA II afferent measured by controlled indentations with a hemispherical probe. Same format as Fig. 8. Reproduced with permission from Johansson and Vallbo (1983).

With respect to the non-myelinated mechanoreceptors, Nordin (1990) found that receptive fields of the low-threshold C-fibres in the glabrous skin of the face (forehead) were similar in size to those of the FA I and SA I afferents on the proximal parts of the hand (85 ± 4 mm2). This suggests that free nerve endings may contribute significantly to tactile sensibility, and that this class of low-threshold mechanoreceptor cannot be ignored in tactile physiology.

1.5. Mechanical thresholds

Psychophysical measurements have shown that the volar aspects of the fingers and the edges of the palm have the highest acuity for tactile detection: when assessed using controlled indentations of the skin the median detection threshold in these areas is 11.2 mm, compared with 36.0 mm in the centre of the palm (Johansson and Vallbo, 1979a). Quantitative studies by Vallbo and Johansson have shown that the mechanical thresholds of each class of afferent are fairly uniform throughout the glabrous area of the hand, but that the rapidly adapting receptors have the lowest thresholds: median indentation thresholds are 9.2 mm for the FA II and 13.8 mm for the FA I afferents

(Johansson and Vallbo, 1979a). SA I afferents have a considerably higher median threshold (56.5 mm), with the threshold for the SA II endings (331 mm) being the highest (Johansson and Vallbo, 1979a). The same is true when force thresholds are measured (using calibrated von Frey hairs): median thresholds are lowest for the FA II (0.54 mN) and FA I (0.58 mN) afferents, and highest for the SA I (1.3 mN) and SA II (7.5 mN) afferents (Johansson et al., 1980).

The higher mechanical threshold of the SA II afferents to punctate stimulation fits with the known properties of Ruffini endings. For instance, mechanoreceptors in the posterior capsule of the cat knee joint, identified as Ruffini organs, respond to strains applied in the plane of the tissue but have very high thresholds to compressive stresses applied perpendicular to the tissue (Eklund and Skogland, 1960; Grigg and Hoffman, 1982; Grigg and Hoffman, 1984); indeed, any responses evoked by compressive forces can be explained by the resultant increases in tensile strain in the immediate receptor environment (Grigg and Hoffman, 1996). Similarly, human joint receptors in the digits, also believed to be Ruffini endings, respond to stresses produced by joint rotation but have very high thresholds to indentation (Burke et al., 1988; Macefield et al., 1990; Macefield, 1995). As noted above, Ruffini endings in human skin respond to planar skin stress (Knibestol and Vallbo, 1970; Knibestol, 1975; Johansson, 1978), and there is evidence to suggest that their sensitivity to stresses applied in the plane of the skin may be higher than that to stresses applied perpendicular to the skin (Macefield et al., 1996).

Von Frey thresholds for the non-myelinated afferents supplying the glabrous skin of the face are similar to the those of the SA I afferents supplying the hand (0.6-2.3 mN; Nordin, 1990), providing further support for a potentially important role of this class of afferent in tactile sensibility.

1.6. Receptor encoding of punctate stimuli

Stimulus-response relations have been calculated for each class of tactile afferent in the glabrous skin of the hand during indentation of the skin with a servo-controlled stimulator (Knibestol, 1973; Knibestol, 1975). For the majority of FA I and FA II afferents the relationship between discharge frequency and indentation velocity is best described by a hyperbolic log tangent function, whereas for others a logarithmic function provides the best fit (Knibestol, 1973). These relationships are illustrated in Fig. 10. For the slowly-adapting afferents, a log tangent function also adequately describes the relation between indentation amplitude and the total number of impulses generated in one second (which includes both the dynamic and static phases of the response), whereas a power function gives the best fit for the majority of afferents when only the mean frequency generated during the static phase is considered (Knibestol, 1975).

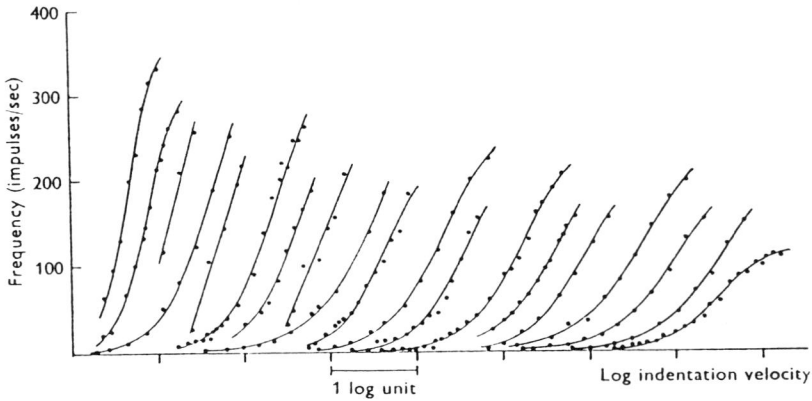

Figure 10. Stimulus-response functions for 19 FA I afferents during controlled indentations. Reproduced with permission from Knibestol (1973).

1.7. Sensitivity to vibration

Sensitivity to vibratory stimuli is not limited to the rapidly-adapting tactile afferents; both classes of slowly-adapting afferent can also follow sinusoidal stimuli up to fairly high frequencies (Knibestol and Vallbo, 1970). However, there is evidence for a differential sensitivity to vibration across each class. Johansson et al. (1982a) used controlled sinusoidal stimulation of identifed receptive fields by perpendicular indentations with a cylindrical probe. Whereas the FA I afferents are more sensitive to low stimulation frequencies (8-64 Hz), the FA II afferents respond preferentially to frequencies above 64 Hz (up to 400 Hz). However, this is only true when indentation amplitudes are small (1 mm); at amplitudes approaching 1 mm the sensitivity of the FA II afferents is reduced towards that of the FA I afferents, with the result being that overall sensitivities to sinusoidal stimulation are similar for either class of rapidly adapting ending. The same is true of the slowly adapting afferents: both the SA I and SA II afferents can respond over a wide range of frequencies (0.5-400 Hz), but with maximal sensitivity at the lowest frequencies; again, responsiveness is similar at higher indentation amplitudes (Johansson et al., 1982a).

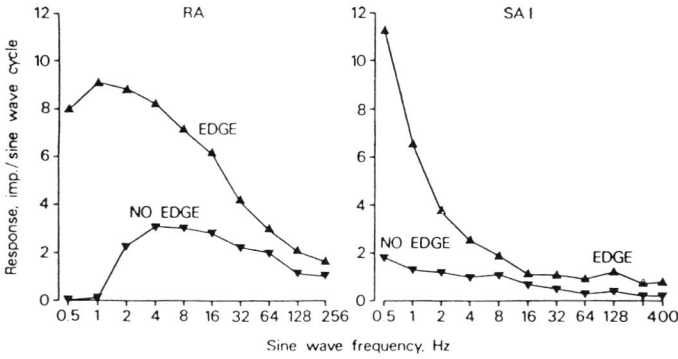

Figure 11. Sensitivity to sinusoidal vibration for an FA I (RA) and SA I afferent when the stimulator probe is enclosed by the receptive field (no edge) or partially covers it (edge). Reproduced with permission from Johansson et al. (1982b).

1.8. Sensitivity to edges

Although the two classes of rapidly adapting afferent have the lowest thresholds to mechanical stimulation, the slowly adapting type I afferent shares a property with the FA I afferents that effectively increases its mechanosensitivity: FA I and SA I afferents are very sensitive to mechanical stimuli applied over the perimeter of their receptive fields. Measured in terms of vibration sensitivity, the number of impulses generated per cycle is greater when the stimulator probe covers the edge of the receptive field than when it is fully enclosed by the field. As illustrated in Fig. 11, this edge sensitivity is particularly pronounced for the SA I afferents (Johansson et al., 1982b).

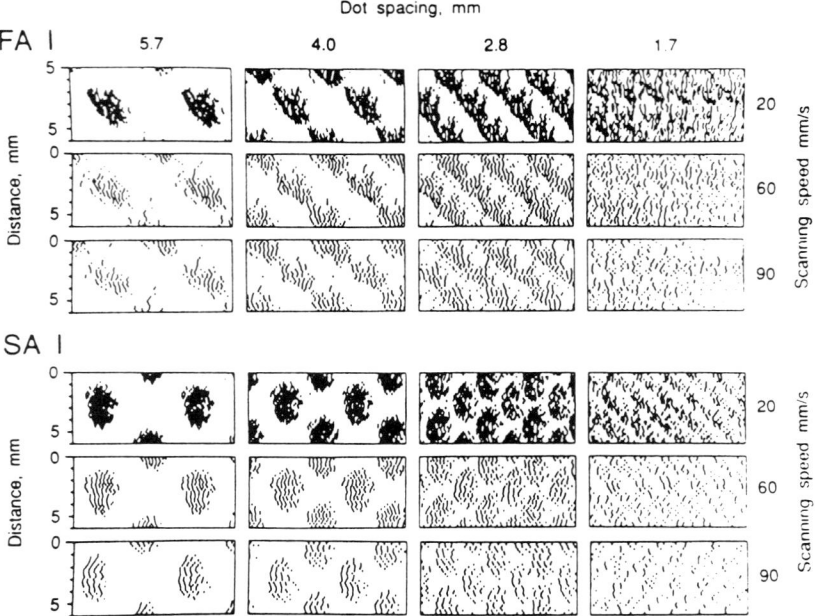

Figure 12. Spatial event plots for an FA I and SAI afferent during application of embossed dot patterns to the skin via a rotating drum. The occurence of a spike is indicated by an event in the raster diagrams. Reproduced with permission from Phillips et al. (1992).

1.9. Encoding of discrete mechanical features

Two reports have documented the capacities of each class of tactile afferent to encode scanned mechnical stimuli in a spatiotemporal pattern of impulses. In the first, standard Braille characters were scanned across the finger pads (Phillips et al., 1990), and in the second embossed dot arrays of differing density were applied at different rates (Phillips et al., 1992). As illustrated in the two-dimensional spatial event plots in Fig. 12, both the SA I and FA I afferent classes can be seen to resolve the dot patterns with high fidelity, down to a spacing of approximately 1.5 mm. Conversely, the maximum resolution provided by the FA II and SA II afferents is only 3.5 mm. These limits of spatial acuity emphasize the importance of both rapidly- and slowly-adapting type I afferents in tactile discrimination (Phillips et al., 1992).

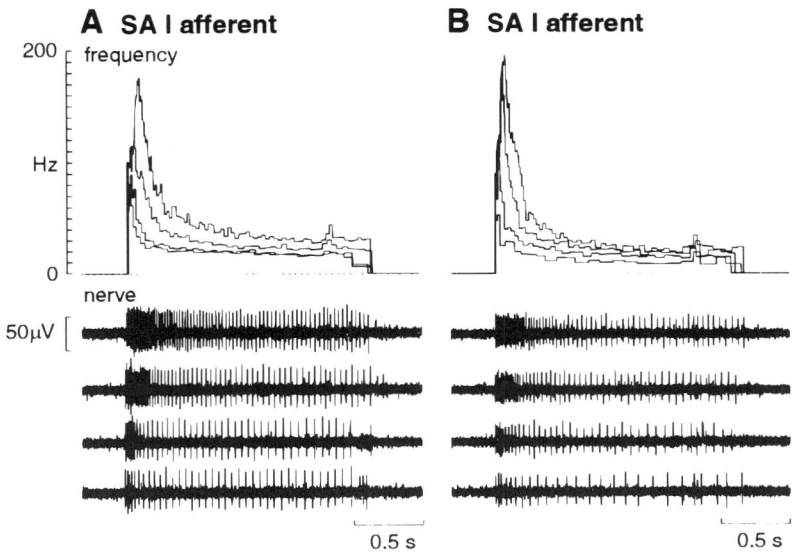

Figure 13. Responses of two SA I afferents in the distal phalanx of the index finger to spheres of different curvature, applied with the same force and rate to the centre of the receptive field. Unpublished observations from Bisley, Macefield and Goodwin.

1.10. Encoding of shape

The capacity of cutaneous afferents to convey haptic information is described by A.W. Goodwin in Chapter 2 of this book. The relationships between curvature of an object in contact with the skin and the discharge of cutaneous receptors in the finger pads of the monkey indicate that the slowly-adapting receptors (equivalent to the SA I class in human skin) are the most important species for conveying information of local curvature to the central nervous system (see Chapter 2). Recently, this has been extended to tactile afferents in human glabrous skin (Bisley, Macefield and Goodwin, unpublished observations). As in the monkey, the SA I afferents in the finger pads show a clear modulation of firing with local curvature. This is shown for two units in Fig. 13, in which four spheres of increasing curvature were applied to the receptive fields on the index finger pad with a constant force and at a constant rate. It can be seen that this increase in discharge frequency with curvature is not limited to the dynamic phase of the response - the mean firing rate during the static phase is also proportional to the curvature. Conversely, for the majority of SA II afferents (which are absent in monkey glabrous skin) or FA I afferents there was no modulation of discharge with curvature, at least when the object contacted the centre of the receptive field.

1.11. Sensitivity to moving tactile stimuli

Recently, the responses of tactile afferents in the glabrous skin of the hand to controlled brush stimuli applied over the receptive fields have been examined (Edin et al., 1995; Essick and Edin, 1995). These data are addressed in detail by G.E. Essick in Chapter 1 of this book. However, it is sufficient to say here that the SA I and SA II, as well as the FA I and FA II afferents all respond to these moving tactile stimuli, and that the discharge patterns of individual receptors differ according to the direction in which the brushing stimuli are applied. It must be noted that these natural, albeit complex, stimuli include both normal and tangential force components: as the brush moves across the field the skin in front is compressed and that behind the brush is stretched, whereas the skin immediately underneath the brush is subjected to indentation (Edin et al., 1995). Interestingly, most of the receptors respond to the stimuli before the brush enters the receptive field.

2. Properties of tactile afferents in human non-glabrous skin

2.1. Receptor types

The hairy skin covers much of the body, and it has been suggested that the properties of mechanosensitive afferents supplying this tissue are probably more representative of the "tactile sensory sheet" than those of the glabrous skin - which is rather more specialized (Vallbo et al., 1995). Three regions of hairy skin will be considered: the dorsal apect of the forearm, the dorsum of the hand, and the face.

In agreement with the types of receptors found in hairy skin of the cat (Burgess et al., 1968), five classes of myelinated tactile afferent have been recorded from the lateral antebrachial cutaneous nerve, which supplies the hairy skin of the human forearm: two types of slowly-adapting afferent (SA I and SA II) that can classified in a similar fashion to those in the glabrous skin, and three types of rapidly-adapting afferent - hair units, field units and Pacinian units (Vallbo et al., 1995). Hair units respond specifically to movements of individual hairs and air puffs onto the receptive field, whereas field units respond to actual skin contact; the behaviour of the SA I, SA II and FA II units is similar to that observed in glabrous skin, with 80% of the SA II endings presenting a low-level background discharge in the absence of stimulation.

The afferent innervation of the skin on the dorsum of the hand, which is supplied by the superficial branch of the radial nerve, is similar to that of the volar surface of the hand: SA I, SA II, FA I and FA II afferents have been identified (; Järvilehto et al., 1976; Järvilehto et al., 1981; Edin and Abbs, 1991). However, differences do exist between the two cutaneous regions. Two studies have reported a high representation of slowly adapting cutaneous afferents (64-68%), and a relative paucity of FA II afferents in the hairy skin of the hand (Järvilehto et al., 1981; Edin and Abbs, 1991). This is quite different to the situation in the glabrous skin, where the dominant species is the FA I afferent (43%) (Johansson and Vallbo, 1979b). A further difference lies in the relative proportions of SA I and SA II afferents: in the glabrous skin there are more of the former (Johansson and Vallbo, 1979b), but Edin and Abbs (1991) found identical representations in the hairy skin. There are also difficulties in differentiating between SA I and SA II units in the hairy skin (Jarvilehto et al., 1981; Edin and Abbs, 1991), whereas the classifications are quite distinct in glabrous skin. Although few endings associated with hairs have been recorded from the radial nerve this may simply reflect the lower density of hairs on the back of the hand compared with that of the forearm.

Microelectrode recordings from the infraorbital nerve have demonstrated that the hairy skin of the human face is innervated by rapidly-adapting and slowly adapting afferents with properties identical to those of the FA I and SA I afferents found in the hand (Johansson et al., 1988b). A distinct population of slowly-adapting afferents that present a very regular discharge characteristic of SA II endings has also been found, although their responsiveness to skin stretch could not be tested. Interestingly, no FA II afferents were encountered; this suggests an absence of Pacinian corpuscles in the human face (Johansson et al., 1988b), and fits with the low sensitivity of the face to high-frequency vibration (Barlow, 1987).

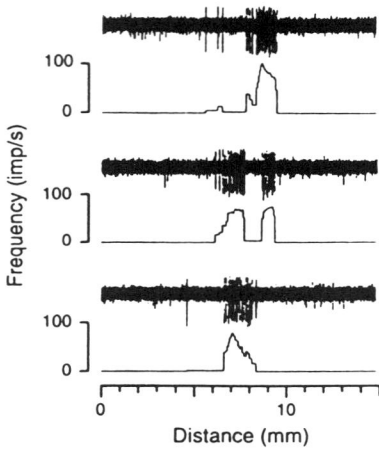

Figure 14. Responses of a non-myelinated afferent in the hairy skin of the forearm to a stimulator probe (4.0 g) moving across the receptive field area. Reproduced with permission from Vallbo et al. (1993).

In addition to these large myelinated afferents, the hairy skin of the forearm is supplied by non-myelinated afferents of low mechanical threshold (Vallbo et al., 1993). These endings can generate peak firing rates of up to 100 Hz during inoccuous mechanical stimulation, such as stroking the skin with a blunt probe across the receptive field (Fig. 14). There is also evidence that these afferents may provide significant innervation of the facial hairy skin; indeed, it was in this region that a single low-threshold mechanosensitive C-fibre was first encountered (Johansson et al., 1988b).

In summary, there are more slowly-adapting and fewer rapidly-adapting tactile afferents in human hairy skin than in glabrous skin. Furthermore, assuming that the forearm is representative of much of the body's cutaneous endowment, the hairy skin is provided with mechanosensitive non-myelinated afferents that, as far as receptive field size and mechanical threshold are concerned, are not so dissimilar to the myelinated SA I afferents.

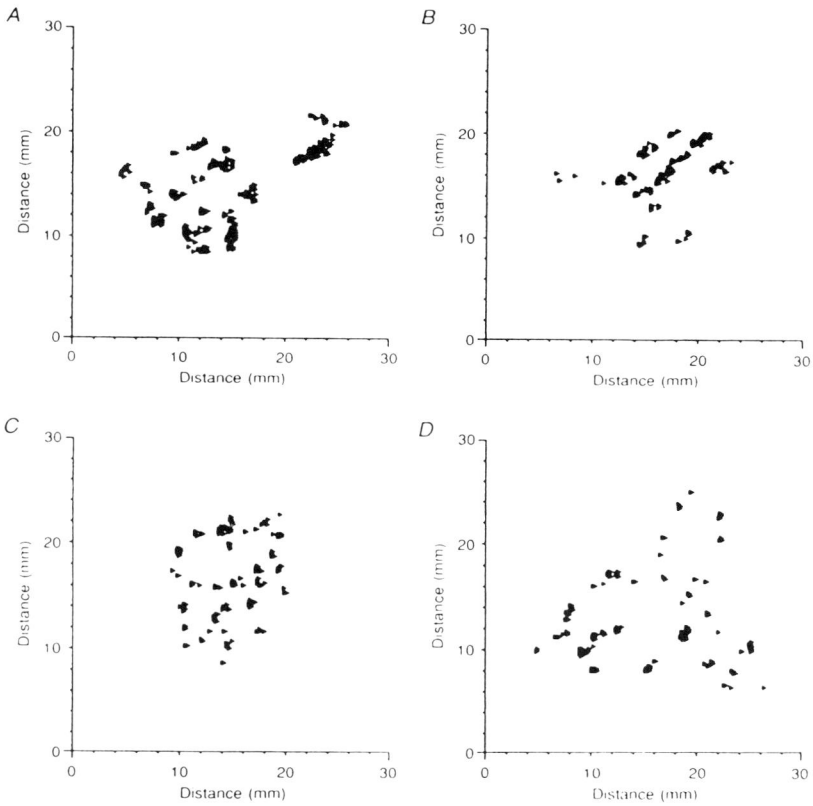

Figure 15. Receptive field maps of four hair units, measured by a stimulator probe scanning an area of skin on the hairy surface of the forearm. Each small symbol represents a spike event. Reproduced with permission form Vallbo et al. (1995).

2.2 Receptive fields

Hair units in the forearm have large ovoid or irregular receptive fields (median lengths of long and short axes 16 x 11 mm) composed of multiple sensitive spots that corresponded to individual hairs (Fig. 15). On average, each afferent innervated 20 hairs (Vallbo et al., 1995). The

field units show a similar arrangement of multiple high sensitivity spots
(10-12) encompassed by a similarly large area (median lengths 13 x 7

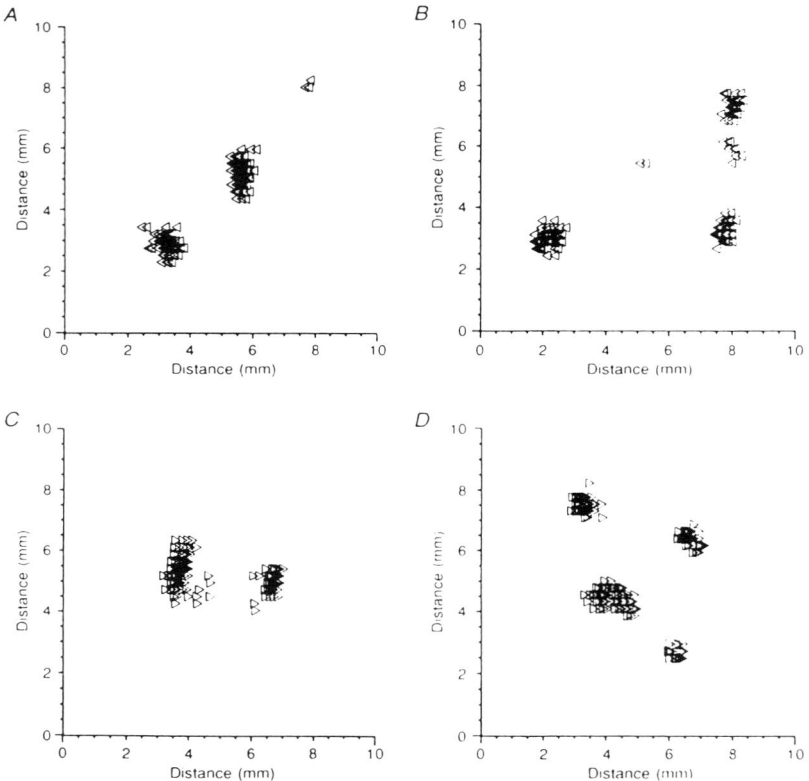

Figure 16. Receptive field maps of four SA I units, measured by a stimulator
probe scanning an area of skin on the hairy surface of the forearm. Each
small symbol represents a spike event. Reproduced with permission from
Vallbo et al. (1995).

mm), although the individual spots were larger and less isolated than
those of the hair units. By contrast, on the dorsum of the hand and the
hairy skin of the face, the rapidly-adapting afferents have small
receptive fields, usually with only a single zone of uniform sensitivity
(Järvilehto et al., 1976; Järvilehto et al., 1981; Johansson et al., 1988b;
Edin and Abbs, 1991).

The receptive fields of the SA I afferents on the forearm consist of 2-4 distinct islands of high sensitivity separated from each other by 3 mm on average (Fig. 16); these spots presumably correspond to the touch domes overlying clusters of Merkel cells (Vallbo et al., 1995). On the dorsum of the hand, most of the SA I fields (like their rapidly-adapting counterparts) consist of a single spot of maximal sensitivity (Järvilehto et al., 1981). Overall, the number of these zones in the hairy skin is lower than in the glabrous skin. Apart from their higher representation in the hairy skin, the SA II afferents are similar to those in the glabrous skin, usually having a single zone of high sensitivity to punctate stimulation.

Like the myelinated SA I endings, the receptive fields of the non-myelinated mechanosensitive afferents in the forearm are composed of 1-6 "hot spots", with a peak separation of 9 mm (Vallbo et al., 1993).

2.3. Mechanical thresholds

Mechanoreceptors in the hairy skin of the forearm are exquisitely sensitive, even more so than those in the glabrous skin: after the hair units the afferents with the lowest threshold to von Frey stimulation are the field units, with a median threshold of 0.1 mN; the SA I and SA II afferents have median thresholds of 0.45 and 1.30 mN, respectively (Vallbo et al., 1995). For comparison, in the glabrous skin of the hand these thresholds are 0.58 mN for the FA I afferents, 1.3 mN for the SA I and 7.5 mN for the SA II afferents (Johansson et al., 1980). Mechanical thresholds of afferents on the dorsum of the hand and on the face are similar to those in the glabrous skin (Johansson et al., 1988b; Edin and Abbs, 1991). The non-myelinated afferents of the forearm have a median threshold of 2.5 mN (Vallbo et al., 1993), comparable to that of the SA I afferents supplying the glabrous skin.

2.4. Receptor encoding of punctate stimuli

Stimulus-response functions have been calculated for FA I, SA I and SA II afferents on the dorsum of the hand (Järvilehto et al., 1976; Järvilehto et al., 1981; Åstrand et al., 1986). The majority of FA I afferents presented negatively accelerating curves when the number of impulses generated was expressed as a function of indentation amplitude (Järvilehto et al., 1981), whereas for the SA I and SA II afferents the curves were best described by power functions (Åstrand et al., 1986). These stimulus-response relationships are similar to those described in the glabrous skin (Knibestol, 1973; Knibestol, 1975).

2.5. Sensitivity to vibration

As in the glabrous skin, the capacity for mechanoreceptors in the hairy skin to follow vibratory stimuli is not limited to the rapidly-adapting afferents: for instance, the range of frequencies over which SA I and SA II afferents are entrained extends from 20 up to 100 Hz for some afferents, whereas FA I afferents can follow up to 200 Hz and FA II afferents 400 Hz (Järvilehto et al., 1976; Järvilehto et al., 1981).

2.6. Sensitivity to moving tactile stimuli

As in the glabrous skin, both rapidly and slowly adapting afferents in the hairy skin of the hand and face respond to light brushing stimuli applied over the skin (Edin et al., 1995; Essick and Edin, 1995). The means by which these afferents encode the mechanical features of moving tactile stimuli are addressed in Chapter 1 of this book.

3. Signalling of finger movements by tactile afferents

3.1. Glabrous skin

The above sections have dealt with the capacities of cutaneous mechanoreceptors to convey sensory information of distinctly tactile qualities. There are appreciable differences in mechanosensitivity among the different receptor types as well as across different skin areas - glabrous and non-glabrous. This is true also when considering the responses of cutaneous afferents to movements of the skin, in which their roles can be seen as proprioceptive. The first quantitative observations on the responsiveness of cutaneous afferents to finger movements were made by Hulliger et al. (1979), who found that 77% of afferents from the glabrous skin of the hand respond to voluntary isotonic movements of the fingers, in which there is no incidental contact with the receptive field (Fig. 17).

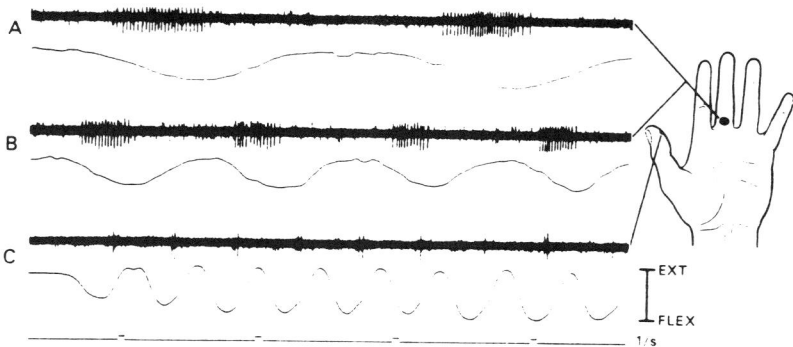

Figure 17. Responses of a SA II afferent located at the base of the middle finger to voluntary movements of the MCP joint (A, B), and of an FA I afferent to movements of the thumb (C). Reproduced with permission from Hulliger et al. (1979).

Specifically, all of the FA II and most of the SA II (94%) afferents respond to cyclic or ramp-and-hold movements, whereas only 57% of the FA I and 66% of the SA I afferents do so. Moreover, whereas the rapidly-adapting afferents fire in either direction of joint rotation, i.e. they respond to movements rather than to joint angle as such, the slowly-adapting afferents often display a preference for a particular direction. Indeed, 81% of the SA II afferents (and 17% of the SA I afferents) display static position sensitivity, in which firing rate is proportional to joint angle (see also Fig. 4 in Knibestol, 1975). It should be emphasized, however, that relatively few of these SA II afferents respond throughout the physiological range of joint rotation; the majority are recruited towards the limits of angular excursion (Hulliger et al., 1979). The same is true when finger movements are brought about passively, with approximately 25% of all cutaneous afferents responding to passive movements about more than one axis of joint rotation (Burke et al., 1988). These results imply that tactile afferents in the glabrous skin of the hand may well provide useful information to the central nervous system about joint movements, but may not provide sufficiently high fidelity signals of joint position across the physiological range.

3.2. Non-glabrous skin

The skin on the palmar surface of the hand is mechanically different to that on the back of the hand. Unlike glabrous skin, the hairy skin is only loosely connected to the subcutaneous tissues, thereby allowing greater stretch and hence greater activation of stretch-sensitive cutaneous afferents (Edin and Abbs, 1991; Edin, 1992; Edin and Johansson, 1995). There do appear to be significant differences in the movement sensitivity of tactile afferents in the non-glabrous and glabrous skin. For instance, 92% of the afferents on the dorsum of the hand responded to finger movements (Edin and Abbs, 1991), whereas only 68% of afferents on the palmar side of the hand responded to passive finger movements (Burke et al., 1988) and 77% responded to active movements (Hulliger et al., 1979).

Moreover, unlike afferents from the glabrous skin of the hand, the majority of afferents on the dorsum respond *throughout* the physiological range of joint rotation, increasing their firing rate as the skin is stretched during flexion of the digital joints. The FA I afferents respond only to movements of the joint over which they are located, but both the SA I as well as SA II afferents are very sensitive to the skin stretch associated with finger movements - whether this is produced by rotation of the nearest joint or by movements of remote digital joints (Fig. 18).

The static sensitivity to stretch is high for both classes of slowly-adapting afferent (Edin, 1992): when measured at an equivalent joint angle, the static sensitivity of the SA I and SA II afferents is similar to that of muscle spindle endings in the long extensors of the fingers, 0.2-0.5 Hz per degree of rotation of the metacarpophalangeal joint (Edin and Vallbo, 1990; Edin, 1992).

Like the hairy skin of the hand, that of the face is also well equipped with receptors for measuring movements of the skin: both rapidly and slowly adapting cutaneous afferents recorded from the infraorbital nerve (Johansson et al., 1988a; Johansson et al., 1988b) respond well to movements of the facial skin associated with phonation and mastication, which supports their provision of proprioceptive information during orofacial movements (Johansson et al., 1988a; Johansson et al., 1988b).

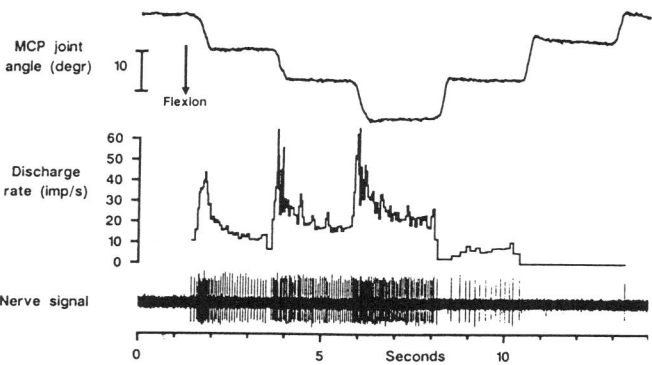

Figure 18. Responses of an SA II afferent located over the MCP joint of the index finger to voluntary movements of the digit about the MCP joint. Reproduced with permission from Edin and Abbs (1991).

4. Signalling of grip and load forces by cutaneous afferents

When holding an object between the fingers and thumb there are two primary forces that act at the skin: a compressive component normal to the skin and a shear component tangential to the skin. The first is brought about by the grip forces exerted by the muscles acting on the digits, the second by the effect of gravity on the held object or any other net force imposed by the object or hand on the object.

Johansson and colleagues have shown that cutaneous afferents in the glabrous skin of the digits are capable of encoding the grip and load forces associated with grasping and lifting an object, and that the information provided by tactile afferents to the central nervous system is of paramount importance in the fine coordination of load and grip forces (Johansson and Westling, 1987a; Johansson and Westling, 1987b; Westling and Johansson, 1987). When grasping and lifting an object between finger and thumb the type I afferents (FA I and SA I) in the tips of the digits respond with high firing rates at the moment of contact and in the early part of the loading phase, during which the grip force increases in parallel with the load force at a rate sufficient to prevent slipping of the object. The FA I and SA I afferents also respond to local mechanical disturbances during the lifting and hold phases, such as incipient or overt slips.

FA II afferents respond to the mechanical transients associated with initial contact and release of the object, and especially the acceleration and deceleration signals related to the start and end of a movement sequence. The SA II afferents generally respond to the grip force during the loading and hold phases of the lift, and also to the tangential loads generated at the skin during the hold phase (Johansson and Westling, 1987a; Johansson and Westling, 1987b; Westling and Johansson, 1987). This ability to respond to force components both normal and tangential to the skin is illustrated in Fig. 19.

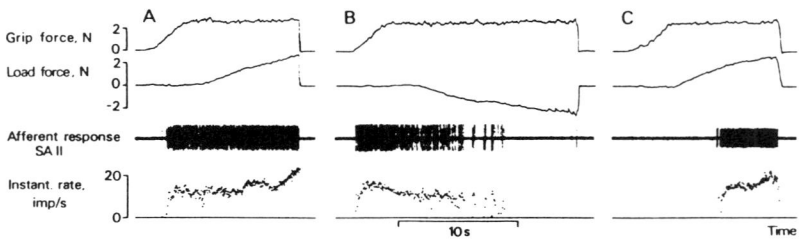

Figure 19. Responses of an SA II afferent to gripping and lifting an object. The subject was instructed to maintain the grip force constant (with visual feedback) while the experimenter produced a load force on the object in the pulling (down) or pushing (up) direction. Reproduced with permission from Westling and Johansson (1987).

Recently, this work was extended to the control of grip force during restraint of objects exerting unpredictable loading or unloading forces (Johansson et al., 1992a; Johansson et al., 1992b; Johansson et al., 1992c; Cole and Johansson, 1993; Macefield et al., 1996; Macefield and Johansson, 1996). In this experimental situation the subject grips a manipulandum between the tips of the thumb and a finger and attempts to prevent escape of the manipulandum from the grasp during unexpected increases or decreases in tangential force. Microneurographic recordings reveal that the FA I, SA I and SA II afferents in the glabrous skin of the digits respond to the shear forces generated between the skin and the manipulandum; the slowly-adapting afferents also respond during the subsequent inceases in grip force that serve to restrain the manipulandum (Macefield et al., 1996). Responses of an FA I afferent to ramp-and-hold tangential loads are illustrated in Fig. 20.

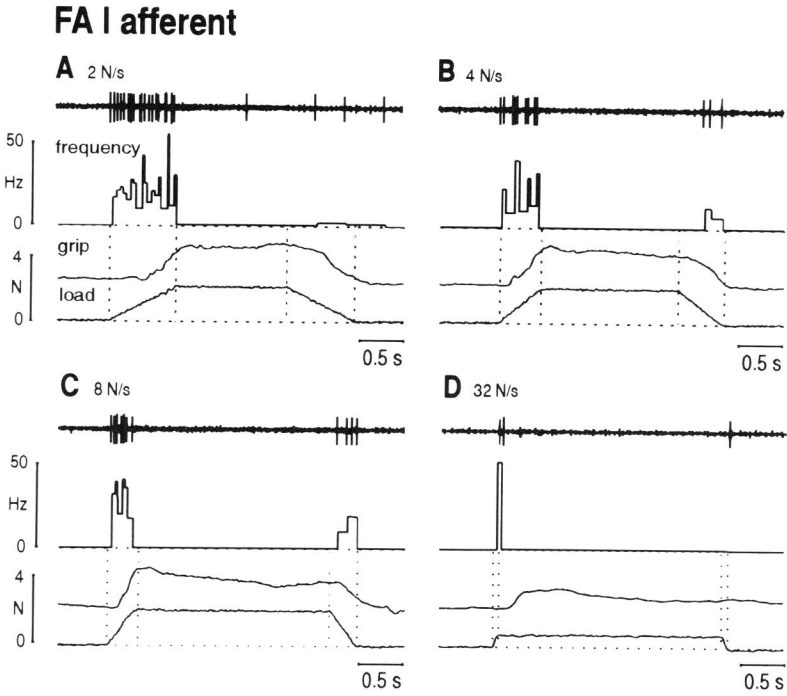

Figure 20. Responses of an FA I afferent during 2 N pulling loads applied to the receptor-bearing digit at different rates via a gripped manipulandum. Unpublished records from Macefield, Häger-Ross and Johansson.

It can be seen that the FA I afferent responds only during the dynamic phases of the stimulus, i.e. during the loading and unloading ramps. Moreover, the ensemble response is not influenced by the subsequent increase in grip force - FA I afferents respond exclusively to the tangential force components in this condition (Macefield et al., 1996).

The responses of an SA I afferent to these same stimuli are shown in Fig. 21. Note that the instantaneous frequency records indicate the receptor's sensitivity to the dynamic and static components of the grip force response. The ensemble response of the SA I afferents is an

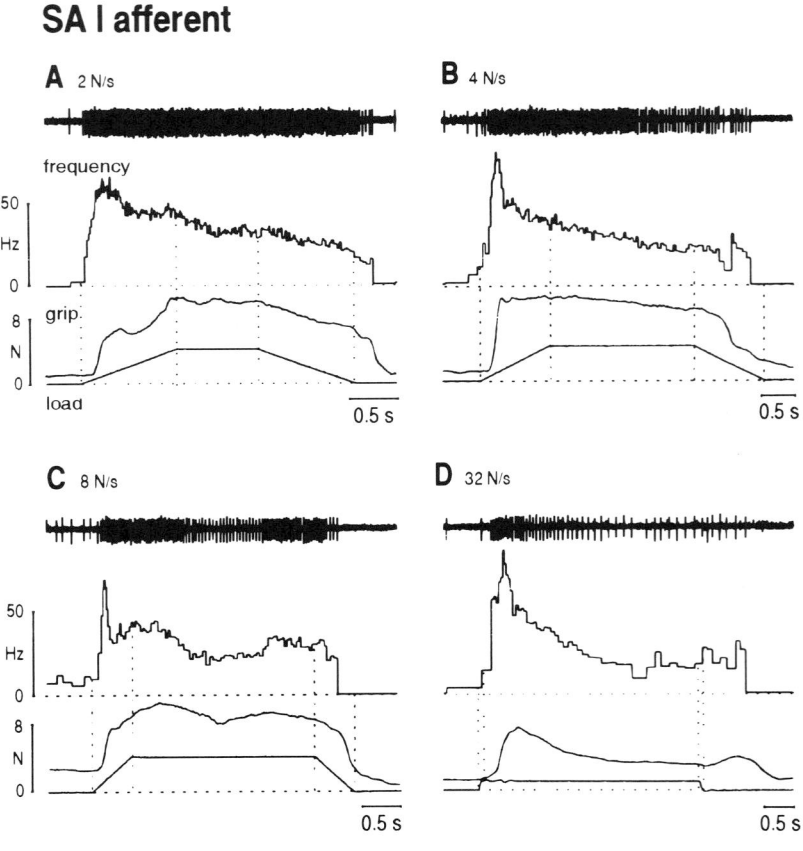

Figure 21. Responses of an SA I afferent during 2 N pulling loads applied to the receptor-bearing digit at different rates via a gripped manipulandum. Unpublished records from Macefield, Häger-Ross and Johansson.

increase in firing during the loading ramp, with a further increase in discharge during the subsequent increase in grip force. SA II afferents behave similarly, but the representation of grip force (normal to the skin) in the ensemble discharge is much smaller than that for the SA I afferents (Macefield et al., 1996).

According to Phillips and Johnson (1981), the Meissner corpuscles (FA I afferents) are primarily sensitive to changes in tensile strain within the local receptor environment. This property would account for their capacity to signal strain changes generated between the skin and a gripped object, such as occur during localized slips of a held object (Johansson and Westling, 1987a) or during the application of tangential loading ramps (Macefield et al., 1996). Conversely, the Merkel cell neurite complexes (SA I afferents) are primarily sensitive to compressive strain (Phillips and Johnson, 1981), which would fit with their high sensitivity to grip force (normal to the skin). As we have seen in Section I, Ruffini endings (SA II afferents) are sensitive to tensile strain over a large area (Eklund and Skogland, 1960; Grigg and Hoffman, 1982; Grigg and Hoffman, 1984; Grigg and Hoffman, 1996), which fits with their high sensitivity to load force (tangential to the skin).

5. Sensory specificity in the tactile system

5.1. Intraneural microstimulation of single axons

Our current understanding of human sensory physiology includes the belief that excitation of a particular class of afferent will produce a particular sensory quality. In the tactile system this concept of sensory specificity has received strong support from the technique of intraneural microstimulation, an extension of microneurography in which controlled current pulses passed down a tungsten microelectrode can be used to determine the input-output (stimulus-sensation) function of a single afferent fibre. This approach was first applied to single cutaneous afferents in conscious human subjects by Torebjörk and Ochoa (1980) and Vallbo (1981). In practice, one uses the microelectrode to locate a single afferent fibre and then stimulates at this intrafascicular site; this allows the experimenter to first identify the afferent and to record its responses to natural stimulation, before determining how the brain utilizes the sensory information evoked by electrical stimulation of this same axon. Alternatively, an intraneural site can be stimulated first and the subject's perceptual responses recorded before the identity of the axon impaled by the microelectrode is known.

Either way, the validity of the technique relies on there being a stimulus level at which an elementary sensation is projected to a small area of skin in the innervation territory, and that this "projected field" and its sensory quality remain constant as the stimulus intensity is increased, until a level is reached at which another distinct sensation and projected field are generated (Torebjörk and Ochoa, 1980; Vallbo, 1981; Ochoa and Torebjörk, 1983; Schady and Torebjörk, 1983; Schady et al., 1983a; Schady et al., 1983b; Vallbo et al., 1984; Torebjörk et al., 1987; Macefield et al., 1990). The quantal nature of the sensations evoked by intraneural stimulation at higher intensities can be explained by the spread of current to adjacent axons, which by virtue of the relatively poor somatotopy within nerve fascicles results in sensory percepts that are referred to sites often remote from the receptive field of the afferent that was excited at the lowest stimulus level. The correspondence between an afferent's receptive field and the projected field (as reported by the subject) is remarkably good, supporting the belief that the same sensory axon that can be recorded from the microelectrode is the same axon that is first stimulated electrically (Schady and Torebjörk, 1983; Schady et al., 1983b; Vallbo et al., 1984; Torebjörk et al., 1987). The ability to selectively stimulate single axons in human nerves has also been proven during stimulation of motor axons, where the criteria for unitary stimulation (uniform EMG and twitch amplitude, with quantal increases in both during stimulation at higher levels) are somewhat easier to measure (Westling et al., 1990; Macefield et al., 1996).

Of relevance here is the observation, raised in Section I, that the microelectrode appears not to cause appreciable propagation block across the nerve fascicle. Indeed, the morphologies of the majority of recorded action potentials indicate that naturally evoked spikes do conduct across the site of impalement (Vallbo, 1976; Inglis et al., 1996), meaning that the microelectrode truly samples neural activity without significantly altering the "local sensory picture". Although at first sight this may appear to be of no consequence during intraneural microstimulation - the impaled axon will be excited by an electrical pulse regardless of whether receptor-generated impulses can pass the impalement site - the interpretation of the single sensory channel by the central nervous system will not be compromised by an altered background afferent inflow.

5.2. Sensations evoked by stimulation of single tactile afferents

Three independent groups of researchers have shown that stimulation of single FA I, FA II and SA I afferents innervating the glabrous skin of the hand evoke elementary sensations of a specific quality (Torebjörk and Ochoa, 1980; Vallbo, 1981; Ochoa and Torebjörk, 1983; Schady and Torebjörk, 1983; Schady et al., 1983a; Schady et al., 1983b; Vallbo et al., 1984; Torebjörk et al., 1987; Macefield et al., 1990). A single pulse delivered to a single FA I afferent can be detected if the subject's attention is directed to it, whereas an SA I afferent requires more impulses and greater attention (Vallbo et al., 1984). This also fits with the lower mechanical threshold of FA I afferents and confirms an earlier interpretation of psychophysical thresholds that subjects can detect a single impulse generated by a single FA I receptor (Johansson and Vallbo, 1979a).

Stimulating a single FA I afferent with a low frequency train generates a percept of intermittent tapping that, as the frequency of stimulation increases, becomes one of flutter or vibration; stimulation of a single FA II afferent with a train of pulses always generates a frequency-dependent perception of mechanical vibration. Percepts of sustained pressure can be evoked by selective stimulation of SA I afferents, the magnitude of which increases with increasing stimulation frequency, although as noted above the generation of a sensation by this class of afferent does appear to be more dependent on the subject's attentional state. Nevertheless, the impulse codes utilized by rapidly and slowly adapting tactile afferents are quite distinct: increasing frequency of stimulation signalling increasing vibration with the former, and increasing pressure with the latter. Stimulation of a single SA II afferent with a train of pulses usually does not elicit a sensation, although there is evidence that some SA II afferents located near the nailbeds may be able to provide meaningful sensations of joint movement (Macefield et al., 1990). Interestingly, microstimulation of a single muscle spindle afferent also fails to generate a sensory percept, yet we know that stimulation of a population of spindle afferents does produce a definite illusion of movement (Macefield et al., 1990); one may speculate that if more than one SA II afferent is activated an illusion of movement would be produced. Of relevance to this point are the results of two recent studies which have shown that illusory movements of the finger joints can be induced by imposed displacements of the dorsal skin of the fingers (without producing overt joint movements), arguing for a significant role of skin strain patterns - as signalled by SA I and SA II afferents in the hairy skin - in the perception of finger position (Edin and Johansson, 1995; Collins and Prochazka, 1995). If this interpretation can be applied to the glabrous skin of the hand then it would be the SA II afferents that are responsible for signalling changes in skin strain.

6. Conclusions

The afferent innervation of human skin is characterized by regional variations in receptor types and densities that indicate specializations of the "tactile sensory sheet". In the glabrous skin of the hand the tips of the digits contain a high proportion of rapidly-adapting afferents (FA I) with small receptive fields, low mechanical thresholds, and a secure transmission to the sensory cortex, whereas on the dorsum of the hand these afferents have a much lower representation. Perhaps the latter region, which contains a high proportion of slowly-adapting afferents of high sensitivity to skin strain (both SA I and SA II), may provide useful kinaesthetic information about finger movements. In the glabrous skin of the hand, the SA I endings appear to be of particular importance in encoding shape, but also in responding with high fidelity to the changes in grip force associated with manipulation of held objects. It would also appear that the glabrous skin of the human hand is unique in possessing SA II afferents (which in other animals are present only in hairy skin), and given their high sensitivity to forces tangential to the skin and poor capacity in spatial discrimination, it is reasonable to conclude that their specific contribution may lie in signalling the load forces encountered during manipulation. In all skin areas the numbers of FA II (Pacinian) afferents is low, but given their large field sizes, low thesholds, exquisite sensitivity to mechanical transients and high security transmission to the sensory cortex, their number need not be so high anyway. The hairy skin of the forearm, which probably typifies the skin of much of the body, has its own specializations: in addition to two classes of very sensitive receptors with large receptive fields (hair units and field units), this region is endowed with non-myelinated mechanosensitive endings of very low threshold. Although these endings, which have also been found in facial skin, may be considered vestigial, it is also reasonable to surmise that they may contribute to the perception of moving tactile stimuli and, by virtue of the after-discharge they characteristically present following a brisk stimulus, to stimulus localization.

References

Åstrand, K., Hämäläinen, H., Aleksandrov, Y.I. and Järvilehto, T. (1986) Response characteristics of peripheral mechanoreceptive units in man: relation to the sensation magnitude and to the subject's task. Electroencephalography and Clinical Neurophysiology, 64, 438-446.

Barlow, S.M. (1987) Mechanical frequency detection thresholds in the human face. Experimental Neurolology, 96, 253-261.

Brink, E.E. and Mackel, R.G. (1993) Time course of action potentials recorded from single human afferents. Brain, 116, 415-432.

Burgess, P.R., Petit, D. and Warren, R.M. (1968) Receptor types in cat hairy skin supplied by myelinated fibres. Journal of Neurophysiology, 31, 833-848.

Burke, D., Gandevia, S.C. and Macefield, G. (1988) Responses to passive movement of receptors in joint, skin and muscle of the human hand. Journal of Physiology, 402, 347-361.

Chambers, M.R., Andres, K.H., Duering, M. and Iggo, A. (1972) The structure and function of slowly-adapting type II mechanoreceptors in hairy skin. Quarterly Journal of Experimental Physiology, 57, 417-445.

Cole, K.J. and Johansson, R.S. (1993) Friction at the digit-object interface scales the sensorimotor transformation for grip responses to pulling loads. Experimental Brain Research, 95, 523-532.

Collins, D.F. and Prochazka, A. (1995) Illusory finger movements evoked by ensemble cutaneous input from the dorsum of the human hand. Society for Neuroscience Abstracts, 21, 1920.

Darian-Smith, I. (1984) The sense of touch: performance and peripheral neural processes. In Darian-Smith, I. (ed) Handbook of Physiology, Section I: The Nervous System, American Physiological Society, Bethesda, 739-788.

Darian-Smith, I. and Kenins, P. (1980) Innervation density of mechanoreceptive fibres supplying glabrous skin of the monkey's index finger. Journal of Physiology, 309, 147-155.

Edin, B.B. (1990) Finger joint movement sensitivity of non-cutaneous mechanoreceptor afferents in the human radial nerve. Experimental Brain Research, 82, 417-422.

Edin, B.B. (1992) Quantitative analysis of static strain sensitivity in human mechanoreceptors from hairy skin. Journal of Neurophysiology, 67, 1105-1113.

Edin, B.B. and Abbs, J.H. (1991) Finger movement responses of cutaneous mechanoreceptors in the dorsal skin of the human hand. Journal of Neurophysiology, 65, 657-670.

Edin, B.B., Essick, G.K., Trulsson, M. and Olsson, K.A. (1995) Receptor encoding of moving tactile stimuli in humans. I. Temporal pattern of discharge of individual low-threshold mechanoreceptors. Journal of Neuroscience, 15, 830-847.

Edin, B.B. and Johansson N. (1995) Skin strain patterns provide kinaesthetic information to the human central nervous system. Journal of Physiology, 487, 243-251.

Edin, B.B. and Vallbo, Å.B. (1990) Dynamic response of human muscle spindle afferents to stretch. Journal of Neurophysiology, 63, 1297-1306.

Eklund, G. and Skogland, S. (1960) On the specificity of the Ruffini like joint receptors. Acta Physiologica Scandinavica, 49, 184-191.

Essick, G.K. and Edin, B.B. (1995) Receptor encoding of moving tactile stimuli in humans. II. The mean response of individual low-threshold mechanoreceptors to motion across the receptive field. Journal of Neuroscience, 15, 848-864.

Ferrington, D.G. (1985) Functional properties of slowly adapting mechanoreceptors in cat footpad skin. Somatosensory Research, 2, 249-261.

Gandevia, S.C., Macefield, G., Burke, D. and McKenzie, D.K. (1990) Voluntary activation of human motor axons in the absence of muscle afferent feedback. The control of the deafferented hand. Brain, 113, 1563-1581.

Grigg, P. and Hoffman, A.H. (1982) Properties of Ruffini afferents revealed by stress analysis of isolated sections of cat knee capsule. Journal of Neurophysiology, 47, 41-54.

Grigg, P. and Hoffman, A.H. (1984) Ruffini mechanoreceptors in isolated joint capsule: responses correlated with strain energy density. Somatosensory Research, 2, 149-162.

Grigg, P. and Hoffman, A.H. (1996) Stretch-sensitive afferent neurones in cat knee joint capsule - sensitivity to axial and compression stresses and strains. Journal of Neurophysiology, 75, 1871-1877.

Hagbarth, K.E. and Vallbo, Å.B. (1968) Discharge characteristics of human muscle afferents during muscle stretch and contraction. Experimental Neurology, 22, 674-694.

Hagbarth, K.E. and Vallbo, Å.B. (1969) Single unit recordings from muscle nerves in human subjects. Acta Physiologica Scandinavica, 76, 321-334.

Hallin, R.G. and Torebjörk, H.E. (1974) Single unit sympathetic activity in human skin nerves during rest and various manoeuvres. Acta Physiologica Scandinavica, 92, 303-317.

Hulliger, M., Nordh, E., Thelin, A.E. and Vallbo, Å.B. (1979) The responses of afferent fibres from the glabrous skin of the hand during voluntary finger movements in man. Journal of Physiology, 291, 233-249.

Inglis, J.T., Leeper, J.B., Burke, D. and Gandevia, S.C. (1996) Morphology of action potentials recorded from human nerves using microneurography. Experimental Brain Research, (in press).

Järvilehto, T., Hämäläinen, H. and Laurinen, P. (1976) Characteristics of single mechanoreceptive fibres innervating hairy skin of the human hand. Experimental Brain Research, 25, 45-61.

Järvilehto, T., Hämäläinen, H. and Soininen, K. (1981) Peripheral neural basis of tactile sensations in man: II. Characteristics of human mechanoreceptors in the hairy skin and correlations of their activity with tactile sensations. Brain Research, 219, 13-27.

Johansson, R.S. (1978) Tactile sensibility in the human hand: receptive field characteristics of mechanoreceptive units in the glabrous skin area. Journal of Physiology, 281, 101-123.

V.G. Macefield

Johansson, R.S., Häger, C. and Bäckström, L. (1992c) Somatosensory control of precision grip during unpredictable pulling loads. III. Impairments during digital anesthesia. Experimental Brain Research, 89, 204-213.

Johansson, R.S., Häger, C. and Riso, R. (1992b) Somatosensory control of precision grip during unpredictable pulling loads. II. Changes in load force rate. Experimental Brain Research, 89, 192-203.

Johansson, R.S., Landström, U. and Lundström, R. (1982a) Responses of mechanoreceptive afferent units in the glabrous skin of the human hand to sinusoidal skin displacements. Brain Research, 244, 17-25.

Johansson, R.S., Landström, U. and Lundström, R. (1982b) Sensitivity to edges of mechanoreceptive afferent units innervating the glabrous skin of the human hand. Brain Research, 244, 27-32.

Johansson, R.S., Riso, R., Häger, C. and Bäckström, L. (1992a) Somatosensory control of precision grip during unpredictable pulling loads. I. Changes in load force amplitude. Experimental Brain Research, 89, 181-191.

Johansson, R.S., Trulsson, M., Olsson, K.A. and Westberg, K.G. (1988a) Mechanoreceptor activity from the human face and oral mucosa. Experimental Brain Research, 72, 204-208.

Johansson, R.S., Trulsson, M., Olsson, K.A. and Abbs, J.H. (1988b) Mechanoreceptive afferent activity in the infraorbital nerve in man during speech and chewing movements. Experimental Brain Research, 72, 209-214.

Johansson, R.S. and Vallbo, Å.B. (1979a) Detection of tactile stimuli. Thresholds of afferent units related to psychophysical thresholds in the human hand. Journal of Physiology, 297, 405-422.

Johansson, R.S. and Vallbo, Å.B. (1979b) Tactile sensibility in the human hand: relative and absolute densities of four types of mechanoreceptive units in glabrous skin. Journal of Physiology, 286, 283-300.

Johansson, R.S. and Vallbo, Å.B. (1983) Tactile sensory coding in the glabrous skin of the human hand. Trends in Neuroscience, 6, 27-31.

Johansson, R.S., Vallbo, Å.B. and Westling, G. (1980) Thresholds of mechanosensitive afferents in the human hand as measured with von Frey hairs. Brain Research, 184, 343-351.

Johansson, R.S. and Westling, G. (1987a) Signals in tactile afferents from the fingers eliciting adaptive motor responses during precision grip. Experimental Brain Research, 66, 141-154.

Johansson, R.S. and Westling, G. (1987b) Significance of cutaneous input for precise hand movements. Electroencephalography and Clinical Neurophysiology Supplement, 39, 53-57.

Knibestol, M. (1973) Stimulus-response functions of rapidly adapting mechanoreceptors in human glabrous skin area. Journal of Physiology, 232, 427-452.

Knibestol, M. (1975) Stimulus-response functions of slowly adapting mechanoreceptors in the human glabrous skin area. Journal of Physiology, 245, 63-80.

Knibestol, M. and Vallbo, Å.B. (1970) Single unit analysis of mechanoreceptor activity from the human glabrous skin. Acta Physiologica Scandinavica, 80, 178-195.

Macefield, G., Gandevia, S.C. and Burke, D. (1990) Perceptual responses to microstimulation of single afferents innervating joints, muscles and skin of the human hand. Journal of Physiology, 429, 113-129.

Macefield, V.G. (1995) The behaviour of cutaneous and joint afferents in the human hand during finger movements. In Ferrell, W.R. and Proske, U. (eds) Neural Control of Movement, Plenum, New York, 73-78.

Macefield, V.G., Fuglevand, A.J. and Bigland-Ritchie, B. (1996) Contractile properties of single motor units in human toe extensors assessed by intraneural motor-axon stimulation. Journal of Neurophysiology, 75, 2509-2519.

Macefield, V.G., Gandevia, S.C., Bigland-Ritchie, B., Gorman, R. and Burke, D. (1993) The firing rates of human motoneurones voluntarily activated in the absence of muscle afferent feedback. Journal of Physiology, 474, 429-443.

Macefield, V.G., Häger-Ross, C. and Johansson, R.S. (1996) Control of grip force during restraint of an object held between finger and thumb: Responses of cutaneous afferents from the digits. Experimental Brain Research, 108, 155-171.

Macefield, V.G. and Johansson, R.S. (1996) Control of grip force during restraint of an object held between finger and thumb: Responses of muscle and joint afferents from the digits. Experimental Brain Research, 108, 172-184.

Macefield, V.G. and Wallin B.G. (1996) The discharge behaviour of single sympathetic neurones supplying human sweat glands. Journal of the Autonomic Nervous System, 61(3), 277-286.

Macefield, V.G., Wallin, B.G. and Vallbo, Å.B. (1994) The discharge behaviour of single vasoconstrictor motoneurones in human muscle nerves. Journal of Physiology, 481, 799-809.

Miller, M.R., Ralston, H.J. and Kasahara, M. (1958) The pattern of cutaneous innervation of the human hand. Americal Journal of Anatomy, 102, 183-217.

Nordin, M. (1990) Low-threshold mechanoreceptive and nociceptive units with unmyelinated (C) fibres in the human supraorbital nerve. Journal of Physiology, 426, 229-240.

Ochoa, J. and Torebjörk, E. (1983) Sensations evoked by intraneural microstimulation of single mechanoreceptor units innervating the human hand. Journal of Physiology, 342, 633-654.

Ochoa, J. and Torebjork, E. (1989) Sensations evoked by intraneural microstimulation of C nociceptor fibres in human skin nerves. Journal of Physiology, 415, 583-599.

Phillips, J.R., Johansson, R.S. and Johnson, K.O. (1990) Representation of braille characters in human nerve fibres. Experimental Brain Research, 81, 589-592.

Phillips, J.R., Johansson, R.S. and Johnson, K.O. (1992) Responses of human mechanoreceptive afferents to embossed dot arrays scanned across fingerpad skin. Journal of Neuroscience, 12, 827-839.

Phillips, J.R. and Johnson, K.O. (1981) Tactile spatial resolution. III. A continuum mechanics model of skin predicting mechanoreceptor responses to bars, edges, and gratings. Journal of Neurophysiology, 46, 1204-1225.

Schady, W.J.L. and Torebjörk, H.E. (1983) Projected and receptive fields: a comparison of projected areas of sensations evoked by intraneural stimulation of mechanoreceptive units, and their innervation territories. Acta Physiologica Scandinavica, 119, 267-275.

Schady, W.J., Torebjörk, H.E. and Ochoa, J.L. (1983a) Peripheral projections of nerve fibres in the human median nerve. Brain Research, 277, 249-261.

Schady, W.J.L., Torebjörk, H.E. and Ochoa, J.L. (1983b) Cerebral localization function from the input of single mechanoreceptive units in man. Acta Physiologica Scandinavica, 119, 277-285.

Simone, D.A., Marchettini, P., Caputi, G. and Ochoa, J.L. (1994) Identification of muscle afferents subserving sensation of deep pain in humans. Journal of Neurophysiology, 72, 883-889.

Torebjörk, E. and Ochoa, J. (1980) Specific sensations evoked by activity in single identified sensory units in man. Acta Physiologica Scandinavica, 110, 445-447.

Torebjörk, H.E. and Ochoa, J.L. (1990) New method to identify nociceptor units innervating glabrous skin of the human hand. Experimental Brain Research, 81, 509-14.

Torebjörk, H.E., Vallbo, Å.B. and Ochoa, J.L. (1987) Intraneural microstimulation in man. Its relation to specificity of tactile sensations. Brain, 110, 1509-29.

Trulsson, M., Johansson, R.S. and Olsson K.A. (1992) Directional sensitivity of human periodontal mechanoreceptive afferents to forces applied to the teeth. Journal of Physiology, 447, 373-389.

Vallbo, Å.B. (1970a) Slowly adapting muscle receptors in man. Acta Physiologica Scandinavica, 78, 315-333.

Vallbo, Å.B. (1970b) Discharge patterns in human muscle spindle afferents during isometric voluntary contractions. Acta Physiologica Scandinavica, 80, 552-566.

Vallbo, Å.B. (1976) Prediction of propagation block on the basis of impulse shape in single unit recordings from human nerves. Acta Physiologica Scandinavica, 97, 66-74.

Vallbo, Å.B. (1981) Sensations evoked from the glabrous skin of the human hand by electrical stimulation of unitary mechanosensitive afferents. Brain Research, 215, 359-363.

Vallbo, Å.B. and Hagbarth, K.-E. (1968) Activity from skin mechanoreceptors recorded percutaneously in awake human subjects. Experimental Neurology, 21, 270-289.

Vallbo, Å.B., Olausson, H., Wessberg, J. and Kakuda, N. (1995) Receptive field characteristics of tactile units with myelinated afferents in hairy skin of human subjects. Journal of Physiology, 183, 783-795.

Vallbo, Å.B., Olausson, H., Wessberg, J. and Norrsell, U. (1993) A system of unmyelinated afferents for innocuous mechanoreception in the human skin. Brain Research, 628, 301-304.

Vallbo, Å.B., Olsson, K.A., Westberg, K.G. and Clark, F.J. (1984) Microstimulation of single tactile afferents from the human hand. Sensory attributes related to unit type and properties of receptive fields. Brain, 107, 727-749.

Westling, G. and Johansson, R.S. (1987) Responses in glabrous skin mechanoreceptors during precision grip in humans. Experimental Brain Research, 66, 128-140.

Westling, G., Johansson, R.S., Thomas, C.K. and Bigland-Ritchie, B. (1990) Measurement of contractile and electrical properties of single human thenar motor units in response to intraneural motor-axon stimulation. Journal of Neurophysiology, 64, 1331-1338.

Yarnitsky, D., Simone, D.A., Dotson, R.M., Cline, M.A. and Ochoa, J.L. (1992) Single C nociceptor responses and psychophysical parameters of evoked pain: effect of rate of rise of heat stimuli in humans. Journal of Physiology, 450, 581-592.

Neural Aspects of Tactile Sensation
J.W. Morley (Editor)
1998 Elsevier Science B.V.

Similarities Between Touch and Vision.

S. S. Hsiao

It has long been speculated that vision and touch may be based on similar neural mechanisms. Both systems rely on inputs from two-dimensional sheets of receptors and both systems are faced with the similar problem of needing to extract from those sensory sheets the rich information that accounts for visual and tactile perception. Recently, there is a growing body of literature emerging from research in both systems that suggests that the similarities go beyond speculation and that analogous pathways in the two system that have similar response properties, are organized in similar ways, and underlie similar aspects of perception.

Although, the mechanisms of sensory transduction in the visual and somatosensory systems are by necessity different in the two systems, it is not obvious whether the neural mechanisms that operate within the two sensory systems are the same or different. In fact, one could propose that it is more likely that the nervous system evolved a common neural mechanism for coding and storing information in the two systems since the sensory goals of the two systems are similar; that is, to recognize and discriminate objects. One teleological reason that one might believe that sensory systems are based on common principles is that sensory integration is integral in everyday life and ordinarily, objects are recognized without confusion whether they are seen or touched and often both senses are required for accurate identification. For example, a ball can be viewed as a round flat surface until it is palpated, and tactually distinguishing a dime from a penny often requires supplementary visual information.

If touch and vision are closely related then what is the neural basis for intersensory cooperation? There are two possibilities. The first is that both systems utilize completely different mechanisms for processing spatial information and that tactile and visual representations of objects are stored in separate locations in the CNS. In this scheme, sensory inputs from the two systems are matched against their individual central representations and the two systems then communicate between each other by some associative mechanism that automatically activates the corresponding representation in the other system (Murray and Mishkin, 1985).   At the other extreme is the possibility that there is a single representation for objects within the central nervous system and that both sensory systems have direct access to that representation.   In this scheme, touch and vision activate the different features of the same central representation through separate ascending pathways. In both of these schemes the primary role of the sensory pathways underlying both vision and touch is to transform the sensory input into the central representations that underlie perception with conflicting information from the two senses being resolved by some hierarchical order whereby the sensory input from one sense dominates the inputs from the other sense. (Cholewiak and Collins, 1991; Lederman et al., 1986; Heller, 1983; Heller, 1982; Lederman and Abbott, 1981). Whether or not perception is based on a common central representations, Craig (1979) has pointed out that if the spatial recognition performance of the two systems is similar then it is most likely that common neural mechanisms are involved in processing spatial information in the two systems.

Although the central neural mechanisms underlying perception are far from being understood, it is clear that at the level of the peripheral inputs (i.e. the skin and retina) that the somatosensory and visual systems are organized into separate parallel pathways that subserve different sensory functions (Shapley, 1995; DeYoe and Van Essen, 1988; Schiller, 1993; Schiller, 1992; Maunsell et al., 1986; Johnson and Hsiao, 1992; Bolanowski et al., 1988; Phillips et al., 1988). The segregation of function begins at the level of the peripheral receptors which are located in the retina and skin and appears to remain fundamentally segregated all the way to the level of perception.   In vision three principal input systems have been reported (Hendry and Yoshioka, 1994), with two of those systems, which will be the focus of this chapter, being well described.   The two systems, called the parvocellular or P system and the magnocellular or M system, receive their inputs from the color-opponent and broad-band ganglia cells of the retina (Schiller and Logothetis, 1990).   The terms parvo- and magnocellular are derived from the projections of the two different kinds of retinal ganglion cells to the parvocellular and magnocellular layers of the lateral geniculate nucleus in the thalamus. These separate pathways, which receive different retinal inputs and have different cortical projections, underlie different aspects of visual perception. In touch, there are four principal input systems: the slowly adapting type I (SAI) system, the rapidly adapting (RA) system, Pacinian (PC) system and the slowly adapting type II (SAII) system that provide the central nervous system with mechanoreceptive information. Of these four input systems, only the SAI and RA afferent pathways innervate the skin with sufficient density to be able to transmit detailed spatial information to the central nervous system. In this chapter I provide evidence that suggests that the SAI and RA afferent pathways appear to play analogous roles in tactile perception that the P and M pathways play in vision. The other two kinds of mechanoreceptive afferents innervate the skin with lower density, are located deep in the skin, and appear to play different roles in tactile perception (see Johnson, 1992 for a review).

The working hypothesis that will be proposed in this chapter is that the P and SAI systems and the M and RA systems play similar roles in vision and touch. The P and SAI systems, which will be referred to as the sustained pathways, have high spatial but low temporal resolving capacities and are devoted to processing information concerning form and texture. In contrast, the M and RA systems, which will be referred to as the transient pathways, have lower spatial but high temporal resolving capacities and are devoted to processing motion across the skin or retina and to processing low frequency temporal information which are termed flutter in the somatosensory system and flicker in the visual system. The intention here is not to suggest that these systems are identical, nor is it to provide an exhaustive review of the literature. Rather, the intention is to compare spatial processing in the two sensory systems in three different ways. The striking results that emerge from such a comparison are that: 1) there are close similarities in the physiological responses of the different afferent types in the two systems; 2) there are close similarities in the way that form, motion and texture are perceived in the two systems; and 3) the sustained and transient pathways in vision and touch are linked to similar aspects of perception. These findings lead to the intriguing possibility that vision and touch may employ similar neural mechanisms at all levels in the central nervous system.

1. Peripheral receptors

For both sensory systems the division of labor begins at the level of the peripheral receptors that densely innervate the skin and retina. In the skin, the SAI and RA afferent populations which receive their inputs from Merkel cells and Meissner corpuscles, respectively, densely innervate the skin on the fingertips suggesting that these afferents are responsible for processing spatial information. The innervation densities of SAI and RA afferent fibers in man and monkey are about 150 RA and about 100 SAI afferents per square cm of skin on the fingertips (Darian-Smith and Kenins, 1980; Johansson and Vallbo, 1979). This corresponds to mean afferent spacings of approximately 0.8 mm for RA afferents and 1.0 mm for SAI afferents. The high innervation densities suggests that both of these afferent types convey to the central nervous system a spatial representation of stimuli presented to the skin. Similarly in vision, the estimated cone spacing is about 0.45 min of arc in the fovea of the retina (O'Brien, 1951). The cones which provide inputs to both the P and M systems are not uniformly distributed across the retina with cells that have transient responses (M like) being relatively more common toward the peripheral retina and cells that have sustained responses (P like) being relatively more common toward the fovea (Gouras, 1968).

The peripheral neurons in both systems respond differently in their adaptive properties, in their sensitivities to spatial and temporal stimuli and their dynamic range or contrast sensitivities. The SAI and P ganglion cells are sensitive to similar features of stimuli as are the RA and M ganglion cells which suggest that these afferent pathways play similar roles in tactile and visual perception. In touch and vision, the SAI and P pathways are the high spatial and low temporal frequency channels. Neurons within these pathways have high spatial acuity and relatively low temporal acuity suggesting that these channels are specialized for resolving the fine spatial details while ignoring the temporal aspects of stimuli. In contrast, the RA and M pathways have lower spatial resolving capacities than do the SAI and P pathways and instead are finely tuned to detect moving or temporal aspects of stimuli.

A comparison of six characteristic properties of these inputs to these separate pathways are listed in table 1. While the fine details such as the absolute receptive field size or the innervation densities are different between the two sensory systems, it is clear that within the somatosensory system the two principal spatial systems correspond in practically all ways to the P and M pathways in the visual system.

The adaptive properties of the peripheral afferents have been extensively studied in both the visual and tactile systems (Darian-Smith, 1984; Mountcastle, 1984; Gouras, 1968). In both

Table 1: Comparison of SAI, RA somatic afferents with Parvo-and Magnocellular cells in the visual system.

Property	SAI	RA	Parvo	Magno
Responses to a sustained stimulus	sustained discharge	Transient discharge	sustained discharge	Transient discharge
Spatial acuity	high	low	high	low
Low Contrast sensitivity	low	high	low	high
Flutter detection	poor	good	poor	good
Receptive field size	small	large	small	large
Motion processing	no	yes	no	yes

systems, the sustained pathways respond with a continuous train of impulses to a steady stimulus. An example of a sustained response in the somatosensory system and schematized response histogram of a color-opponent cell is shown in figure 1. These sustained neurons are characterized by having an initial burst of activity that coincides with the onset of the stimulus, that rapidly declines to a sustained impulse rate proportional to the intensity of the stimulus. The response at the offset of the stimulus appears to be different between the two systems in that while the SAI afferents stop firing after the stimulus is removed, the P ganglion cells appear to show a transient decrease in firing rate below the background firing rate (see figure 1). The onset and sustained responses of neurons in the transient pathways also appears to be similar (see figure 1). Again, both the RA and M systems respond with a rapid burst of activity to the onset of the stimulus, however for both afferent types the activity rapidly declines to the background firing rate of the neuron. The offset responses for the two systems appear to differ, with the RA afferents responding with a brief burst of impulses at the offset of the stimulus and the broad-band system showing a transient decrease in impulse rate.

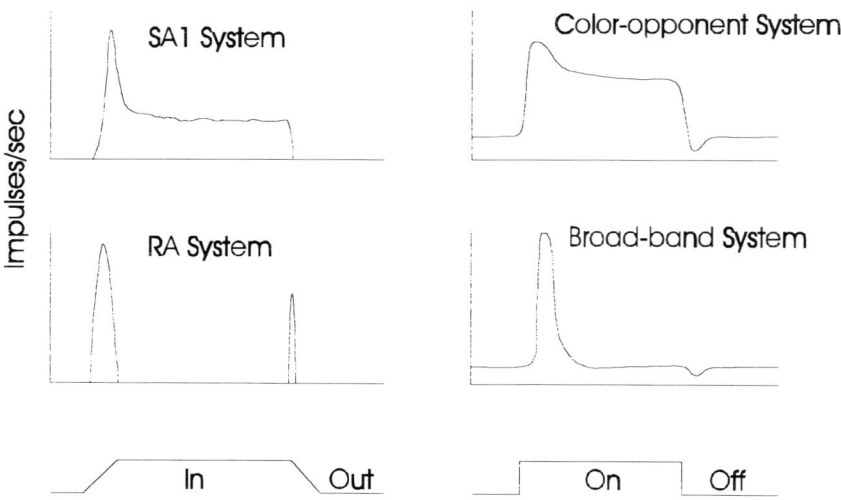

Figure 1 Comparison of the adaptive properties of afferents in the somatic and visual system. The SAI and RA responses are the mean impulse rates evoked by a 1 mm diameter probe indented 500 micron into the skin. The visual responses are adapted from Schiller and Logothetis (1990).

1.1 Receptive fields and spatial and temporal acuity

A significant distinction between the two processing streams is that the sustained systems have smaller receptive fields and a higher capacity to resolve spatial information than do the transient systems (Croner and Kaplan, 1995; Phillips et al., 1988; Johnson and Lamb, 1981).

Receptive field sizes of the SAI and RA afferents are consistent within each class, with SAI afferents having mean diameters of about 2-3 mm and RA afferents having significantly larger fields that average about 4-5 mm (Johnson and Lamb, 1981). SAI afferents also have a higher spatial resolving capacity than RA afferents. Phillips and Johnson (1981a) investigated the responses of SAI and RA afferents to embossed gratings that were pressed into the skin without horizontal scanning. They found that the SAI afferents were highly sensitive to the spatial structure of the gratings and clearly resolved gratings that had gap widths 0.5 mm wide. (see figure 2). In contrast, the RA afferents were insensitive to the fine spatial structure of the stimuli, and a

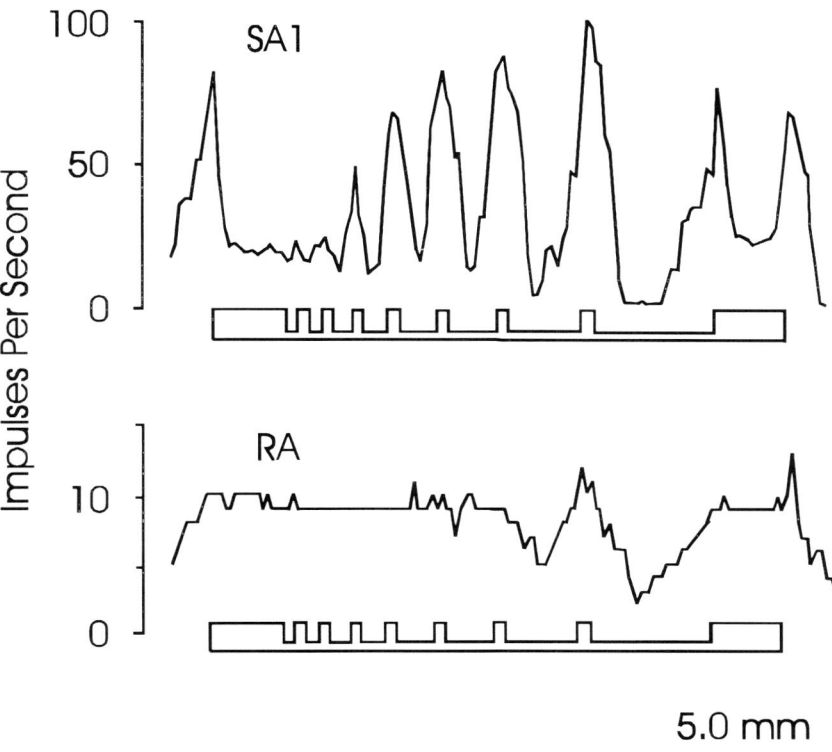

Figure 2. Responses of SAI and RA afferents to an aperiodic grating composed of 0.5 mm wide bars separated by gaps of various widths. The grating pattern was pressed 1.0 mm into the monkey's fingertip at various positions relative to the neuron's receptive field. (Adapted from Phillips and Johnson 1981).

majority of the afferents were unable to resolve gratings with gaps 3 to 5 mm wide. Consistent with these findings, the ability of the SAI afferents to resolve spatial details corresponds closely to the minimum psychophysical gap detection thresholds which will be discussed in detail later.

Many studies have shown that the RA afferents are acutely tuned for low frequency vibrations (see Darian-Smith, 1984 and Mountcastle 1984 for a review). Briefly, studies that investigated the sensitivity of the peripheral afferents to vibrations show that when low frequency vibrations are presented to the receptive fields of the peripheral SAI, RA, and PC afferents of monkeys using a punctate probe, the vibratory tuning curves for these afferents are "U" shaped functions, with RA afferents having minimum absolute and entrainment thresholds of about 20 microns near 30 Hz, SA afferents having significantly higher thresholds (about 120 microns at 30 Hz) and PC afferents having minimum thresholds near 200 Hz. Mountcastle and his colleagues (Talbot et al., 1968; Mountcastle et al., 1972) also showed that only the RA afferents had absolute and entrainment thresholds that could account for human psychophysical detection thresholds at low vibratory frequencies (below about 100 Hz), which they termed the sense of flutter to contrast it with the sensation of high frequency vibrations, which are coded by the PC afferents.

Studies in the visual system show that neurons in the P system have smaller receptive fields than neurons in the M system, and that neurons in the P system are more sensitive to high spatial frequencies than are neurons in the M system (see Croner and Kaplan 1994). The mean central radii of P ganglion cells near the fovea are less than 0.05 degrees, while M ganglion cells have radii that are about 4 times greater. Figure 3 shows a schematic adapted from Schiller and Logothetis (1990) illustrating the relative processing capacities of the color-opponent and broad-band channels for spatial and temporal information. Although there is some overlap between the two systems, the P system is more effective at processing high spatial frequencies while the M system is more effective at processing high temporal frequencies.

These studies strongly suggest that the SAI and P systems are high spatial acuity systems and are well suited for detecting the fine surface features of objects while the RA and M systems are tuned to detect low spatial but high temporal features of stimuli and as such are well suited for detecting tactile flutter, visual flicker, or motion of stimuli.

1.2 Contrast sensitivity

Studies on which afferent types are primarily responsible for detecting contrast sensitivity in the tactile system show mixed results. Lamotte and his colleagues have shown that at low contrasts, the ability to detect a small asperity on a smooth background can only be conveyed by RA afferent responses (Srinivasan et al., 1990; LaMotte and Whitehouse, 1986; Johansson and LaMotte, 1983). They showed that RA afferents are highly sensitive to asperities below 20 microns, while the SA and PC afferents only responded when the asperity height was greater than 30 microns. Furthermore, they show that increases in

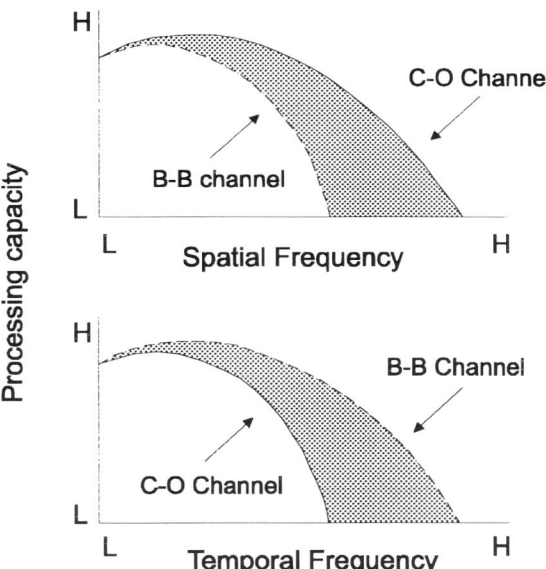

Figure 3. Schematic showing the relative spatial and temporal processing capacities for the color-opponent and broad-band systems in the visual system. H and L stand for high and low. (Adapted from Schiller and Logothetis, 1990)

dot height results in a rapid increase in firing rate in the RA afferents, illustrating that these afferents have high gain sensitivity to low contrast stimuli.   However, recent studies indicate that RA afferent responses tend to saturate at large contrasts above 200 microns.   At these contrast levels the roles of the SAI and RA afferents reverse, with SAI afferents being highly sensitive to changes in pattern height and RA afferents being unaffected.   Increasing the height from 280 to 620 microns results in a 44% increase in the firing rate of SAI afferents, while identical changes in height only minimally affected (10% increase) the response of the RA afferents (Blake et al., 1997).

There have been multiple studies showing the differences in contrast gain for the magno- and parvocellular systems (see Croner and Kaplan, 1995 and Shapley, 1990 for a review). Briefly, these studies have shown that when contrast gain is compared between the P and M systems, the populations form two completely separate clusters, with the P cells having significantly lower contrast gains than M cells, and with P cells becoming unresponsive to gratings when the mean luminance drops below about 0.1 cd/m2. M cells also become less sensitive but remain responsive into the scotopic range. Croner and Kaplan (1995) show that the contrast gains for both kinds of cells are constant across the visual field, with the M cells having, on average, a gain that is six times greater than that of P cells. Their data also indicate that at high contrasts the M cells tend to show saturated responses with increases in contrast while the P cells show increased sensitivity with increased contrast.

2. Psychophysical studies

Evidence from studies of acuity, form processing and roughness perception indicate that the psychophysics of touch and vision and are similar. When stimuli are scaled to the primary afferent spacings in the retina and fingertip there is strong evidence that spatial acuity is determined by the spacings of the peripheral afferents, and that when stimuli are scaled to the innervation density of the receptors the perception of spatial form in the two systems is similar (Johnson and Phillips, 1981; Phillips and Johnson, 1981a). Figure 4 shows the results of a series of psychophysical experiments that investigated the tactile spatial resolving capacities of human subjects. In these studies, subjects were required to either detect the presence of a varying sized gap in a smooth surface, determine the orientation of a grating that varied in spatial frequency, or identify embossed letters that varied in height. In each case there was a systematic increase in subject performance as the stimulus element increased in size indicating that the discrimination was based on spatial rather than nonspatial cues. In figure 4 the psychometric functions for each of these tasks are normalized to the fundamental element width, which corresponds to the gap size, bar width, or one-fifth the letter height. This study showed that the psychometric threshold for tactile spatial acuity is about 1.0 mm, which is approximately the spacing of SAI receptors in the fingertip (see above). In each case the threshold corresponds to the minimum number of primary afferent receptors necessary to resolve the critical details of the stimulus. Thus for gaps and gratings this corresponds to resolving the gap from the flat surface or the gap from the bar and for the letter stimuli this corresponds to the ability to resolve the minimum details of a typical letter. For letters this corresponds to a minimum resolution of 5 receptors to resolve the three horizontal bars and the gaps within a typical letter such as the letter E. Because RA afferents were unable to resolve the spatial details of grating with spatial elements spaced 3 to 5 mm apart (see above), it is clear that SAI's and not RA's are responsible for fine spatial tactile form processing. Phillips et al. (1983) reported similar findings in their studies in which they compared the visual and tactile systems in letter discrimination tasks and concluded that like in touch, the limits of visual letter recognition ability is determined by the innervation density of neurons innervating the retina (see below).

Several experiments have directly compared vision and touch in form processing tasks. Owen and Brown (1970) compared the perception of tactile surfaces and black-and-white figures that contained identical forms that varied in complexity. They showed that the forms were judged to be nearly identically complex regardless of the input modality and concluded that there is a perceptual world that exhibits invariance regardless of the input modality. Phillips et al. (1983) showed similar findings in their studies comparing tactile and visual letter recognition abilities. They tested subject recognition performance using embossed and visual letters scaled so that the letter

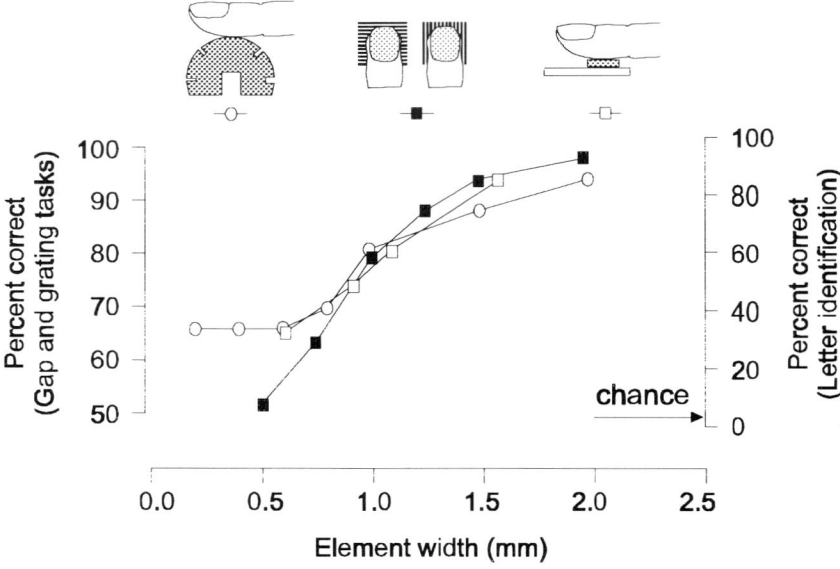

Figure 4. Human performance in tactile gap detection, grating orientation, and letter recognition tasks. The abscissa represents the gap size in the detection task, the bar width in the grating orientation task and one-fifth the letter height in the letter recognition task. The left and right axis are scaled to align change and perfect performance in the three tasks with threshold is defined to be the element size midway between these two levels. (Adapted from Johnson and Phillips, 1981).

heights would span approximately the same number of receptors in the two systems (i.e. embossed letters that range from 3.0 to 8.0 mm high and visual letters ranging from 1.5 to 3.0 mm of arc). Phillips et al., (1983) found that static or scanned modes of presentation yielded similar recognition performance and that letter heights that spanned about five receptor spacings in both sensory systems yielded similar levels of performance (see figure 5). In figure 5, letters that were easy to recognize tactually such as the letters I, L J, T and A were also the letters that were the most accurately recognized visually and similarly letters that were often misidentified in vision such as B, S, R, and G were also misidentified in touch. In addition, they found that the patterns of letter confusions that were made in the two modes of presentation were highly similar (figure 6). For example, V was often called a Y and R was often misidentified to be an A in both systems. The patterns of confusions although similar were not identical suggesting that there may be differences in the way that information is represented in both systems. The patterns of confusions that were made in the tactile experiments were predictable based on the visual similarities of the letters. Thus, for example, the letter B was often called a D and the letter T was most often misidentified as the letter Y. Similar findings have been demonstrated by Craig (1979, 1981) and Loomis (1981, 1982) using vibrotactile and embossed letter patterns. Loomis (1982) concluded from his studies, in which he compared visual and tactile confusion matrices, that visual recognition of low-pass filtered characters is, to a first approximation, a model of tactile letter recognition. In both systems, performance decreased as the complexity of the letters increased supporting the results found by Owen and Brown (1970).

In both the study by Phillips et al. (1983) and in a more recent tactile letter recognition study by Vega-Bermudez et al. (1991), subjects were given no training or feedback during the entire experiment and were instructed simply to respond with one of the letters of the alphabet. Because subjects had no previous training in this task, they must have based their tactile responses on previously stored visual representations of letters. These results strongly support the notion that there is a close correspondence between the neural mechanisms underlying spatial processing in the tactile and visual system, and that tactile representations must somehow have access to central visual representations of spatial form. Vega-Bermudez et al. (1991) also showed that there is no difference between active and passive scanning of the letters, that changes in scanning velocity had minimal effects on tactile letter recognition, and that there is a small but significant learning effect when subjects are repeatedly tested. Over multiple repeated trials, subjects showed an increase in performance of about 20%, illustrating that, since the subjects were well acquainted with letters of the alphabet, the learning must have occurred within the tactile system.

There have been many studies that have compared tactile and visual roughness (for example, Lederman et al., 1986; Heller, 1982;

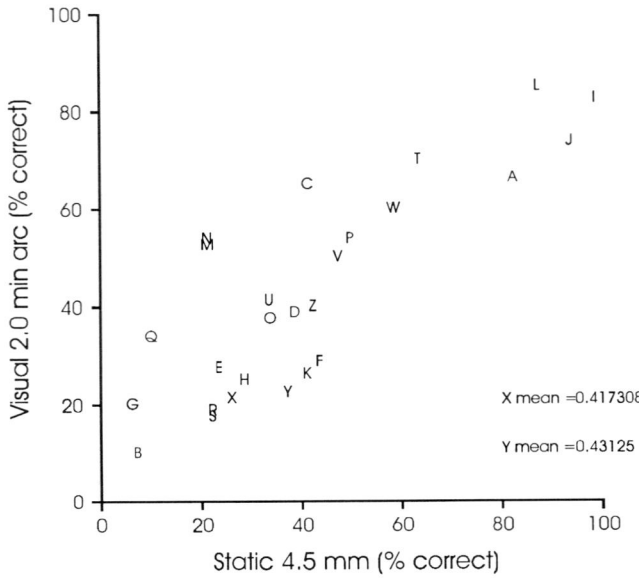

Figure 1.   Comparison of human visual and tactile letter recognition
performance. The letter heights were scaled so that the mean percent correct
in both sensory modalities was about 50%. Letters within the graph represent
the mean correct identification for that letter. (Adapted from Phillips, Johnson
and Browne, 1983).

Lederman and Abbott, 1981). In a direct comparison of tactile and
visual roughness perception, Heller (1982) found that subjects made
comparable judgments in texture discrimination when asked to
compare the relative smoothness of various surfaces when these
surfaces were either viewed or touched alone, or when both senses
were used. Similarly, Lederman and her colleagues have shown that
visual and tactile judgments of surface roughness yielded nearly
identical subjective magnitude estimates (Lederman et al., 1986). She
showed that sandpaper with increasing grit both felt and appeared to
be of comparable smoothness. These results indicate that roughness,
as well as form, may have similar attributes in the tactile and visual
systems. Whether other aspects of the tactile and visual perception
such as flicker, flutter, and motion are the same in the two systems
have not been studied.

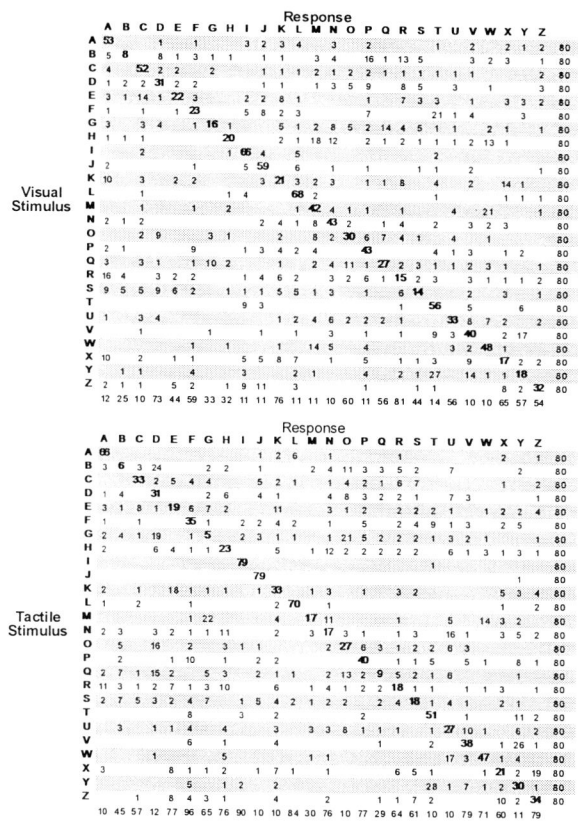

Figure 2. Confusion matrices constructed from the pooled data from four subjects for visual and static tactile letter recognition tasks. The visual letters (2.0 min high) and the tactile letter heights (4.5 mm high) resulted in similar performance levels (43% visual and 42% tactile). The right hand column shows the number of time each letter was presented and the bottom row shows the number of times subjects responded with that letter. (Adapted from Phillips, Johnson and Browne, 1983).

3. Linking the different pathways to perception

One of the principal goals in neurophysiology is to link observed neural responses with perception. The problem is difficult and can be approached from at least two different directions. The first is to train animals to perform a behavioral task and observe whether there are changes in the behavior when single populations of neurons are either selectively activated or deactivated. One way of selecting a population of neurons, which is the approach that was used in the visual studies that will be described, is to identify neural structures, such as the LGN, where the different populations of cells can be selectively lesioned. The animal's behavior is then tested before and after the lesions and one can then infer from the changes in the animal's performance whether the lesioned pathway was responsible for the original behavior. The second approach, which was originally employed by Mountcastle (Mountcastle et al., 1972; Talbot et al., 1968), is to first carry out psychophysical studies in humans or animals using a stimulus that evokes the desired behavior, and to follow those studies with neurophysiological studies in monkeys (or humans if possible) using the identical stimulus. The goal from these studies is to infer from the neural population response the neural basis for the behavior, that is, to understand which neural populations are responsible for the behavior and how the information is represented within the population of neurons. This is the approach that was used in most of the somatosensory studies that will be discussed.

3.1 Studies Linking SAI afferents to Tactile Form and texture perception

There is now strong evidence that the SAI system is responsible for tactile spatial form, shape and texture perception (see Johnson, 1992 for a review and Goodwin et al., 1996; LaMotte et al., 1996; Goodwin et al., 1991 for a discussion of the coding of tactile shape). The evidence for SAI coding for two-dimensional spatial form comes from three lines of experimentation. The first, which has already been discussed in some detail, is studies which show that the SAI afferents account for fine spatial acuity. The relevant results from those studies are that the SAI afferents are able to resolve gaps as small as 0.5 mm and this corresponds to the limits measured in psychophysical experiments. The RA afferents are unable to resolve gaps that are more than five times wider. So the conclusion from these studies is that for statically placed stimuli, only the SAI afferents have sufficient acuity to account for the fine spatial acuity that is observed in human psychophysical studies (Johnson and Phillips, 1981; Phillips and Johnson, 1981a).

The second line of evidence comes from studies that investigated tactile letter recognition using the Optacon, which is a vibrotactile pattern stimulator that consists of 144 vibrating pins arranged in a two dimensional array. Gardner and Palmer (1989) have shown that while the Optacon vigorously activates the RA (and PC) afferents, it does not activate the SAI afferents. Presumably, the reason for this is that the vibrating pins do not indent the skin and generate a strain field with sufficient magnitude to activate the Merkel cells. When the Optacon is employed in psychophysical letter recognition experiments, the results are consistent with what would be predicted based on the grating studies. That is, the threshold height for recognizing vibrating letters increases from around 5 mm when the finger is directly in contact with the embossed letters to 20 mm when only the RA afferents are activated (Phillips et al., 1983; Loomis, 1980; Craig, 1979; Bliss, 1969). These results show that although the RA afferents do not code for fine spatial form, they do provide a crude representation that is accessible by the central form processing mechanisms.

The responses of the SAI and RA neurons have been extensively examined in neurophysiological studies at the peripheral and central levels using embossed letters as stimuli. If SAI or RA afferents are responsible for fine form perception, then there must exist at all levels in the pathways leading to perception a neural representation of the stimulus that accounts for the psychophysical data. Figure 7 shows the response of typical peripheral SAI and RA afferents to the letters A-D, 6.0 mm, high scanned at two different velocities across the fingertip of a macaque monkey. Both afferent types respond with images that are isomorphic to the letter stimuli, and both afferents are relatively unaffected by changes in the scanning velocity or contact force (20-80 gms). Within each population, all of the afferents in the periphery respond in a similar way indicating that there is an isomorphic representation of spatial form at the peripheral level (Phillips et al., 1990; Phillips et al., 1988). Although the peripheral responses have the same overall shape as the stimulus letters, they are distorted in two fundamental ways. First, the responses are blurred relative to the stimulus letters. Second, both features that lie parallel to the direction of scanning and internal structural features of the letters are less well preserved than are leading and trailing features. For example, in figure 7, the leading bar in the letter B is enhanced while the trailing and

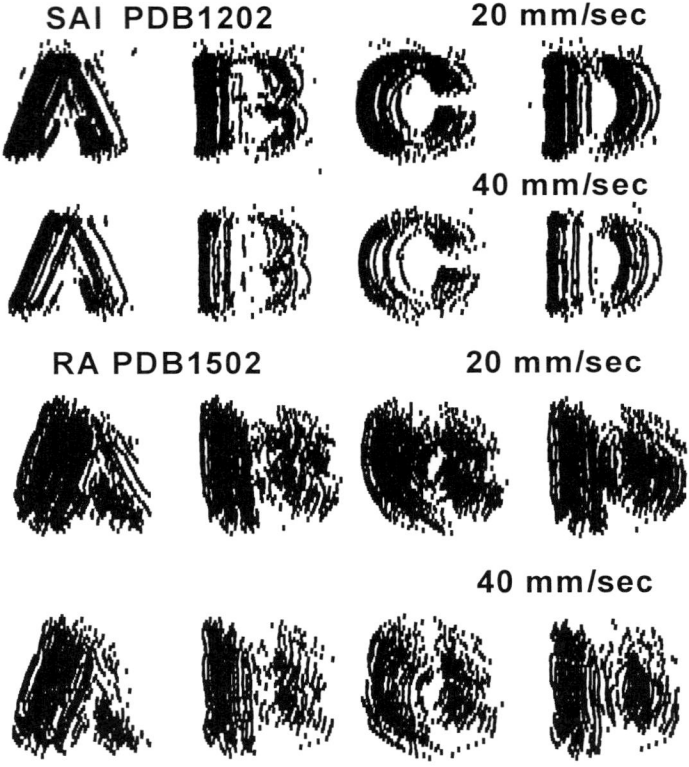

Figure 1. Response of typical SAI and RA afferent fibers to embossed letters scanned at 20 and 40 mm/sec across the neuron's receptive field. The stimulus letters were 6.0 mm high and were raised 500 microns above the background surface. The letters were mounted on a drum that repeatedly scanned the patterns across the animal's restrained fingertip. Initially no part of the drum was in contact with the neuron's receptive field, however after every rotation of the drum, the pattern was shifted 200 microns so that after about 75 scans the entire pattern scanned across the receptive field. Each black tick represents the occurrence of an action potential.

horizontal features are either poorly represented or are missing completely. The loss in internal features due to scanning is clearly reflected in the psychophysical results that were shown earlier and may account for the differences in the tactile and visual confusion matricies. For example in the confusion matrix shown in Figure 6, there is a response asymmetry for the letters B and D. Subjects often reported the letter B as being the letter D, but rarely reported D as being B. The explanation for this asymmetry can be found in the lack of response by SAI (and RA) afferents to internal features that are parallel to the scanning direction supporting the view that there is a direct link between the neural responses of the SAI and RA afferents and behavior.

Although the SAI afferents preserve more of the details and provide a more acute image of the letters, the image provided by the RA afferents is more acute than what would be expected based on the results of the grating study described earlier (Phillips and Johnson, 1981a). The likely explanation is that since RA afferents are highly sensitive to indentation velocity (Pubols, Jr. and Pubols, 1976) and movement across the skin, scanning causes these afferents to modulate their responses in concert with the spatial pattern. Because both afferent types were well modulated by scanned patterns, the peripheral RA response does not provide evidence to rule out the RA system in playing a role in form discrimination.

A significant difference, however, does exist in the sustained and transient cortical responses to scanned letters. In contrast to what is observed in the peripheral afferent response, neurons in area 3b respond to embossed letters with a wide variety of neural responses (see figure 8). Some neurons have isomorphic responses, like those observed in the periphery, while other neurons have nonisomorphic responses. Because almost all of the neurons in area 3b that respond with high spatiotemporal structure tend to show sustained responses, and neurons that respond with weak or no structure tend to be RA- or PC-like, these results support the notion that the sustained and not the transient pathway is involved in form processing (Phillips et al., 1988; Hsiao et al., 1996).

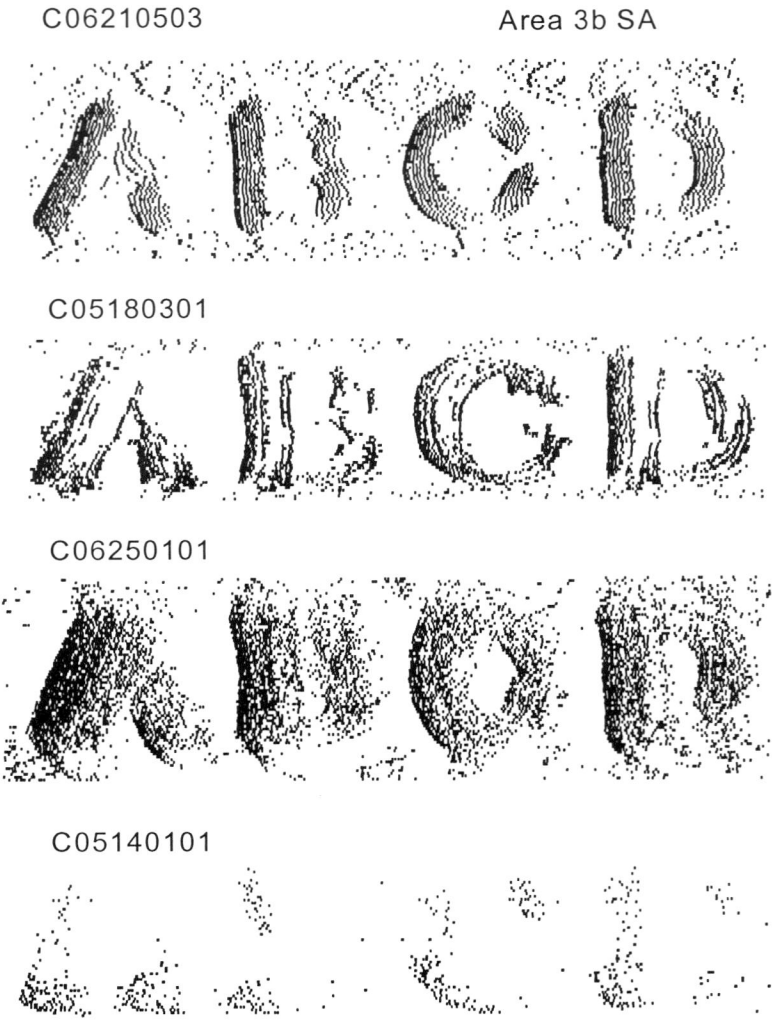

Figure 2. Response of four slowly adapting neurons to embossed letters of the alphabet in area 3b of the somatosensory cortex of an awake macaque monkey. See figure 7 for details.

4. Texture perception

Tactile texture perception is in general poorly understood. It is a multidimensional perception of surface features that includes the sensations of roughness, hardness, leatheriness, slipperiness, etc. Of these, the sensations of roughness-smoothness and hardness-softness have been studied in some detail, and recent results have shown that discrimination along both of these dimensions is mediated by activity in the SAI afferent population. These results are surprising since it had been previously proposed that texture perception is based on the relative neural activity among the SAI, RA and PC afferents (Johnson, 1983). However, recent studies have indicated that the perception of texture when the finger is in direct contact with the surface is most likely based solely on the SAI afferent population response (Blake et al., 1994; Johnson and Hsiao, 1994; Hsiao et al., 1993).

There have been many studies on the neural mechanisms underlying roughness perception (Blake et al., 1994; Burton and Sinclair, 1994; Johnson and Hsiao, 1994; Smith, 1994; Johnson et al., 1996; Goodwin, 1993; Hsiao et al., 1993; Connor and Johnson, 1992; Connor et al., 1990; Lederman, 1985) to name a few). Three recent studies (Blake et al., 1994; Connor and Johnson, 1992; Connor et al., 1990) have shown that when surfaces are scanned directly with the finger, roughness is based on the spatial variation in firing rates between the SAI afferents innervating the skin. The three studies employed the identical experimental design and differed only in the surfaces that were used. The experimental design consisted of using the identical surfaces in neurophysiological experiments in monkeys and psychophysical experiments in humans, and inferring from the neurophysiological recordings the physiological basis for the observed human performance. In the neurophysiological experiments, recordings were performed in the peripheral nerves of anesthetized animals while the stimulus surfaces were scanned across the receptive fields of a large number of SAI, RA, and PC afferents. In the psychophysical experiments, human subjects were required to scan their fingertips across the identical surfaces and to report their subjective magnitude estimates of the roughness of the surfaces. The assumption is then made that if the peripheral afferent types and their respective innervation densities are similar between the two species, then the physiological data from the monkeys can be used to model the afferent information upon which humans based their psychophysical judgments.

The first two experiments (Connor and Johnson, 1992; Connor et al., 1990) were designed to investigate the peripheral neural code underlying roughness perception. Once the coding mechanisms were understood, then a third experiment was performed to test whether roughness is conveyed by the SAI or RA afferent response. The first experiment (Connor et al., 1990) used surfaces with dots that varied in spacing (1.2 to 6.2 mm) and diameter (0.5, 0.7, and 1.0 mm). A large number of hypothetical coding mechanisms were tested such as mean firing rate, variance in firing rate, short-term temporal variation, and local spatial variation. No coding mechanism based on combined information from the three afferent types accounted for the psychophysical data. However, coding measures based on firing rate variation, especially the local spatial variation in firing rate among the SAI afferents, accounted well for the psychophysical behavior. The second study (Connorand Johnson, 1992) addressed the question of whether the coding is based on spatial (across fiber) or temporal (within fiber) variation in firing rates. The surfaces that were used consisted of dots in which the spacing varied either in the along (scanning direction) or across direction. If roughness is based on temporal variation then decreasing the spacing of the dots in the across direction while leaving the spacing constant in the along direction should have opposite effects on spatial and temporal codes. The reasoning is that as the dots move farther apart in the across dimension, fewer and fewer afferents should be activated by the pattern since some afferents that were activated when the dots were closely spaced will now fall between the dots resulting in a decrease in the mean temporal variation in firing rate in the afferent population. In contrast, the total spatial variation should increase since fibers in the across direction will show minimal variation when the dots are close together and maximal variation when the dots are far apart. The psychophysical and neurophysiological results clearly supported the hypothesis that roughness is based on local spatial variation in firing rates.

Although there were strong indications from the first two studies that roughness is based on the activity in the SAI afferent response (Johnson and Hsiao, 1994), a third study (Blake et al., 1997) was designed to directly test whether roughness is based on the SAI or the RA population response, or some combination of the two. In this study, dot patterns were used with dots that varied in diameter and in height, since the first study had shown that SAI and RA responses were differentially affected by dot diameter, and a second study had shown that there are dramatic differences in the way that the SAI and RA afferents respond to pattern height. The results from this third study indicate that while both afferents were affected by changes in dot diameter only the SAI afferents were significantly affected by changes in dot height. These results matched precisely what was observed in the psychophysical experiments with the perceived roughness of the pattern increasing with the spatial variation of the SAI, not RA, population response. The RA responses appeared to saturate for dot heights greater than about 2-300 microns whereas the SAI afferents continued to increase their firing rates as dot height increased to 680 microns.

SAI afferents are also critical for the perception of softness. Srinivasan and LaMotte (1995, 1996) investigated the ability of human subjects to discriminate surfaces that varied in softness while subjects actively and passively palpated the surfaces. They showed that a subject's ability to discriminate the softness of a surface, which is normally quite good, was practically eliminated if the cutaneous afferent responses were blocked with a local anesthetic. This clearly indicated that softness is based on the activity in either the SAI or RA responses. These authors then showed that only the SAI afferents showed a consistent relationship between their firing rate and the compliance of the surfaces. In particular, they showed that when compliant surfaces are pressed into the skin at constant rate, only the SAI afferents showed decreased responses as the surfaces became softer. Under identical circumstances, RA afferents were unaffected by changes in the compliance of the surfaces.

Despite all of these findings, to make the broad statement that all tactile texture perception is coded by the SAI population, is at present, not reasonable since all aspects of texture have yet to be studied. For instance the perception of surface texture when holding a tool is most likely not coded by the activity in the SAI afferent population. When holding a tool, the pressures and forces remain constant at the skin surface and thus the SAI afferent response would be unable to differentiate between surfaces that were at the end of the tool. In this case it is highly likely that surface roughness would be coded by vibrations that are transmitted from the working end of the tool to the hand (Katz, 1989) which most likely is coded by the responses of the PC afferents due to their high sensitivity to vibration. However, at present it is reasonable to assume that when the finger is in direct contact with a surface, the gross textural features of the surface are coded by the SAI system.

5. Motion and temporal coding.

There are many studies that have investigated temporal coding mechanisms in the somatosensory system. As described earlier, the response of the RA afferents have been clearly linked to the sensation of low frequency vibrations termed flutter (Mountcastle et al., 1972). The results from recent studies suggests that in addition to flutter, the RA afferents are responsible for the sensation of tactile motion. Gardner and Sklar (1994) used the Optacon in a study of motion processing in the somatosensory system. In their study, a moving stimulus was simulated on subjects' fingertips by activating successive pins of the stimulus array with various time delays. The results indicate that subjects have clear perception of stimulus motion that correlates well with patterns being "scanned" across the skin at velocities ranging from 30 to 120 mm/sec. Since the Optacon does not activate SAI afferents, these results imply that motion is coded by the RA (and possibly PC) afferent responses. However, recent studies using a brush stimulus indicate that all of the afferents, particularly the SAII's may play a role in motion perception (Edin et al., 1995).

Additional evidence that the RA afferents are important for detecting movement across the skin comes from experiments investigating surface slip. Srinivasan et al. (1987, 1990) showed that the detection of slip between the skin and a surface depends on activity in the RA or PC afferents. In these studies, surfaces with asperities 2 microns high were shown to activate RA afferents alone. When these surfaces were scanned across their skin, subjects reported the sensation of a dot moving across the skin. When surfaces that only activated the PC afferents were used, subjects reported that they felt a diffuse vibration instead of a moving pattern suggesting that motion is coded by the RA and not the PC afferents. Similarly, Johansson and Westling (1987) (Johansson and Westling, 1987; Westling and

Johansson, 1987) studied the activity evoked in the peripheral afferents during the different phases of gripping an object. From these studies, they concluded that the SAI afferents, which had an ongoing discharge, were providing information about the spatial features of the object. On the other hand, the RA afferents, which responded during the initial phases of the grip and during periods of microslip, were providing information to the CNS related to the coefficient of friction between the skin and the surface.

Together, these studies strongly indicate that the RA afferents are critical for detecting small movements across the skin, however, they do not necessarily rule out the SAI afferents as being unimportant, since there have been no studies in which only the SAI afferents were activated in a motion detection task.

6. Coding of form and movement by the P and M systems

There is a large volume of literature that deals with the parallel flow of information through the visual system. One recent study, in particular, has shown that the P and M systems have distinct roles in visual perception (Schiller et al., 1990a; Schiller et al., 1990b). The key in this study was that Schiller and his colleagues were able to measure how the behavior of animals changed after making selective lesions of either the M or P pathways. For these experiments they trained rhesus monkeys to do either a detection or a discrimination task in which an array of stimuli were presented around a central fixation point, and the animals learned to make a single saccade to the stimulus pattern that was different from the other patterns. The stimuli were designed to vary along a number of perceptual dimensions, including contrast sensitivity, color, brightness, spatial frequency, form, texture, depth, flicker, and motion (see Schiller et al., 1990 for details of this study). The P and M systems were then selectively disrupted by making small ibotenic-acid lesions into either the parvocellular or magnocellular portions of the lateral geniculate nucleus. After recovering from the surgery, the animals were retested intensively for 1-3 months using the battery of stimulus patterns. The effect of the lesion on the animal's behavior could be directly assessed by comparing the performance rate with the target stimulus either within or outside the lesioned portion of the visual field. The results from these studies are summarized in table 2. Lesions of the parvocellular layers left the monkeys with severe losses in the ability to make discriminations of texture, color, fine pattern and fine stereopsis. Animals with lesions of the magnocellular layers had severe losses in the perception of flicker and motion. These studies strongly suggest that the P and M systems play different roles in visual perception that are analogous to the roles that the SAI and RA systems play in tactile perception. The P system like the SAI system is responsible for the perception of texture and fine form, whereas the M system, like the RA system, is responsible for flicker and motion.

7. Discussion

This chapter describes evidence suggesting that tactile and visual processing are based on similar neural mechanisms. The emerging evidence in both systems is that at the primary input level, inputs are segregated into separate but overlapping spatial channels with one channel specialized for processing high spatial frequencies with low temporal resolution and another channel is specialized for processing low spatial and high temporal information. Furthermore, the different pathways in the two systems have analogous response properties and are responsible for similar aspects of sensory perception. These results lead to two conjectures as to how the sensory systems process information and are organized. The first is that physiological properties of the peripheral afferents that the two systems have in common must be critical for perception since these pathways underlie similar aspects of perception. The second is that, if the organizational principles in the two sensory systems are similar at the peripheral level then similar organizational principles may also apply in the central nervous system.

Neurons in the the sustained pathways share the common properties of having sustained responses to a steady input, small receptive fields, high spatial acuity, low sensitivity to low contrast stimuli, and high sensitivity to high contrast stimuli. These neurons also have low sensitivity to moving or temporally dynamic stimuli. Small peripheral receptive fields are necessary for a system to have high spatial resolution since increasing receptive field size results in a spreading-out or blurring of the input. The high spatial acuity that is observed in the sustained pathways is augmented in both vision and

Table 2. Deficit magnitudes in visual performance following lesions of the parvocellular or magnocellular layers of the LGN. Dashes indicate no deficit. (Adapted from Schiller and Logothetis, 1990)

Function		Parvo lesion	Magno lesion
Color vision		Severe	-
Texture perception		Severe	-
Pattern perception		Severe	-
Shape perception	Fine	Severe	-
	Coarse	mild	-
Brightness perception		-	-
Coarse Scotopic vision		-	-
Contrast sensitivity	Fine	Severe	-
	Coarse	mild	-
Stereopsis	Fine	Severe	-
	Coarse	Pronounced	-
Motion perception		-	Severe
Flicker perception		-	Severe

touch by peripheral mechanisms that enhance the sensitivity of these neurons to edges placed within the receptive fields. In the somatosensory system, SAI afferents respond much more vigorously to edges than to flat surfaces (Phillips and Johnson, 1981a). Phillips and Johnson (1981b) inferred that the edge enhancement seen in these afferents can be attributed to skin mechanics rather than being neural in origin. In vision, the P ganglion cells also show enhanced responses to edges which in this case is due to these cells having center-surround receptive field structures.

The sustained responses and low sensitivity to slow moving stimuli, which are characteristic of neurons in the SAI and P pathways, is also reasonable for a pathway that is responsible for form and texture perception. Sustained responses allow for images to persist for long periods which commonly occurs when stimuli are either fixated or grasped. It is also reasonable that these neurons have low sensitivity to low contrast stimuli and moving stimuli, since this protects these systems from responding to the minute details of objects (e.g. small asperities related to the frictional forces) or movement of the object, and leaves them free for discriminating gross spatial features instead.

Neurons within the transient pathways have high sensitivity to high temporal frequencies and to moving stimuli and are relatively insensitive to the spatial features of stimuli. Their larger receptive fields accounts for their inability resolve the fine spatial details. It also allows these receptor types to integrate information over a larger surface area, which accounts partially for their increased sensitivities to motion.

Whether the similarities in touch and vision go beyond the peripheral level is speculative. Cortical studies indicate that in both touch and vision there are two parallel processing streams that originate in the primary somatosensory and visual cortices (Maunsell and Newsome, 1987; Murray and Mishkin, 1984; Mishkin et al., 1983). The ventrally oriented processing streams are termed the "what" pathways and are considered to be the pathways that underlie tactile and visual form and texture perception. Neurons along these pathways have properties that are somewhat like the P and SAI peripheral inputs. The other processing streams, which are termed the dorsal or "where" pathways, appear to inherit their properties from the transient pathways. Neurons along this pathway, especially those in the visual system respond well to motion. Recent studies in the visual system have shown that these pathways are not entirely segregated into M and P processing streams, since lesions in area V4 of the "what" pathway only mildly disrupt form processing (Ferrera et al., 1994; Schiller and Lee, 1994). Neurons in V4 are also well modulated by inputs from the M pathway illustrating that the complete segregation of function, observed in the peripheral responses is lost along this ventral stream. Whether a similar segregation of function or a similar degree of divergence and convergence occurs in the somatosensory system is not known.

Although the two systems appear to be quite similar in many respects, there are many differences that indicate that visual processing is significantly more difficult and requires more neural machinery than touch. Not only is the organization of striate cortex more complex than primary somatosensory cortex but the number of identified visual and somatosensory cortical areas differ by 3 to 4 orders of magnitude (Felleman and Van Essen, 1991). If touch and vision are similarly organized, then what accounts for this huge discrepancy? One possibility is that visual processing has the increased task of needing color information which is absent in touch, and additionally needs to transform two-dimensional retinal images of overlapping objects into three-dimensional internal representations. In contrast, object recognition in the somatosensory system is simpler since it requires that the skin be in direct contact with the object and the three dimensional structure of objects comes directly from the conformation of the fingers and hand. Future studies are required to resolve many of these issues.

Acknowledgements

I would like to thank Paul Fitzgerald and Alexander Twombly for their comments and inputs to this paper. The work reported in this paper was supported by NIH Grant NS34086.

References

Blake DT, Hsiao SS, Johnson KO (1994) Neural basis for tactile roughness perception: The relative contributions of slowly adapting and rapidly adapting afferents. Society for Neuroscience Abstracts, 20, 1387(Abstract)

Blake DT, Hsiao SS, Johnson KO (1997) Slowly and rapidly adapting mechanoreceptive responses to raised and depressed scanned patterns: effects of width, height and a raised surround. (submitted)

Bliss JC (1969) A relatively high resolution reading aid for the blind. IEEE Transactions, Man-Machine Sys 10,1-9.

Bolanowski SJ, Gescheider GA, Verrillo RT, Checkosky CM (1988) Four channels mediate the mechanical aspects of touch. Journal of the Acoustical Socociety of America, 84, 1680-1694.

Burton H, Sinclair RJ (1994) Representation of tactile roughness in thalamus and somatosensory cortex. Canadian Journal of Physiology and Pharmacology, 72, 546-557.

Cholewiak RW, Collins AA (1991) Sensory and physiological bases of touch. In: The Psychology of Touch (Heller MA, Schiff WR eds), pp 23-60. Hillsdale, N.J. Lawrence ErlbaumAssociates.

Connor CE, Hsiao SS, Phillips JR, Johnson KO (1990) Tactile roughness: Neural codes that account for psychophysical magnitude estimates. Journal of Neuroscience, 10, 3823-3836.

Connor CE, Johnson KO (1992) Neural coding of tactile texture: Comparisons of spatial and temporal mechanisms for roughness perception. Journal of Neuroscience, 12, 3414-3426.

Craig JC (1979) A confusion matrix for tactually presented letters. Perception and Psychophysics, 26, 409-411.

Craig JC (1981) Tactile letter recognition. International Journal of Rehabilitation Research, 400-402.

Croner LJ, Kaplan E (1995) Receptive fields of P and M ganglion cells across the primate retina. Vision Research, 35, 7-24.

Darian-Smith I (1984) The sense of touch: Performance and peripheral neural processes. In: Handbook of Physiology. Sec. 1. Vol. III Sensory Processes (Darian-Smith I, Mountcastle VB, Brookhart JM eds), pp 739-788.

Darian-Smith I, Kenins P (1980) Innervation density of mechanoreceptive fibers supplying glabrous skin of the monkey's index finger. Journal of Physiology, 309, 147-155.

DeYoe EA, Van Essen DC (1988) Concurrent processing streams in monkey visual cortex. Trends in Neuroscience, 11, 219-226.

Edin BB, Essick GK, Trulsson M, Olsson KA (1995) Receptor encoding of moving tactile stimuli in humans. I. Temporal pattern of discharge of individual low-threshold mechanoreceptors. Journal of Neuroscience, 15, 830-47.

Felleman DJ, Van Essen DC (1991) Distributed hierarchial processing in the primate cerebral cortex. Cerebral Cortex, 1, 1-47.

Ferrera VP, Nealey TA, Maunsell JH (1994) Responses in macaque visual area V4 following inactivation of the parvocellular and magnocellular LGN pathways. Journal of Neuroscience, 14, 2080-2088.

Gardner EP, Palmer CI (1989) Simulation of motion on the skin. I. Receptive fields and temporal frequency coding by cutaneous mechanoreceptors of Optacon pulses delivered to the hand. Journal of Neurophysiology, 62, 1410-1436.

Gardner EP, Sklar BF (1994) Discrimination of the direction of motion on the human hand: a psychophysical study of stimulation parameters. Journal of Neurophysiology, 71, 2414-2429.

Goodwin AW, John KT, Marceglia AH (1991) Tactile discrimination of curvature by humans using only cutaneous information from the fingerpads. Experimental Brain Research, 86, 663-672.

Goodwin AW (1993) The code for roughness. Comparing psychophysical measurement in humans with neural recordings in monkeys points to a neural code for tactile roughness. Current Biology, 3, 378-379.

Goodwin AW, Browning AS, Wheat HE (1996) Representation of the shape and contact force of handled objects in populations of cutaneous afferents. In: Somesthesis and the Neurobiology of the Somatosensory Cortex (Franzen O, Johansson RS, Terenius L eds), pp 137-146. Basel: Birkh@user Verlag.

Gouras P (1968) Identification of cone mechanisms in monkey ganglion cells. Journal of Physiology, 533-547.

Heller MA (1982) Visual and tactual texture perception: Intersensory cooperation. Perception and Psychophysics, 31, 339-344.

Heller MA (1983) Haptic dominance in form perception with blurred vision. Perception 12, 607-613.

Hendry SHC, Yoshioka T (1994) A neurochemically distinct third channel in the macaque dorsal lateral geniculate nucleus. Science 264, 575-577.

Hsiao SS, Johnson KO, Twombly IA (1993) Roughness coding in the somatosensory system. Acta Psychologica (Amsterdam), 84, 53-67.

Hsiao SS, Johnson KO, Twombly IA, DiCarlo JJ (1996) Form processing and attention effects in the somatosensory system. In: Somesthesis and the Neurobiology of the Somatosensory Cortex (Franzen O, Johansson R, Terenius L eds), pp 229-247. Switzerland: Birkhauser Verlag Basel.

Johansson RS, LaMotte RH (1983) Tactile detection thresholds for a single asperity on an otherwise smooth surface. Somatosensory and Motor Research, 1, 21-31.

Johansson RS, Vallbo AB (1979) Tactile sensibility in the human hand: Relative and absolute densities of four types of mechanoreceptive units in glabrous skin. Journal of Physiology, 286, 283-300.

Johansson RS, Westling G (1987) Signals in tactile afferents from the fingers eliciting adaptive motor responses during precision grip. Experimental Brain Research, 66, 141-154.

Johnson KO, Hsiao SS, Blake DT (1996) Linearity as the basic law of psychophysics: Evidence from studies of the neural mechanisms of roughness magnitude estimation. In: Somesthesis and the Neurobiology of the Somatosensory Cortex (Franzen O, Johansson R, Terenius L eds), pp 213-228. Switzerland: Birkhauser Verlag Basel.

Johnson KO (1983) Neural mechanisms of tactual form and texture discrimination. Federation Proceedings, 42, 2542-2547.

Johnson KO, Hsiao SS (1992) Neural mechanisms of tactual form and texture perception. Annual Review of Neuroscience, 15, 227-250.

Johnson KO, Hsiao SS (1994) Evaluation of the relative roles of slowly and rapidly adapting afferent fibers in roughness perception. Canadian Journal of Physiology and Pharmacology, 72, 488-497.

Johnson KO, Lamb GD (1981) Neural mechanisms of spatial tactile discrimination: Neural patterns evoked by braille-like dot patterns in the monkey. Journal of Physiology, 310, 117-144.

Johnson KO, Phillips JR (1981) Tactile spatial resolution: I. Two-point discrimination, gap detection, grating resolution, and letter recognition. Journal of Neurophysiology, 46, 1177-1191.

Katz D (1989) The World of Touch. Hillsdale, NJ: Erlbaum (originally published in 1925).

LaMotte RH, Lu C, Srinivasan MA (1996) Tactile neural codes for shapes and orientations of objects. In: Somesthesis and the Neurobiology of the Somatosensory Cortex (Franzen O, Johansson RS, Terenius L eds), pp 113-122. Basel: Birkh@user

LaMotte RH, Whitehouse JM (1986) Tactile detection of a dot on a smooth surface: Peripheral neural events. Journal of Neurophysiology, 56, 1109-1128.

Lederman SJ (1985) Tactual roughness perception in hand:a psychophysical assessment of the role of vibration. Experimental Brain Research, Supplement 10, 77-92.

Lederman SJ, Thorne G, Jones B (1986) Perception of texture by vision and touch: Multidimensionality and intersensory integration. Journal of Experimental Psychology [Human Perception and Performance] 12, 169-180.

Lederman SJ, Abbott SG (1981) Texture perception: Studies of intersensory organization using a discrepancy paradigm, and visual versus tactual psychophysics. Journal of Experimental Psychology [Humand Perception and Performance] 7, 902-915.

Loomis JM (1980) Interaction of display mode and character size in vibrotactile letter recognition. Bulletin of the Psychonomic Society, 16, 385-387.

Loomis JM (1981) On the tangibility of letters and Braille. Perception and Psychophysics, 29, 37-46.

Loomis JM (1982) Analysis of tactile and visual confusion matrices. Perception and Psychophysics, 31, 41-52.

Maunsell JH, Sandell JH, Schiller PH (1986) Functions of the on and off channels of the visual system. Nature, 322, 824-825.

Maunsell JH, Newsome WT (1987) Visual processing in monkey extrastriate cortex. Annual Review of Neuroscience, 10, 363-401.

Mishkin M, Ungerleider LG, Macko KA (1983) Object vision and spatial vision: Two cortical pathways. Trends in Neuroscience, 6, 414-416.

Mountcastle VB, LaMotte RH, Carli G (1972) Detection thresholds for stimuli in humans and monkeys: Comparison with threshold events in mechanoreceptive afferent nerve fibers innervating the monkey hand. Journal of Neurophysiology, 35, 122-136.

Mountcastle V B (1984) Central nervous mechanisms in mechanoreceptive sensibility. In: Handbook of Physiology. Sept 1. Vol III Sensory processes (Darian-Smith I, Mountcastle VB eds), pp 789-878.

Murray EA, Mishkin M (1984) Relative contributions of SII and area 5 to tactile discrimination in monkeys. Behavioural Brain Research, 11, 67-83.

Murray EA, Mishkin M (1985) Amygdalectomy impairs crossmodal association in monkeys. Science, 228, 604-606.

O'Brien B (1951) Vision and resolution in the central retina. Journal of the Optometrical Society of America, 882-894.

Owen DH, Brown DR (1970) Visual and tactual form complexity: A psychophysical approach to perceptual equivalence. Percept Psychophys 7, 225-228.(Abstract)

Phillips JR, Johnson KO, Browne HM (1983) A comparison of visual and two modes of tactual letter resolution. Perception and Psychophysics, 34, 243-249.

Phillips JR, Johnson KO, Hsiao SS (1988) Spatial pattern representation and transformation in monkey somatosensory cortex. Proceedings of the National Academy of Science, USA, 85, 1317-1321.

Phillips JR, Johansson RS, Johnson KO (1990) Representation of braille characters in human nerve fibers. Experimental Brain Research, 81, 589-592.

Phillips JR, Johnson KO (1981a) Tactile spatial resolution: II. Neural representation of bars, edges, and gratings in monkey afferents. Journal of Neurophysiology, 46, 1192-1203.

Phillips JR, Johnson KO (1981b) Tactile spatial resolution: III. A continuum mechanics model of skin predicting mechanoreceptor responses to bars, edges, and gratings. Journal of Neurophysiology, 46, 1204-1225.

Pubols BH, Jr., Pubols LM (1976) Coding of mechanical stimulus velocity and indentation depth by squirrel monkey and raccoon glabrous skin mechanoreceptors. Journal of Neurophysiology, 39, 773-787.

Schiller PH, Logothetis NK, Charles ER (1990a) Role of the color-opponent and broad-band channels in vision. Visual Neuroscience, 5, 321-346.

Schiller PH, Logothetis NK, Charles ER (1990b) Functions of the colour-opponent and broad-band channels of the visual system [see comments]. Nature, 343, 68-70.

Schiller PH (1992) The ON and OFF channels of the visual system. [Review]. Trends in Neuroscience, 15, 86-92.

Schiller PH (1993) The effects of V4 and middle temporal (MT) area lesions on visual performance in the rhesus monkey. Visual Neuroscience, 10, 717-746.

Schiller PH, Lee K (1994) The effects of lateral geniculate nucleus, area V4, and middle temporal (MT) lesions on visually guided eye movements. Visual Neuroscience, 11, 229-241.

Schiller PH, Logothetis NK (1990) The color-opponent and broad-band channels of the primate visual system. [Review]. Trends in Neuroscience, 13, 392-398.

Shapley R (1995) Parallel neural pathways and visual function. In: The Cognitive Neuronsciences (Gazzaniga MS ed), pp 315-324. Cambridge, Massachusetts.London, England: The MIT Press.

Shapley RM (1990) Visual sensitivity and parallel rentinocortical channels. Annual Review of Psychology, 41, 635-658.

Smith AM (1994) Some shear facts and pure friction related to roughness discrimination. Canadian Journal of Physiology and Pharmacology, 72, 583-590.

Srinivasan MA, Whitehouse JM, LaMotte RH (1987) Detection of slip: Peripheral neural coding. Neuroscience, 7, 1682-1697.

Srinivasan MA, Whitehouse JM, LaMotte RH (1990) Tactile detection of slip: Surface microgeometry and peripheral neural codes. Journal of Neurophysiology, 63, 1323-1332.

Srinivasan MA, LaMotte RH (1996) Tactual discrimination of softness: Abilities and mechanisms. In: Somesthesis and the Neurobiology of the Somatosensory Cortex (Franzen O, Johansson RS, Terenius L eds), pp 123-136. Basel: Birkh@user Verlag.

Talbot WH, Darian-Smith I, Kornhuber HH, Mountcastle VB (1968) The sense of flutter-vibration: Comparison of the human capactiy with response patterns of mechanoreceptive afferents from the monkey hand. Journal of Neurophysiology, 31, 301-334.

Vega-Bermudez F, Johnson KO, Hsiao SS (1991) Human tactile pattern recognition: Active versus passive touch, velocity effects, and patterns of confusion. Journal of Neurophysiology, 65, 531-546.

Westling G, Johansson RS (1987) Responses in glabrous skin mechanoreceptors during precision grip in humans. Experimental Brain Research, 66, 128-140.

Neural Aspects of Tactile Sensation
J.W. Morley (Editor)
© 1998 Elsevier Science B.V. All rights reserved

Functional Organization of the Somatosensory Cortex in the Primate

A.B. Turman, J.W. Morley and M. J. Rowe

At the beginning of the twentieth century, clinical observations on vascular and traumatic lesions in the region of the postcentral gyrus in human patients led to the suggestion that this region of the cerebral cortex had a role in somatic sensation, in particular, in the sensations of touch and kinaesthesia (Head and Holmes,1911). However, in such observations the clinical presentation could differ markedly from case-to-case and the extent of a lesion could not always be well-defined in post-mortem studies. From these initial observations it was not clear whether a single large cortical area or indeed multiple cortical regions in this part of the brain were responsible for the processing of somatosensory inputs. It was not until the introduction of the evoked potential method of cortical mapping (Gerard et al., 1936; Marshall et al., 1937) that it became evident that there was more than a single area of cerebral cortex related to the somatosensory system. These electrophysiological studies, which were conducted on various species, including cats and monkeys, showed that the short-latency, positive-going surface potentials recorded in response to brief tactile stimulation of the body surface or to electrical stimulation of peripheral nerves occurred in at least two separate locations on the contralateral hemisphere (Adrian, 1940, 1941; Marshall et al., 1941; Woolsey, 1943; Woolsey et al., 1942; Woolsey and Fairman, 1946). These two areas were designated the first (primary) somatosensory area (SI) and the second somatosensory area (SII), on the basis of the chronology of their discovery. Later studies in a number of species revealed further distinct cortical areas that receive somatic sensory inputs. Furthermore, as recording and mapping techniques became more refined and accurate some of these areas were found to contain multiple representations of the body surface. In this chapter we will summarize the connectivity and functional organization of the cortical areas that have been implicated in the processing of somatic sensory information in the primate.

1. Cortical somatosensory areas

1.1 Primary somatosensory area

The primary somatosensory area, SI, is located in the postcentral gyrus in primates, including humans, and consists of cytoarchitectural areas 3, 1 and 2 of Brodmann, arranged in a rostrocaudal sequence (Brodmann, 1909) (Fig. 1). Area 3 is further subdivided into area 3a and area 3b; area 3b has a typical six-layered appearance with densely packed small granular cells in lamina IV, whereas area 3a contains large pyramidal cells in layer V, similar to the motor areas of cortex, in addition to inner granular layers (Merzenich et al., 1978). Within the SI area the contralateral body surface is represented somatotopically, with the area of cortex devoted to a particular body part being approximately proportional to the innervation density of that body part (Woolsey, 1958). In this representation the caudal regions of the body

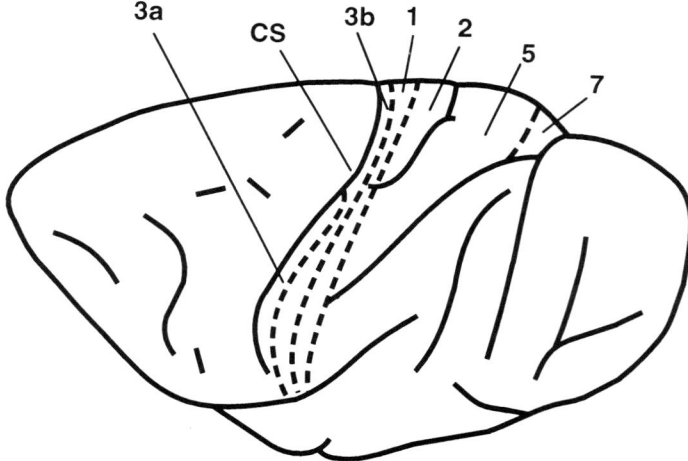

Figure 1. Diagramatic representation of cytoarchitecturaly distinct regions of the primary somatosensory cortex (areas 3a, 3b, 1 and 2) and the posterior parietal cortex (areas 5 and 7) in the macaque monkey. CS: central sulcus (modified from Brodmann 1909).

are represented medially and the more rostral regions represented progressively more laterally. In recent years, studies using microelectrode mapping techniques reported that in monkeys separate somatotopic representations exist in each of the three cytoarchitectural areas of the primary somatosensory cortex, with the separate somatotopically organized body representations in areas 3b and 1 being roughly mirror reversals of each other (Merzenich et al., 1978, 1981; Kaas, 1983; Kaas et al., 1979, 1981a, 1981b; Nelson et al., 1980; Sur et al., 1982; Felleman et al., 1983; Pons et al., 1985). In addition, these studies have shown that there are correlations between cytoarchitectural divisions of the primary somatosensory cortex and the functional properties of neurons within these areas. The neurons in areas 3b and 1 respond to cutaneous stimulation, whereas neurons in areas 3a and 2 are driven largely by deep receptors located in muscles and joints. The response properties of neurons are reported to change sharply at cytoarchitectural borders.

In areas 3b and 1, almost all neurons respond to activation of cutaneous receptors (Dreyer et al., 1975; Hyvärinen and Poranen, 1978a; Hyvärinen et al., 1980; Tanji and Wise, 1981; Felleman et al., 1983; Cusick et al., 1989). However, functional differences exist between neurons in the two areas. First, there appear to be proportionally more neurons with slowly adapting (SA) response properties in area 3b than in area 1 (Paul et al., 1972a; Sur, 1980). Second, neurons that are responsive to high frequency vibrotactile stimuli, and appear to be driven predominantly by Pacinian corpuscle (PC) receptor input, are found in area 1 but not in area 3b (Paul et al., 1972b; Hyvärinen and Poranen, 1978b). Third, neurons in area 3b have small receptive fields while neurons in area 1 display relatively large, discontinuous receptive fields, and in some cases respond selectively to a particular direction of stimulus motion on the skin (Hyvärinen and Poranen, 1978a; Sur, 1980).

Neurons in area 3a receive inputs arising principally in muscle spindle afferents (Phillips et al., 1971; Schwartz et al., 1973; Yumiya et al., 1974; Tanji, 1975; Heath et al., 1976; Hore et al., 1976; Wise and Tanji, 1981), with the majority of neurons responsive to passive movement of a limb in a particular direction (Wise and Tanji, 1981). However, some area 3a neurons, especially in the digit representations, respond to cutaneous stimuli (Tanji and Wise, 1981; Wise and Tanji, 1981).

In area 2, most neurons are activated by deep receptors located in muscles and joints (Burchfiel and Duffy, 1972; Schwartz et al., 1973; Iwamura and Tanaka, 1978; Iwamura et al., 1980; Soso and Fetz, 1980; Gardner and Costanzo, 1981; Pons et al., 1985). However, a substantial number of neurons in this area (in particular, in the area of hand representation) receive cutaneous inputs (Kaas et al., 1979; Nelson et al., 1980; Pons et al., 1985). Most of these neurons respond to more complex forms of tactile stimuli and show sensitivity to the movement, orientation and direction of tactile stimuli (Hyvärinen and Poranen, 1978a; Iwamura and Tanaka, 1978).

1.2. Second somatosensory area and neighboring cortical regions

The second somatosensory area, SII, has been identified in a great variety of mammalian species. In primates, it is buried deep in the superior bank of the Sylvian fissure and the somatotopic representation of the body is oriented from the head anteriorly to the tail posteriorly (for review see Burton, 1986). This anteroposterior orientation of the body representation is approximately perpendicular to the mediolateral body representation in the primary somatosensory cortex. The border between SI and SII is a congruent border where similar receptive fields from superior regions of the head and the nose are aligned on each side of the border (see Burton, 1986). It has been suggested by Krubitzer and Kaas (1990) that the SII area in monkeys be referred to as the "SII region" as it contains several cytoarchitectural subregions that include SII-proper, the parietal ventral (PV) area, area 7b, the retroinsular (Ri) area, the posterior auditory (PA) area, the granular (Ig) and dysgranular (Id) insular fields (Robinson and Burton, 1980a, b; Friedman et al., 1986) and the ventral somatic (VS) area (Cusick et al., 1989).

Almost all neurons in SII-proper are responsive to inputs arising from contralateral hairy and glabrous skin. The majority of neurons are classified as rapidly adapting (RA) to touch or movement of hairs (Robinson and Burton, 1980b; Cusick et al., 1989), with a small number responding to high frequency vibrotactile stimulation. In addition, some SII neurons can be activated from receptors located in muscles and joints (Robinson and Burton 1980c). Recently, it was suggested by Krubitzer et al. (1995) that within this region of macaque monkeys there are two complete body representations. They proposed that the term SII be maintained for the caudal field and to use the term parietal ventral (PV) for the rostral field. This suggestion was based on the description of similar divisions in other mammals, including the marmoset monkey (Krubitzer and Kaas, 1992).

Other cytoarchitectural areas in the "SII region" of primates that contain neurons responsive to somatic sensory stimuli include area 7b, the retroinsular area (Ri), and the granular (Ig) and dysgranular (Id) insular fields.   Area 7b has a crude somatotopic organization and contains neurons with large and often bilateral receptive fields responsive, predominantly, to somatic inputs as well as neurons receiving convergent inputs from somatic and visual sources (Robinson and Burton, 1980a, b). Together with area 7a, this region has been considered a visual association centre and may be involved in visually guided behavior (Mountcastle et al., 1975; Yin and Mountcastle, 1978). The response properties of neurons in Ri are similar to those in the SII region. They respond almost exclusively to cutaneous stimuli, the majority being rapidly adapting (RA) to light touch and low frequency vibrations (20-80 Hz), while a small number are most responsive to higher frequency vibration (over 100 Hz) which suggests that they receive inputs from Pacinian corpuscles (Robinson and Burton, 1980a, b). The response properties of neurons in the granular (Ig) and dysgranular (Id) insular fields have not yet been extensively studied, but the available evidence suggests that the neurons in Ig have receptive field properties similar to those in SII-proper (Robinson and Burton, 1980b). In the owl monkey (a New-World monkey), the region extending laterally from SII on the lower bank of the lateral sulcus has been termed the ventral somatic (VS) area and contains neurons activated exclusively by contralateral cutaneous stimuli (Cusick et al., 1989). The response properties of neurons in this area also resemble those in the SII-proper area.   Although recently Krubitzer and colleagues (1995) recorded from the insular region in the macaque monkey and also used the term VS for this functional region, it is still not clear to what extent this area corresponds to traditionally identified cytoarchitectural areas in other studies.

2. Anatomical Connections of Cortical Somatosensory Areas

2.1. Thalamic connections

The ventral posterior nucleus (VP) is the major source of thalamic input to somatosensory cortical areas (for review see Jones, 1985). The VP nucleus is divided into two subnuclei; the ventral posterior medial nucleus (VPM) and the ventral posterior lateral nucleus (VPL). Input to these nuclei differs, with the trigeminal lemniscus terminating in VPM and the medial lemniscus in VPL (see Jones, 1985). Although most neurons in the VP nucleus are activated from receptive fields on the contralateral side of the head and the body (Mountcastle and Henneman, 1949, 1952; Rose and Mountcastle, 1952; Gordon and Manson, 1967; Bombardieri et al., 1975) reports have been made of both bilateral (Gaze and Gordon, 1954) and ipsilateral input (Gordon and Manson, 1967). The VP nucleus is organized somatotopically, such that in VPM the mouth and face are represented mediolaterally and in VPL the hindlimb and tail representations are located laterally (for review see Welker, 1973; Jones, 1985) while the distal parts of the limbs are represented ventrally and the trunk and proximal parts of limbs dorsally (for review see Jones, 1985).

Most neurons in the VP nucleus have relatively small receptive fields and can be activated by different forms of somatic stimuli, including light touch, pressure, movement of joints and manipulation of tendons and muscles (for review see Mountcastle, 1984). The medial lemniscus fibres, which are bundled according to functional modality and place, project upon mode- and place-specific "modular" groups of thalamic neurons, resulting in a differential distribution of neurons of different functional classes within the VP nucleus (Rose and Mountcastle, 1952). However, it has still not been clearly established how these "modular" groups are distributed within the VP nucleus and how they project upon different cytoarchitectural areas of the somatosensory cortex (for review see Mountcastle, 1984). Since the introduction of microelectrode recording techniques various studies have identified several zones within the VP nucleus that contain neurons with different functional properties. Vertical electrode penetrations into the nucleus initially record neurons that respond only to stimulation of deep tissues (for a distance of 300-500 μm) and then pass into a larger area that contains neurons responsive to stimulation of cutaneous receptive fields (Poggio and Mountcastle, 1963; Loe et al., 1977; Pollin and Albe-Fessard, 1979). Further advances of microelectrodes into ventral or posteroventral parts of the VPL (the region identified as ventral posterior inferior nucleus, VPI) encounter neurons that respond preferentially to high frequency vibrotactile stimulation (Dykes et al., 1981). As this sequence of modality representation can also be observed in microelectrode penetrations that traverse the nucleus horizontally (Friedman and Jones, 1981), it has been suggested that the VP nucleus has a central "core" region containing modules of neurons that respond to cutaneous stimuli and a surrounding "shell" region with neurons responsive to activation of receptors located in deep tissues (Fig. 2). It has also been suggested that the "dorsal capping zone" (the shell region) contains at least one further body representation, the "ventral posterior superior" nucleus (VPS) that is largely responsive to deep inputs (Cusick et al., 1985). The ascending projections from each module in the VP nucleus are to small, restricted zones of the somatosensory cortical areas and each module in turn receives a point-to-point descending projection from that same cortical area (Jones et al., 1979; Jones et al., 1982).

The central core region of the VP nucleus in monkeys projects to somatosensory cortical areas 3b and 1, and the shell region to areas 3a and 2 (Fig.2; for review see Kaas and Pons, 1988). However, the presence of thalamic projections to multiple cortical areas and their functional role still remains unclear. Although earlier anatomical labeling studies by Jones and his colleagues (Jones et al., 1979; Jones, 1983) argued that virtually no VP neurons project to more than one cytoarchitectural area within the primary somatosensory cortex, an increasing number of recent studies suggest otherwise. An anatomical labelling study in squirrel monkeys showed that approximately 20% of neurons in the core region project to both areas 3b and 1, and up to 40% in the shell region project to areas 3a and 2 (Cusick et al., 1985). In another labeling study it was reported that area 3b receives connections from at least three thalamic nuclei, the VPS nucleus, the VPI nucleus and the anterior pulvinar (aPul) (Cusick and Gould, 1990). The aPul has also been reported to have projections to areas 3a and 2 (Cusick et al., 1985; Pons and Kaas, 1985). Although thalamic

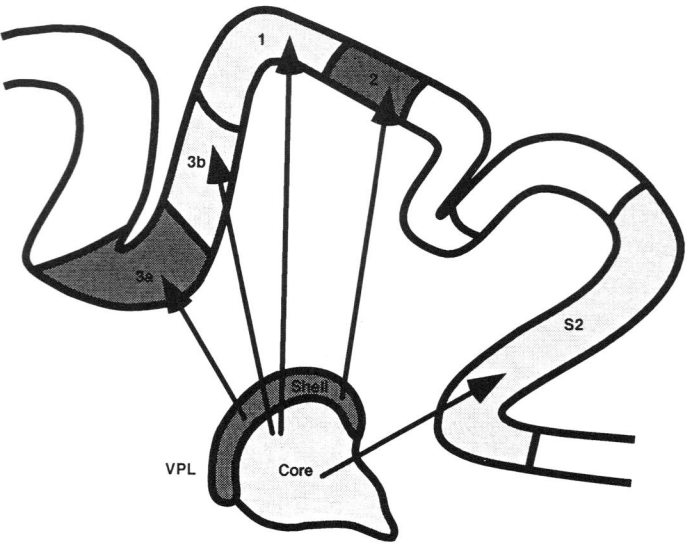

Figure 2. Illustration of direct thalamocortical projections to SI and SII from
the shell and core regions of the ventral posterior lateral nucleus (VPL).
The core region of VPL projects to cytoarchitectural areas 3b and 1, while
the shell region projects to areas 3a and 2. The diagram illustrates a
projection from the core region of VPL to SII, however there has been some
disagreement about the source of thalamic input to SII (see text) (modified
from Jones and Friedman 1982).

projections to multiple areas of the primary somatosensory cortex have
also been shown in *newborn* macaque monkeys (Darian-Smith et al.,
1990b), no double-labeled thalamic neurons were found in *adult*
macaques following injections into multiple cortical areas (Darian-
Smith et al., 1990a). This has led to the suggestion that there may be
differences in the thalamocortical projections to somatosensory areas in
New- and Old-World monkeys.

The SII area of the cortex also receives direct anatomical
projections from thalamic nuclei. However, there is continuing debate
on the issue of thalamic nuclei that project to SII. In the macaque
monkey, although the VPL, the central nucleus and the posterior
nucleus (PO) have been reported to project to SII (for review see Jones,
1985), Friedman and Murray (1986) suggested that on the basis of

retrograde labeling and degeneration studies the VPI nucleus is the major source of input to the SII area. More recently, a similar suggestion was also made for thalamic projections to SII in marmoset monkeys (Krubitzer and Kaas, 1992). However, while most neurons in the VPI nucleus respond to inputs presumably arising from Pacinian corpuscle receptors (Dykes et al., 1981), there appear to be few neurons in the SII area of the macaque monkey that selectively respond to high frequency vibrotactile stimulation (Robinson and Burton, 1980b), in contrast to the cat (Ferrington and Rowe, 1980; Fisher et al., 1983). It is unlikely therefore that VPI is the *major* source of input to SII in the primate.

2.2. Corticocortical connections

One major difference between most primates and non-primate species is in the body representation *within* the primary somatosensory cortex. While the SI region of the cortex in macaque monkeys and other New-World monkeys such as the owl monkey, squirrel monkey and the cebus monkey contains two separate and complete cutaneous body representations in areas 3b and 1 (Merzenich et al., 1978; Sur et al., 1982; Felleman et al., 1983), in prosimians and other mammals such as rats, rabbits, cats, racoons and tree shrews there appears to be only a single body representation within SI (for review see Kaas, 1983). This difference in the number of cortical representations within SI between primate and non-primate species may be attributable to the divergent lines taken by these species in the course of evolution (Kaas, 1983, 1987).

There are extensive intracortical connections between cortical fields of the primary somatosensory cortex (3a, 3b, 1 and 2) and between different somatosensory areas (SI and SII) in all primate species studied   (for review see Kaas and Pons, 1988).   These connections are organized in highly specific patterns and the somatotopic order is preserved so that any two areas within the one hemisphere representing the same body part are interconnected. This pattern of intrinsic reciprocal connections has led to the suggestion that there are feedforward and feedback information processing schemes of the type described originally in the visual cortex (Felleman and Van Essen, 1991).   In the *feedforward* scheme the corticocortical projections to other cortical areas originate predominantly from superficial layers (layers II and III), although in some cases include neurons located in deep layers. The terminations of these axons in the target field is preferentially in layer IV and to some extent in layer III. In *feedback* patterns, the majority of projection fibres arise from deep layers of the cortex and terminate predominantly in both superficial and deep layers, avoiding layer IV.

Within the primary somatosensory cortex of primates there are projections from area 3b to area 1 and return projections from area 1 to area 3b (Fig. 3) (Jones et al., 1978; Vogt and Pandya, 1978; Cusick et al., 1985). As projections from area 3b terminate predominantly in layers III and IV of area 1 and the neurons in area 1 in turn project mainly to layer I of area 3b, it has been suggested that area 3b is the major source of cutaneous tactile information for area 1, and that area 1 sends feedback information to area 3b (for review see Kaas, 1983). It has been argued that this hierarchical scheme is further supported by evidence accumulated from lesion studies on awake behaving animals (Randolph and Semmes, 1974; Carlson, 1981). In monkeys, selective ablations of area 3b results in an inability to learn tactile discriminations in a variety of different tasks. However, lesions in area 1 result in impairment of the monkey's ability to discriminate surface texture (hard-soft and smooth-rough) while no apparent effects are observed in the ability to judge shapes. These results are interpreted as being consistent with a hierarchical information processing scheme between areas 3b and 1.

Areas 3b and 1 project caudally to area 2, which also sends projections back to these areas (Fig. 3). The pattern of connectivity between these areas also led to the suggestion that a feedforward scheme operates for the flow of information from area 3b and 1 to area 2 (Pons and Kaas, 1986). Areas 1 and 2 are reciprocally connected with area 3a and the motor areas of the precentral gyrus, and also project caudally to area 5 (Jones et al., 1978; Vogt and Pandya, 1978; Friedman et al., 1986; Pons and Kaas, 1986). Furthermore, areas 1 and 2 send projections rostrally to the supplementary motor area (SMA),

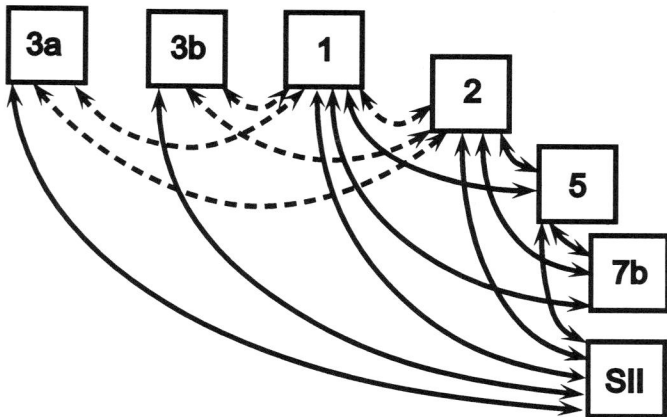

Figure 3. Diagram of corticocortical connections between cytoarchitecturally distinct regions of the primary somatosensory cortex (SI), posterior parietal cortex and the second somatosensory cortex (SII). Dashed lines indicate corticocortical connections between cytoarchitectural areas that comprise SI, and solid lines indicate connections between SI areas and the posterior parietal cortex and SII.

which is reciprocally linked with areas 5 and 4 (Bowker and Coulter, 1981). This pattern of connectivity with the SMA is thought to play a role in the initiation of movements (for review see Mountcastle, 1984). Surgical lesions in area 2 result in impairment of monkey's ability to discriminate the shape and size of objects (Randolph and Semmes, 1974; Carlson, 1981) while injections of the GABA agonist, muscimol (which is used to inactivate the area of cortex in the region of injection) impair temporarily the coordination of finger movements (Hikosaka et al., 1985). Therefore, it appears that processing within area 2 is dependent upon the integration of inputs from several sites such as the thalamus, and intracortical areas which include area 3a, from which information about finger and limb position is presumably derived, and 3b and 1, from which information may come about texture and spatial details of objects.

The SII area of monkeys is also reciprocally connected to areas of the primary somatosensory cortex (Fig. 3) (Jones et al., 1978; Friedman et al., 1986; Pons and Kaas, 1986). Additional reciprocal connections of SII are with areas 4, 7b, the retroinsular area (Ri) and the granular (Ig) and dysgranular (Id) insular fields (Friedman et al.,

1986). Based on results obtained from anatomical labeling studies, a hierarchical scheme for the processing of somatosensory information through a series of caudo-ventrally directed cortical fields was proposed by Friedman et al. (1986). This pathway, which includes areas 3a, 3b, 1, 2, 5, 7b, SII, Ig, Id and the connections to the amygdaloid complex and the hippocampal formation, was suggested to play a role in tactile learning and memory (Friedman et al., 1986). In a detailed review by Felleman and Van Essen (1991) on the connectivity patterns in the cerebral cortex of primates 13 cytoarchitectural areas of somatosensory and motor regions were assigned to one of nine levels of a proposed sensorimotor hierarchy. However, considerable inconsistency remains over the dominant direction of information flow among these areas, for example, between area 7b and SII (Friedman et al., 1986; Neal et al., 1987; Andersen et al., 1990).

3. Parallel/serial networks for processing within somatosensory cortex

The demonstration of direct thalamocortical projections to both SI and SII has led to the suggestion that somatosensory pathways at the thalamocortical level were organized in parallel (Jones and Friedman, 1982; Fisher et al., 1983; Burton, 1984; Burton and Kopf, 1984; Friedman and Murray, 1986; Landry et al., 1987). However, a serial scheme that resembles the strongly serial or hierarchical scheme proposed for the processing of visual information at thalamocortical levels (Van Essen and Maunsell, 1983) was also proposed for the somatosensory system (Mishkin, 1979). This proposed serial scheme was later elaborated by Friedman et al. (1986) and Felleman and Van Essen (1991) based on the pattern of anatomical connectivity. The first physiological evidence for the existence of the serial model was provided by Pons and colleagues (Pons et al., 1987), who reported that 6-8 weeks after *surgical* removal of the hand representation in the primary somatosensory cortex of macaque monkeys, there was an abolition of responsiveness in the topographically corresponding region of SII. In contrast, surgical ablation of the SII area had no detectable effect on the responsiveness of neurons in the primary somatosensory cortex (Pons et al., 1987). This pattern of inactivation was later confirmed for the macaque by Burton et al. (1990) and also observed by Garraghty et al. (1990) (again following surgical ablation of SI) for the SII area in the marmoset monkey, suggesting that the thalamocortical projections involving SI and SII in primates are organized in a strictly serial or hierarchical functional order.

These findings in primates, based on the disappearance of SII responsiveness following surgical ablation of SI, are in striking contrast to the experimental evidence in support of the *parallel* processing hypothesis in a diverse range of mammals including the cat (Burton and Robinson, 1987; Manzoni et al., *1979*; Turman et al., *1992*), the rabbit (Murray et al., *1992*), the tree shrew and the prosimian galago (Garraghty et al., *1991*), as SI inactivation brought about by a variety of methods in these species has little effect on tactile responsiveness within the SII area of cortex. These reported differences between primate and non-primate species led to the view that there are fundamental differences between simian primates and other eutherian mammals, including the prosimians, in the organization of thalamocortical systems for tactile processing (Garraghty et al., 1991; Murray et al., 1992; Turman et al., 1992). However, the ablation procedure used for SI inactivation in the primate studies of SII responsiveness (Burton et al., 1990; Garraghty et al., 1990; Pons et al., 1987, 1992) is clearly irreversible and does not permit examination of responsiveness in an *individual* SII neuron *before, during* and *after* SI inactivation, as was possible in our non-primate studies with the rapidly reversible procedure for SI inactivation based on localized cortical cooling (Murray et al., 1992; Turman et al., 1992). In addition, the ablation technique could entail complications, such as vascular impairment, that may jeopardize areas of cortex beyond the bounds of the actual ablation.

On account of these reservations about SI inactivation based on surgical ablation, there has been a recent re-investigation of the extent to which SI and SII are organized *in parallel* or *in series* for tactile processing in the marmoset monkey, one of the primate species that was the subject of the surgical ablation studies (Zhang et al., 1996; Rowe et al., 1996). This re-investigation was undertaken with the SI inactivation bought about with the rapidly reversible procedure based upon localized cortical cooling. The results demonstrated *first*, that SII evoked potentials were never abolished by SI inactivation and remain unaffected in a similar proportion of cases as is found in the cat

(Turman et al., 1992), and *second*, that > 90% of SII neurons retained some responsiveness to tactile stimulation.

Although some reductions in SII responsiveness were observed in association with SI inactivation these were also observed for a proportion of SII neurons in the cat and rabbit when SI was inactivated in these species (Turman et al., 1995; Murray et al., 1992), and in all three species there is evidence that the reductions were not attributable to part of the peripheral input to SII traversing a serial path from the thalamus via SI. Instead, it appears that the reduction reflects a loss of background facilitatory influence from SI operating as a tonic influence mediated via the intracortical connections that link SI and SII (Alloway and Burton, 1985; Friedman, 1983; Jones and Powell, 1968; 1969; Manzoni et al., 1979; Zarzecki, 1986). There are several arguments that point to the loss of background facilitation from SI as the explanation (Turman et al., 1992; Murray et al., 1992; Zhang et al., 1996; Rowe et al., 1996). *First*, as SI was cooled, there was a progressive increase in latency and slowing in the time course of the SI responses prior to the disappearance of SI activity, but no comparable delay in SII responses, whether evoked potential or single neuron, as might be expected if they were entirely, or in part, dependent upon inputs coming via SI. If SI were placed earlier than SII in a strict hierarchical scheme of thalamocortical processing, then any slowing or delay in SI neuron responses should inevitably lead to a similar delay in SII responses.

The *second* argument in favour of disfacilitation is that SI inactivation failed to bring about a tightening in the phaselocking of SII impulse activity to vibrotactile stimuli. If the tactile inputs to SII were to come from both a direct path from the thalamus and an indirect path via SI, then SI inactivation might be expected to eliminate some temporal differences in the inputs to the SII neurons and result in a tightening of phase-locking in the SII responses to vibrotactile stimulation. As this did not happen for SII neurons in the marmoset (Zhang et al., 1996), or for SII neurons in the cat (Turman et al., 1992; Rowe et al., 1996), it appears that the fall in responsiveness reflects a removal of facilitation, rather than removal of a component of peripheral input that traverses a serial path via SI.

The *third* argument in support of the disfacilitation interpretation is that SII neurons showing a reduction in responsiveness in association with SI inactivation often displayed a reduction in their background activity (Turman et al., 1992; Zhang et al., 1996; Rowe et al., 1996). Furthermore, for some SII neurons, the decline in background activity could account entirely or predominantly for the reduction in the response to the peripheral tactile stimulus. If this background activity is endogenous to SII, and not generated by afferent drive that comes via SI, it would mean that the SI inactivation removes a source of facilitation that operates as a background influence on SII responsiveness (Turman et al., 1992; Zhang et al., 1996; Rowe et al., 1996).

Based on these arguments, it appears that the reductions observed in SII responsiveness in the marmoset, as in the cat (Turman et al., 1992) and rabbit (Murray et al., 1992), are attributable to removal of background facilitation that operates over the intracortical linkage from SI to SII. Therefore, in the marmoset, as well as in the cat, rabbit, tree shrew and galago, tactile signals that reach SII from skin may come entirely or predominantly over a direct pathway from the thalamus that is organized in parallel with that to SI. (Zhang et al., 1996; Rowe et al., 1996).

These recent findings in the marmoset necessitate a revision of the hypothesis that tactile processing at thalamocortical levels in simian primates is based on a strict *serial* scheme in which signals are conveyed from the thalamus to SI, and thence to SII (Garraghty et al., 1990; Pons et al., 1987, 1992). The explanation for the differences between the findings, based on cooling-induced inactivation of SI (Zhang et al., 1996; Rowe et al., 1996) and those based on surgical ablation is not altogether clear. However, the ablation may have set up an injury discharge in corticocortical neurons that project from SI to the topographically-related area of SII which may have lead to changes in extracellular ion concentrations, in particular, K^+ ion accumulation, and accommodation block of neurons within this region of SII (Zhang et al., 1996; Rowe et al., 1996).

The results of this recent study (Zhang et al., 1996; Rowe et al., 1996) show unequivocally that there is substantial direct thalamic input to SII in the marmoset, and therefore refute any hypothesis for a serial organization of SI and SII as a generalized attribute of the primate somatosensory cortex. Increasingly, it appears that schemes that place SII at a higher processing level than SI (or its component cytoarchitectural zones, 3a, 3b 1 and 2) in a hierarchical (or serial) network for tactile processing (Pons et al., 1987, 1992; Garraghty et al., 1990; Felleman and van Essen, 1991) are no longer valid for most mammals, including primate and non-primate species.

4. Summary

There has been much controversy over the organization of the thalamocortical somatosensory networks, in particular, over the issue of serial versus parallel processing in the SI and SII cortical areas. Several reports have established unequivocally that, in non-primate species, there is a parallel scheme of processing in which inputs from the thalamus reach both SI and SII independently. However, observations that SII responsiveness was abolished in primate species (the macaque and marmoset monkey) following the surgical ablation of the SI area, led to the interpretation that in primates a serial scheme operated in which tactile information is conveyed from the thalamus to SI and thence to SII via intracortical connections (Pons et al., 1987; Garraghty et al., 1990). However, re-investigation of this issue in the marmoset monkey in experiments based on reversible inactivation of SI by means of localized cooling (Zhang et al., 1996; Rowe et al., 1996) show unequivocally that there is substantial direct thalamic input to SII in the marmoset.

In conclusion, it appears that there is no longer justification for the hypothesis that there are fundamental differences in terms of serial and parallel organization of the SI and SII areas, between simian primates *in general*, and other mammals (Murray et al., 1992; Turman et al., 1992, 1995). Whether the serial scheme can be confirmed for the *macaque* monkey, the other primate in which SI ablation was reported to abolish SII responsiveness (Burton et al., 1990; Garraghty et al., 1990; Pons et al., 1987, 1992), will depend upon the application of refined reversible methods such as cooling, for SI inactivation, and quantitative evaluation of SII responsiveness in association with the reversible inactivation procedures.

References

Adrian, E.D. (1940) Double representation of the feet in the sensory cortex of the cat. Journal of Physiol, 98, 16-18P.

Adrian, E.D. (1941) Afferent discharges to the cerebral cortex from peripheral sense organs. Journal of Physiology, 100, 159-191.

Alloway, K.D. and Burton, H. (1985) Homotypical ipsilateral cortical projection between somatosensory areas I and II in the cat. Neuroscience, 14, 15-35.

Andersen, R.A., Asanuma, C., Essick, G. and Siegel, R.M. (1990) Cortico-cortical connections of anatomically and physiologically defined subdivisions within the inferior parietal lobule. Journal of Comparitive Neurology, 296, 65-113.

Bombardieri, R.A., Jr., Johnson, J.I.and Campos, G.B. (1975) Species differences in mechanosensory projections from the mouth to the ventrobasal thalamus. Journal of Comparitive Neurology, 163, 41-64.

Bowker, R.M. and Coulter, J.D. (1981) Intracortical connectivities of somatic sensory and motor areas: multiple cortical pathways in monkeys. In: Cortical Sensory Organization. Multiple Somatic Areas; edited by C.N. Woolsey. Clifton, NJ: Humana Press, pp. 205-242.

Brodmann, K. (1909) Vergleichende Lokalisationslehre der Grosshirnrinde. Leipzig: Barth.

Burchfiel, J.L. and Duffy, F.H. (1972) Muscle afferent input to single cell in primate somatosensory cortex. Brain Research, 45, 241-246.

Burton, H. (1984) Corticothalamic connections from the second somatosensory area and neighboring regions in the lateral sulcus of macaque monkeys. Brain Research, 309, 367-372.

Burton, H. (1986) Second somatosensory cortex and related areas. In: Cerebral Cortex. Sensory-Motor Areas and Aspects of Cortical Connectivity; edited by E.G. Jones and A. Peters. New York: Plenum Press, Vol. 5, p. 31-98.

Burton, H. and Kopf, E.M. (1984) Connections between the thalamus and the somatosensory areas of the anterior ectosylvian gyrus in the cat. Journal of Comparitive Neurology, 224, 173-205.

Burton, H. and Robinson, C.J. (1987) Responses in the first and second somatosensory cortical in cats during transient inactivation of the other ipsilateral area with lidocaine hydrochloride. Somatosensory Research, 4, 215-236.

Burton, H., Sathian, K. and Dian-Hua, S. (1990) Altered responses to cutaneous stimuli in the second somatosensory cortex following lesions of the postcentral gyrus in infant and juvenile macaques. Journal of Comparitive Neurology, 291, 395-414.

Carlson, M. (1981) Characteristics of sensory deficits following lesions of Brodmann's areas 1 and 2 in the postcentral gyrus of *Macaca mulatta*. Brain Research, 204, 424-430.

Cusick, C.G. and Gould, H.J., III (1990) Connections between area 3b of the somatosensory cortex and subdivisions of the ventroposterior nuclear complex and the anterior pulvinar nucleus in squirrel monkeys. Journal of Comparitive Neurology, 292, 83-102.

Cusick, C.G., Steindler, D.A. and Kaas, J.H. (1985) Corticocortical and collateral thalamocortical connections of postcentral somatosensory cortical areas in squirrel monkeys: a double-labeling study with radiolabeled wheatgerm agglutinin and wheatgerm agglutinin conjugated to horseradish peroxidase. Somatosensory Research, 3, 1-31.

Cusick,C.G.,Wall,J.T.,Felleman,D.J.andKaas, J.H.(1989)Somatotopic organization of the lateral sulcus of Owl monkeys: Area 3b, S-II, and a ventral somatosensory area. Journal of Comparitive Neurology, 282, 169-190.

Darian-Smith, C., Darian-Smith, I. and Cheema, S.S. (1990a) Thalamic projections to sensorimotor cortex in the macaque monkey: Use of multiple retrograde fluorescent traces. Journal of Comparitive Neurology, 299, 17-46.

Darian-Smith, C., Darian-Smith, I.and Cheema, S.S.(1990b) Thalamic projections to sensorimotor cortex in the newborn macaque. Journal of Comparitive Neurology, 299, 47-63.

Dreyer, D.A., Loe, P.R., Metz, C.B. and Whitsel, B.L. (1975) Representation of head and face in postcentral gyrus of the macaque. Journal of Neurophysiology, 38, 714-733.

Dykes, R.W., Sur, M., Merzenich, M.M., Kaas, J.H. and Nelson, R.J. (1981) Regional segregation of neurons responding to quickly adapting, slowly adapting, deep and Pacinian receptors within thalamic ventroposterior lateral and ventroposterior inferior nuclei in the squirrel monkey *(Saimiri sciureus)*. Neuroscience, 6, 1687-1692.

Felleman, D.J., Nelson, R.J., Sur, M. and Kaas, J.H. (1983) Representations of the body surface in areas 3b and 1 of postcentral parietal cortex of cebus monkey. Brain Research, 268, 15-26.

Felleman, D.J. and Van Essen, D.C. (1991) Distributed hierarchical processing in the primate cerebral cortex. Cerebral Cortex, 1, 1-47.

Ferrington, D.G. and Rowe, M.J. (1980) Differential contributions to coding of cutaneous vibratory information by cortical somatosensory areas I and II. Journal of Neurophysiology, 43, 310-331.

Fisher, G.R., Freeman, B. and Rowe, M.J. (1983) Organization of parallel projections from Pacinian afferent fibers to somatosensory cortical areas I and II in the cat. Journal of Neurophysiology, 49, 75-97.

Friedman, D.P. (1983) Laminar patterns of termination of cortico-cortical afferents in the somatosensory system. Brain Research, 273, 147-151.

Friedman, D.P. and Jones, E.G. (1981) Thalamic input to areas 3a and 2 in monkeys. Journal of Neurophysiology, 5, 59-85.

Friedman, D.P. and Murray, E.A. (1986) Thalamic connectivity of the second somatosensory area and neighboring somatosensory fields of the lateral sulcus of the macaque. Journal of Comparitive Neurology, 252, 348-373.

Friedman, D.P., Murray, E.A., O'Neill, J.B. and Mishkin, M. (1986) Cortical connections of the somatosensory fields of the lateral sulcus of macaques: evidence for a corticolimbic pathway for touch. Journal of Comparitive Neurology, 252, 323-347.

Gardner, E.P. and Costanzo, R.M. (1981) Properties of kinesthetic neurons in somatosensory cortex of awake monkeys. Brain Research, 214, 301-319.

Garraghty, P.E., Florence, S.L., Tenhula, W.N. and Kaas, J.H. (1991) Parallel thalamic activation of the first and second somatosensory areas in prosimian primates and tree shrews. Journal of Comparitive Neurology, 311, 289-299.

Garraghty, P.E., Pons, T.P. and Kaas, J.H. (1990) Ablations of areas 3b (SI proper) and 3a of somatosensory cortex in marmosets deactivate the second and parietal ventral somatosensory areas. Somatosensory and Motor Research, 7, 125-135.

Gaze, R.M. and Gordon, G. (1954) The representation of cutaneous sense in the thalamus of the cat and monkey. Quarterly Journal of Experimental Physiology, 39, 279-304.

Gerard, R.W., Marshall, W.H. and Saul, L.J. (1936) Electrical activity to the cat's brain. Archives of Neurology and Psychiatry, 36, 675-735.

Ghosh, S., Murray, G.M., Turman, A.B. and Rowe, M.J. (1994) Corticothalamic influences on transmission of tactile information in the ventroposterolateral thalamus of the cat: effect of reversible inactivation of somatosensory cortical areas I and II. Experimental Brain Research, 100,276-286.

Gordon, G. and Manson, J.R. (1967) Cutaneous receptive fields of single nerve cells in the thalamus of the cat. Nature, 215, 597-599.

Head, H. and Holmes, G. (1911) Sensory disturbances from cerebral lesions. Brain, 34, 102-254.

Heath, C.J., Hore, J. and Phillips, C.G. (1976) Inputs from low threshold muscle and cutaneous afferents of hand and forearm to areas 3a and 3b of baboon's cerebral cortex. Journal of Physiology, 257, 199-227.

Hikosaka, O., Tanaka, M., Sokamoto, M. and Iwamura, Y. (1985) Deficits in manipulative behaviors induced by local injections of muscimol in the first somatosensory cortex of the conscious monkey. Brain Research, 325, 375-380.

Hore, J., Preston, J.B., Durkovic, R.G. and Cheney, P.D. (1976) Responses of cortical neurons (areas 3a and 4) to ramp stretch of hindlimb muscles in the baboon. Journal of Neurophysiology, 39, 484-500.

Hyvärinen, J. and Poranen, A. (1978a) Movement-sensitive and direction and orientation-selective cutaneous receptive fields in the hand area of the post-central gyrus in monkeys. Journal of Physiology, 283, 523-537.

Hyvärinen, J. and Poranen, A. (1978b) Receptive field integration and submodality convergence in the hand area of the post-central gyrus of the alert monkey. Journal of Physiology, 283, 539-556.

Hyvärinen, J., Poranen, A., and Jokinen, Y. (1980) Influence of attentive behavior on neuronal responses to vibration in primary somatosensory cortex of the monkey. Journal of Neurophysiology, 43, 870-882.

Iwamura, Y. and Tanaka, M. (1978) Postcentral neurons in hand region of area 2: their possible role in the form discrimination of tactile objects. Brain Research, 150, 662-666.

Iwamura, Y., Tanaka, M. and Hikosaka, O. (1980) Overlapping representation of fingers in the somatosensory cortex (area 2) of the conscious monkey. Brain Research, 197, 516-520.

Johnson, J.I. (1990) Comparative development of somatic sensory cortex. In: Cerebral Cortex. Comparative Structure and Evolution of Cerebral Cortex. Vol. 8B, part II, edited by E.G. Jones and A. Peters. New York: Plenum Press, pp. 335-449.

Jones, E.G. (1983) Lack of collateral thalamocortical projection to fields of the first somatic sensory cortex in monkeys. Experimental Brain Research, 52, 375-384.

Jones, E.G. (1985) The Thalamus. New York: Plenum Press.

Jones, E.G., Coulter, J.D. and Hendry, S.H.C. (1978) Intracortical connectivity of architectonic fields in the somatosensory, motor and parietal cortex of monkeys. Journal of Comparitive Neurology, 181, 291-348.

Jones, E.G. and Friedman, D.P. (1982) Projection pattern of functional components of thalamic ventrobasal complex on monkey somatosensory cortex. Journal of Neurophysiology, 48, 521-544.

Jones, E.G., Friedman, D.P. and Hendry, S.H.C. (1982) Thalamic basis of place- and modality-specific columns in monkey somatosensory cortex: a correlative anatomical and physiological study. Journal of Neurophysiology, 48, 545-568.

Jones, E.G. and Powell, T.P.S. (1968) The ipsilateral cortical connections of the somatic sensory areas in the cat. Brain Research, 9, 71-94.

Jones, E.G. and Powell, T.P.S. (1969) Connections of the somatic sensory cortex of the rhesus monkey.   I. Ipsilateral cortical connections. Brain, 92, 477-502.

Jones, E.G., Wise, S.P. and Coulter, J.D. (1979) Differential thalamic relationships of sensory-motor and parietal cortical fields   in monkeys. Journal of Comparitive Neurology, 183, 833-881.

Kaas, J.H. (1983) What, if anything, is SI?   Organization of first somatosensory area of cortex. Physiological Review, 63, 206-231.

Kaas, J.H. (1987) The organization of neocortex in mammals: implication for theories of brain function.   Annual Review of Psychology, 38, 129-151.

Kaas, J.H., Nelson, R.J., Sur, M., Lin, C.-S. and Merzenich, M.M. (1979) Multiple representation of the body within primary somatosensory cortex of primates. Science, 204, 521-523.

Kaas, J.H., Nelson, R.J., Sur, M. and Merzenich, M.M. (1981a) Organization of somatosensory cortex in primates. In: The Organization of the Cerebral Cortex; edited by F.O. Schmitt, F.G. Worden, G. Adelman and S.G. Dennis.   Cambridge, MA: MIT Press, pp. 237-261.

Kaas, J.H. and Pons, T.P. (1988) The somatosensory system of primates, In: Comparative Primate Biology. Vol. 4: Neurosciences, edited by H.P. Steklis and J. Erwin New York, Liss. pp.421-468.

Kaas, J.H., Sur, M. and Wall, J.T. (1981b) Modular segregation in monkey somatosensory cortex. Trends in Neuroscience, 10, 13-14.

Krubitzer, L. A. and Kaas, J.H. (1990) The organization and connections of somatosensory cortex in marmosets. Journal of Neuroscience, 10, 952-974.

Krubitzer, L.A. and Kaas, J.H. (1992) The somatosensory thalamus of monkeys: Cortical connections and a redefinition of nuclei in marmosets. Journal of Comparitive Neurology, 319,123-140.

Krubitzer, L., Clarey, J., Tweedale, R., Elston, G. and Calford, M. (1995) A redefinition of somatosensory areas in the lateral sulcus of macaque monkeys. Journal of Neuroscience, 15, 3821-3839.

Landry, P., Diadori, P., Leclerc. S. and Dykes, R.W. (1987) Morphological and electrophysiological characteristics of somatosensory thalamocortical axons studied with intra-axonal staining and recording in the cat. Experimental. Brain Research, 65, 317-330.

Loe, P.R., Whitsel, B.L., Dreyer, D.A. and Metz, C.B. (1977) Body representation in ventrobasal thalamus of Macaque: a single unit analysis. Journal of Neurophysiology, 40, 1339-1357.

Manzoni, T., Caminiti, R., Spidalieri, G. and Morelli, E. (1979) Anatomical and functional aspects of the associative projections from somatic areas SI to SII. Experimental Brain Research, 34, 453-470.

Marshall, W.H., Woolsey, C.N. and Bard, P. (1937) Cortical representation of tactile sensibility as indicated by cortical potentials. Science, 85, 388-390.

Marshall, W.H., Woolsey, C.N. and Bard, P. (1941) Observations on cortical somatic sensory mechanisms of cat and monkey. Journal of Neurophysiology, 4, 1-24.

Merzenich, M., Kaas, J.H., Sur, M. and Lin, C-S. (1978) Double representation of the body surface within cytoarchitectonic areas 3b and 1 in "SI" in the owl monkey *(Aotus trivirgatus)*. Journal of Comparitive Neurology, 181,41-73.

Merzenich, M.M., Sur, M., Nelson, R.J. and Kaas, J.H. (1981) Organization of the S-I cortex: multiple cutaneous representations in areas 3b and 1 of the owl monkey. In: Cortical Sensory Organization. Multiple Somatic Areas; edited by C.N. Woolsey. Clifton, NJ: Humana Press, Vol. 1, pp. 47-66.

Mishkin, M. (1979) Analogous neural models for tactual and visual learning. Neuropsychologia, 17, 139-151.

Mountcastle, V.B. (1984) Central nervous mechanisms in mechanoreceptive sensibility. In: Handbook of Physiology. The Nervous System. Sensory Processes; edited by I. Darian-Smith. Bethesda, MD: Am. Physiol. Soc., Vol. 3, part 2, pp. 789-878.

Mountcastle, V.B. and Henneman, E. (1949) Pattern of tactile representation in thalamus of cat. Journal of Neurophysiology, 12, 85-100.

Mountcastle, V.B. and Henneman, E. (1952) The representation of tactile sensibility in the thalamus of the monkey. Journal of Comparitive Neurology, 97, 409-440.

Mountcastle, V.B., Lynch, J.C., Georgopoulos, A., Sakata, H. and Acuna, C. (1975) Posterior parietal association cortex of the monkey: command functions for operations within extrapersonal space. Journal of Neurophysiology, 38, 871-908.

Murray, G.M., Zhang, H.Q., Kaye, A.N., Sinnadurai, T., Campbell, D.H. and Rowe, M.J. (1992) Parallel processing in rabbit first (SI) and second (SII) somatosensory cortical areas: Effects of reversible inactivation by cooling of SI on responses in SII. Journal of Neurophysiology, 68,703-710.

Neal, J.W., Pearson, R.C.A. and Powell, T.P.S. (1987) The corticocortical connections of area 7b, PF, in the parietal lobe of the monkey. Brain Research, 419, 341-346.

Nelson, R.J., Sur, M., Felleman, D.J. and Kaas, J.H. (1980) Representations of the body surface in postcentral parietal cortex of *Macaca fascicularis*. Journal of Compartive Neurology, 192, 611-643.

Paul, R.L., Goodman, H. and Merzenich, M. (1972a) Alterations in mechanoreceptor input to Brodmann's areas 1 and 3 of the postcentral hand area of *Macaca mulatta* after nerve section and regeneration. Brain Research, 36, 1-19.

Paul, R.L., Merzenich, M. and Goodman, H. (1972b) Representation of slowly and rapidly adapting cutaneous mechanoreceptors of the hand in Brodmann's areas 3 and 1 of Macaca mulatta. Brain Research, 36, 229-249.

Phillips, C.B., Powell, T.P.S. and Wiesendanger, M. (1971) Projection from low-threshold muscle afferents of the hand and forearm to area 3a of baboon's cortex. Journal of Physiology, 217, 419-446.

Poggio, G.F. and Mountcastle, V.B. (1963) The functional properties of ventrobasal thalamic neurons studied in unanesthetised monkeys. Journal of Neurophysiology, 26, 775-806.

Pollin, B. and Albe-Fessard, D. (1979) Organization of somatic thalamus in monkeys with and without section of dorsal spinal tracts. Brain Research, 173, 431-449.

Pons, T.P., Garraghty, P.E., Cusick, C.G. and Kaas, J.H. (1985) The somatotopic organization of area 2 in macaque monkeys. Journal of Comparitive Neurology, 241, 445-466.

Pons, T.P., Garraghty, P.E., Friedman, D.P. and Mishkin, M. (1987) Physiological evidence for serial processing in somatosensory cortex. Science, 237, 417-420.

Pons, T.P., Garraghty, P.E. and Mishkin, M. (1992) Serial and parallel processing of tactual information in somatosensory cortex of rhesus monkeys. Journal of Neurophysiology, 68, 518-527.

Pons, T.P. and Kaas, J.H. (1985) Connections of area 2 of somatosensory cortex with the anterior pulvinar and subdivisions of the ventroposterior complex in macaque monkeys. Journal of Comparitive Neurology, 240, 16-36.

Pons, T.P. and Kaas, J.H. (1986) Corticocortical connections of area 2 of somatosensory cortex in macaque monkeys: a correlative anatomical and electrophysiological study. Journal of Comparitive Neurology, 248, 313-335.

Randolph, M. and Semmes, J. (1970) Behavioral consequences of selective subtotal ablations in the postcentral gyrus of *Macaca mulatta*. Brain Research, 70, 55-70.

Robinson, C.J. and Burton, H. (1980a) Organization of somatosensory receptive fields in cortical areas 7b, retroinsula, postauditory and granular insula of *M. fascicularis*. Journal of Comparitive Neurology, 192, 69-92.

Robinson, C.J. and Burton, H. (1980b) Somatic submodality distribution within the second somatosensory (SII), 7b, retroinsula, postauditory, and granular insula cortex of *M. fascicularis*. Journal of Comparitive Neurology, 192, 93-108.

Rose, J.E. and Mountcastle, V.B. (1952) The thalamic tactile region in rabbit and cat. Journal of Comparitive Neurology, 97, 441-489.

Rowe, M.J., Turman, A.B., Murray, G.M. and Zhang, H.Q. (1996) Parallel processing in somatosensory areas I and II of the cerebral cortex. In: Somesthesis and the Neurobiology of the Somatosensory Cortex; edited by O. Franzén, R. Johansson and L. Terenius. Basel, Birkhäuser Verlag, pp. 197-212.

Schwartz, D.W., Deecke, L. and Fredrickson, J.M. (1973) Cortical projection of group I muscle afferents to areas 2, 3a, and the vestibular field in the rhesus monkey. Experimental Brain Research, 17, 516-526.

Soso, M.J. and Fetz, E.E. (1980) Responses of identified cells in postcentral cortex of awake monkeys during comparable active and passive joint movements. Journal of Neurophysiology, 43, 1090-1110.

Sur, M. (1980) Receptive fields of neurons in areas 3b and 1 of somatosensory cortex in monkeys. Brain Research, 198, 465-471.

Sur, M., Nelson, R.J. and Kaas, J.H. (1982) Representation of the body surface in cortical areas 3b and 1 of squirrel monkeys: comparisons with other primates. Journal of Comparitive Neurology, 211, 177-192.

Tanji, J. (1975) Activity of neurons in cortical area 3a during maintenance of steady postures by the monkey. Brain Research, 88, 549-553.

Tanji, J. and Wise, S.P. (1981) Submodality distribution in sensorimotor cortex of the unanesthetized monkey. Journal of Neurophysiology, 45, 467-481.

Turman, A.B., Ferrington, D.G., Ghosh, S., Morley, J.W. and Rowe, M.J. (1992) Parallel processing of tactile information in the cerebral cortex of the cat: effect of reversible inactivation of SI on responsiveness of SII neurons. Journal of Neurophysiology, 67, 411-429.

Turman, A.B., Morley, J.W., Zhang, H.Q. and Rowe, M.J. (1995) Parallel processing of tactile information in the cerebral cortex of the cat: effect of reversible inactivation of SII on responsiveness of SI neurons. Journal of Neurophysiology, 73, 1063-1075.

Van Essen, D.C. and Maunsell, J.H.R. (1983) Hierarchical organization and functional streams in the visual cortex. Trends in Neuroscience, 6, 370-375.

Vogt, B.A. and Pandya, D.N. (1978) Cortico-cortical connections of somatic sensory cortex (areas 3, 1 and 2) in the rhesus monkey. Journal of Comparitive Neurology, 177, 179-191.

Welker, W.I. (1973) Principals of organization of the ventrobasal complex in mammals. Brain and Behavioural Evolution, 7, 253-336.

Wise, S.P. and Tanji, J. (1981) Neuronal responses in sensorimotor cortex to ramp displacements and maintained positions imposed on hindlimb of the unanesthetized monkey. Journal of Neurophysiology, 45, 482-500.

Woolsey, C.N. (1943) Second" somatic receiving areas in the cerebral cortex of cat, dog and monkey. Federation Proceedings, 2, 55.

Woolsey, C.N. (1958) Organization of somatic sensory and motor areas of the cerebral cortex. In: Biological and Biochemical Bases of Behavior; edited by H.F. Harlow and C.N. Woolsey. Madison: Univ. of Wisconsin Press, pp. 63-81.

Woolsey, C.N. and Fairman, D. (1946) Contralateral, ipsilateral and bilateral representation of cutaneous receptors in somatic areas I and II of cerebral cortex of pig, sheep and other animals. Surgery, 19, 684-702.

Woolsey, C.N., Marshall, W.H. and Bard, P. (1942) Representation of cutaneous tactile sensibility in the cerebral cortex of the monkey as indicated by evoked potentials. Bulletin of the Johns Hopkins Hospital, 70, 339-341.

Yin, T.C.T. and Mountcastle, V.B. (1978) Mechanisms of neural integration in the parietal lobe for visual attention. Federation Proceedings, 37, 2251-2257.

Yuan, B., Morrow, T.J. and Casey, K.L. (1985) Responsiveness of ventrobasal thalamic neurons after suppression of SI cortex in the anaesthetized rat. Journal of Neuroscience, 5, 2971-2978.

Yumiya, H., Kubota, K. and Asanuma, H. (1974) Activities of neurons in area 3a of the cerebral cortex during voluntary movements in the monkey. Brain Research, 78, 169-177.

Zarzecki, P. (1986) Functions of corticocortical neurons of somatosensory, motor, and parietal cortex. In: Cerebral Cortex. Vol. 5: Sensory-Motor areas and Aspects of Cortical connectivity. edited by E.G. Jones and A. Peters. New York: Plenum, pp. 185-215.

Zhang, H.Q., Murray, G.M., Turman, A.B., Mackie, P.D., Coleman, G.T. and Rowe, M.J. (1996) Parallel processing in cerebral cortex of the marmoset monkey: effect of reversible SI inactivation on tactile responses in SII. Journal of Neurophysiology, 76, 3633-3655.

Representation of Tactile Functions in the Somatosensory Cortex

Y. Iwamura

The hand is an organ to explore and perceive the outer world at the same time. Sensory signals required for movement control and perception originate from both surface and deep tissues, of not only the hand and fingers but also the arm and shoulders. We also use a single hand or both hands, depending on the situation. Thus the spatial and temporal patterns of the sensory signals during the performance must be enormously complex. How these complex patterns of sensory signals during actual hand use are processed in the somatosensory cortex has been our primary concern. In this chapter I review the results of our single neuronal studies done in awake monkeys.

1. Hierarchical information processing in the somatosensory cortex

Sensory signals from the skin and deep tissue project to the cortex via the dorsal column and spinothalamic pathways. Classical studies in this field have emphasized the fidelity with which response properties of peripheral receptors are reproduced by single units in the somatosensory thalamic relay as well as the primary cortical receiving area, SI (areas 3, 1, and 2)(c.f. Mountcastle 1984). This notion has been accepted as a basis of later studies in the somatosensory cortices. This was in marked contrast to the visual system. Since the earlier work by Hubel and Wiesel (1959) at the single neuronal level, it has been recognized that systematic and hierarchical processing of sensory information starts at the earliest stage of the cortex and proceeds gradually and continuously to the higher level.

The hierarchical organization in information processing was suggested in the somatosensory cortical system by Duffy and Burchfiel (1971), Sakata (1971) and Sakata et al (1973). They found in area 5, neurons with complex receptive field properties such as those receiving converging inputs from different body parts or different submodalities. Neither group of investigators however, mentioned that the integration starts within SI. Several years later, Hyvarinen and Poranen (1978) proposed a hierarchical scheme of information processing within SI based on their single neuron study in the postcentral digit region of alert monkeys. Iwamura and Tanaka (1978a) described receptive fields of many neurons in area 2 covering multiple digit skin, and that among them exist neurons with response preference to the presence of an edge. They also described neurons which were activated only when the monkey actively grasped a wooden ball or a cube(rectangular or edged objects). Iwamura et al (1980) further showed that because of the receptive field convergence, representations of different digits are not strictly separated but rather they overlap with each other in the caudal part of the gyrus starting in area 1. Iwamura and Tanaka (1978b,c) also found differences in the receptive field size and shape among different cytoarchitectonic areas of the cat somatosensory cortex. The hierarchical scheme was incompatible with the concept of multiple, independent and parallel representations of the body in four different cytoarchitectonic subareas proposed at that time (Kaas et al 1979).

1.2 Anatomical background of hierarchical processing in the postcentral gyrus

Anatomical organization of the primate postcentral gyrus has been described in terms of several different cytoarchitectonic nomenclatures (Campbell 1905, Brodmann 1909, Economo and Koskinas 1925, Vogt and Vogt 1919, von Bonin and Bailey 1947), and there has been a common understanding that cytoarchitectonic characteristics change rostrocaudally. Powell and Mountcastle (1959a) described differences among the subdivisions in detail, stating that area 3 is a typical koniocortex with granular cells while in areas 1 and 2 the morphological characteristics change gradually to the homotypical parietal association cortex showing increase in the number of large pyramidal cells in layers III and V. They emphasized that the divisions between the three cytoarchitectonic areas are regions of gradients of morphological change, rather than sharp lines. This impression has been confirmed and shared by many investigators of this region and thus the idea of dividing the postcentral gyrus into four cytoarchitectonic subdivisions appeared inappropriate.

2. Physiological correlates of the anatomical gradients along the rostrocaudal axis of the postcentral gyrus

Powell and Mountcastle (1959b) studied receptive field properties of single neurons in lightly anesthetized monkeys and found that in the rostrocaudal dimension of the postcentral gyrus there is a gradient change in the representation of submodalities which they thought to be paralleled by the gradual morphological changes in the same region described above. However, Hyvarinen and Poranen (1978) showed that there is not much change in the rates of deep neurons nor skin neurons among the different cytoarchitectonic subdivisions. On the other hand, as describedalready, Hyvarinen and Poranen (1978) and Iwamura and Tanaka (1978) found that the receptive field size increases toward the posterior part of the gyrus. The increase in the size of the receptive field of neurons in areas 1 and 2 of monkeys have been confirmed repeatedly since then (McKenna et al., 1982, Darian-Smith et al., 1984, Pons et al., 1985, Iwamura et al., 1983a,b; 1985a,b,c,d; Gardner 1988, Wannier et al., 1991, Ageranioti-Belanger and Chapman 1992). More recently Iwamura et al. (1993) demonstrated the rostrocaudal gradients in the complexity of neuronal receptive field properties in the postcentral gyrus of alert monkeys in a quantitative manner, based on the large body of data which have been accumulated in a series of experiments described below. A total of 2656 neurons were recorded from 6 hemispheres of 4 monkeys, in various portions of the postcentral gyrus including the deep rostral bank of the intraparietal sulcus. Among them, 1979 neurons were responsive to somatosensory stimulation, and their submodality types were identified.

2.1 Neurons with composite type receptive field increase caudally

The smallest receptive fields of the SI neurons are those confined to one segment or single joint of a digit (Iwamura et al 1983a). We defined the unitary receptive fields as follows: the distal skin area covering either one of the digits (I-V), the hand dorsum (D) and the radial (R) or ulnar (U) half of the palmar skin. The forearm skin (F) is the only proximal skin area encountered in the present study. Deep submodality neurons were classified into the digit type (I-V), the wrist (W), the forearm muscles (F), the elbow (E) and the shoulder joint (S) manipulation types (Iwamura et al 1993).

Large receptive fields found in areas 1 and 2 cover multiple digits, both digits and the palmar skin, both sides of the palmar skin etc. We classified all the neurons into single locus type and multi-loci (composite) type in terms of their receptive fields. Figure 1 is a scatter map showing distribution of skin neurons of single-locus type (small

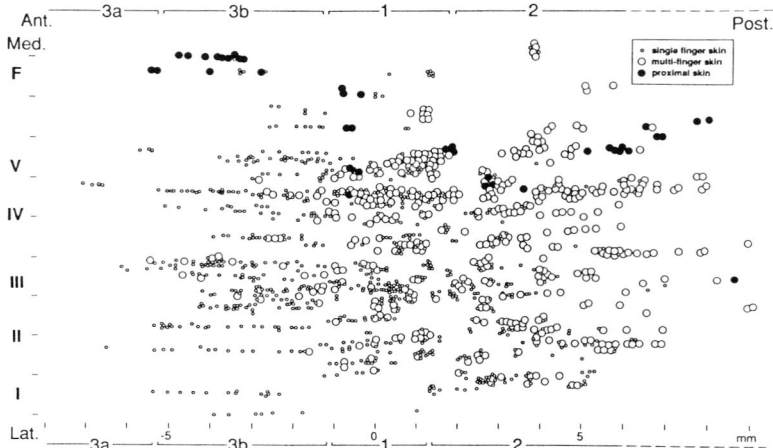

Figure 1. A scatter map on the unfolded postcentral gyrus showing differential distribution of skin neurons. The abscissa represents the anterior-posterior distance along the gyrus and the ordinate represents the medio-lateral distance. In the abscissa the reference point (zero) was taken at the corner of the postcentral sulcus. Letters along the ordinate indicate approximate sites of representation of digits (I-V) or forearm (F) in area 3. Small open circles: neurons with single digit (location) type receptive fields. Large open circles: neurons with composite type receptive fields. Filled circles: neurons representing the forearm skin. All the skin submodality neurons are included which responded to light touch, specific types of skin stimulation, hair bending and nails. Data from six hemispheres of four monkeys are superimposed. (Adapted from Iwamura et al 1993).

open circles) or neurons of composite type (large open circles). Neurons with single locus type receptive fields are the great majority in area 3b. The single-locus type neurons were found in area 2 up to 7 mm caudally from the reference point, but they were mostly in the lateral part. In contrast, composite type neurons increased rapidly in area 1 and were dominant in further caudal regions. This tendency was remarkable in the medio-caudal part where neurons representing the forearm skin (filled circles) were intermingled. Skin neurons were rare in area 3a.

A similar tendency was found in deep submodality neurons (Figure 2) which responded to manipulation of either single (small open circles) or multiple joints (large open circles). In general the deep

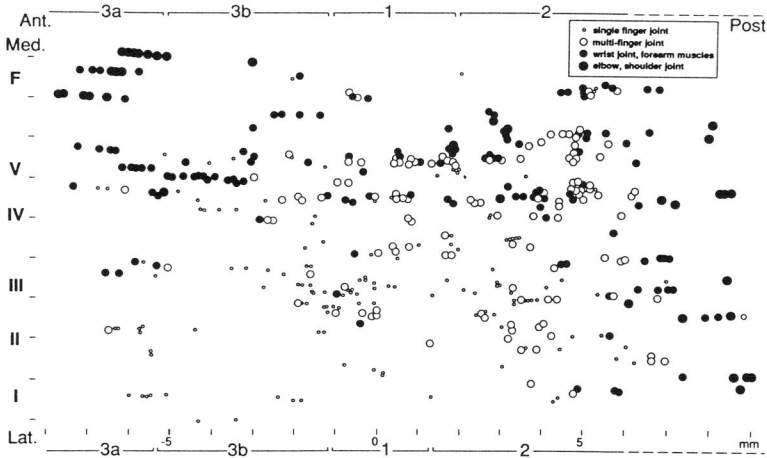

Figure 2. A scatter map on the unfolded postcentral gyrus showing distribution of deep submodality neurons. The same arrangement as in Figure 1. Small open circles: neurons with single digit type receptive fields. Large open circles: neurons with composite type receptive fields. Small filled circles: neurons representing the wrist joint or forearm muscles. Large filled circles: neurons representing the elbow or shoulder joint. All the deep submodality neurons are included which responded to joint manipulation, tapping or pressing of joints or muscles. Data from six hemispheres of four monkeys, same as those in Figure 1. (Adapted from Iwamura et al 1993).

submodality neurons were less numerous than the skin neurons in the digit region. They were particularly sparse in the lateral part where receptive fields of skin neurons were on the thumb or the index finger. The composite type dominated the single locus type in the mediocaudal part, similar to the skin submodality neurons.

2.2 The presence of proximal neurons in the digit region

Figure 2 also shows that neurons responding to manipulation of the proximal joints or muscles (filled circles), which were found in the rostral region only medially to the digit representation, invaded the digit region in its caudalmost part. In the rostral region the proximal neurons were mostly of single locus type with receptive fields localized to the forearm. In the caudal region, receptive fields of the proximal neurons were mostly of multi-loci type. Sometimes the shoulder joint was involved. The presence of neurons with proximal receptive fields is interpreted as reflecting the fact that the proximal and distal parts of the forelimb work together in hand manipulation and reaching actions.

2.3 Rostrocaudal increase in the complexity of receptive fields

To evaluate the differences along the rostrocaudal axis of the digit region, we performed a quantitative analysis based on combining the data obtained from six hemispheres, regardless of small mediolateral or individual variations in the width of the postcentral gyrus. The ratios of neurons with different characteristics were calculated in each medio-lateral strip of 1 mm width along the entire rostrocaudal axis of the map. Figure 3 shows how the composite type receptive fields increase toward the caudal part of the gyrus. Only those neurons of which receptive fields involve the digits were taken into account here. Neurons with receptive fields of a single locus type comprised the overwhelming majority in the rostral part (areas 3a and 3b), but its ratio started to decrease at the border between areas 3b and 1. Neurons with a two-loci type receptive field started to increase within area 3b near its border to area 1; neurons with a three-or-more-loci type receptive field started to increase more caudally in area 1 and further in area 2.

3. Comparison of submodalities among different cytoarchitectonic subdivisions

Submodalities of neurons collected in the entire postcentral gyrus were compared among different cytoarchitectonic subdivisions. The cytoarchitectonic borders between areas 3 and 1 and areas 1 and 2 were determined in each histological section according to the criteria

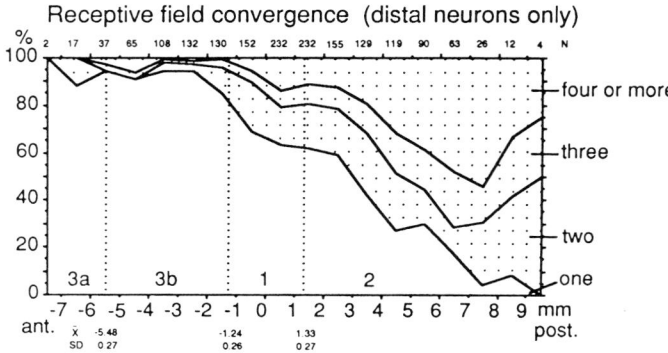

Figure 3 The increase in the receptive field convergence toward the caudal part of the gyrus. Dotted lines indicate borders between two successive cytoarchitectonic subareas. They were the mean of those which were determined in each histological section. Only neurons representing the distal part were included. Data from six hemispheres of four monkeys. (Adapted from Iwamura et al 1993).

described by Powell and Mountcastle (1959a) and the recording site of the individual neuron was assigned to one of the subdivisions. The caudal border of area 2 is unclear cytoarchitectonically. Physiologically, however, we found changes of neural activity in several features at around 6 mm caudal from the reference point. The changes include the increase in the number of neurons responding to proximal forelimb manipulation and the increase in the number of neurons whose receptive field positions or submodalities were undetermined and those of which activity was influenced by visual stimulation. We take these observations to indicate the presence of a functional subdivision in the caudalmost part of the postcentral gyrus and labeled this region as "d(deep) IPS".

Figure 4 summarizes the results of classification of postcentral neurons in terms of submodality. They indicate that the skin submodality neurons were the majority in areas 3b and 1, that the deep submodality neurons were similarly distributed except in area 3a, that a small number of neurons responding to both skin and deep stimuli were found mostly in area 2, and that submodality-unspecified neurons increase posteriorly, reaching as many as 60% in the dIPS.

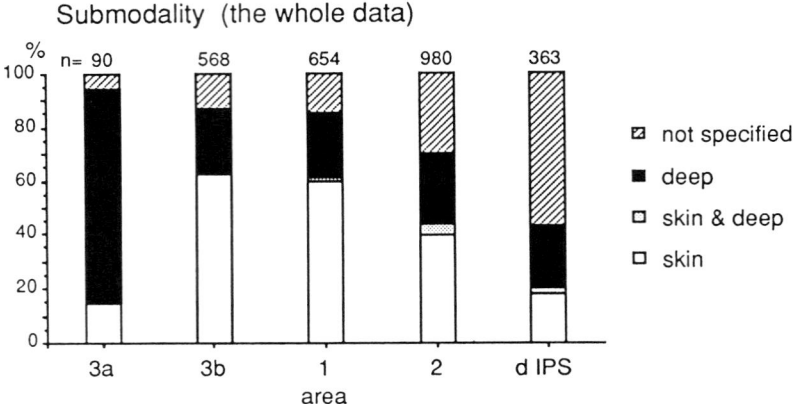

Figure 4. Comparison of the ratio of submodality populations. Data from six hemispheres of four monkeys. dIPS, deep part of intraparietal sulcus. (Adapted from Iwamura et al 1993).

The ratio of skin neurons was largest in area 3b and decreased gradually toward the caudal part of the gyrus. The ratio of deep submodality neurons was largest in area 3a, as expected from previous studies (Jones and Porter 1980). It was around 20% in area 3b, and was kept almost constant caudally from there. The results are at variance with the results of Powell and Mountcastle (1959b) who reported that more than 90% of area 2 neurons were of deep submodality. The discrepancy may be explained by the regional difference: Powell and Mountcastle recorded neurons in the more medial part of the gyrus, (those with receptive fields on the forearm, arm, shoulder, trunk, thigh, leg, and very few on digits), while we recorded neurons in the digit region which may be a highly specified area for cutaneous inputs.

4. A hierarchical model and a columnar hypothesis

The present results are in accordance with the hierarchical scheme, i.e. within the postcentral gyrus sensory information is processed from the primary sensory receiving stage to integrative and more associative stages (Iwamura et al 1983b). This model fits better to the gradual shift of the cytoarchitectonic characteristics along the rostrocaudal axis of the postcentral gyrus. Rich intrinsic corticocortical connections are demonstrated within the postcentral gyrus starting with area 3b and projecting to areas 1 and 2 (Jones and Powell 1970; Jones, 1975; Jones and Burton, 1976; Vogt and Pandya, 1978; Jones et al., 1978; Shanks and Powell, 1981; Kunzle, 1978). These cortico-cortical connections may be the main route of inputs to area 2 and further caudal region, area 5, where many neurons did not respond to stimulation of the periphery. The direct thalamic inputs from the shell of the ventroposterior nuclei (Jones and Friedman, 1982; Pons and Kaas, 1985) to area 2 may be subordinate, if not minimal, among various inputs to area 2. On the other hand, additional inputs from thalamic association nuclei such as anterior pulvinar reach area 2 as well as areas 5 and 7 (Pons and Kaas, 1985). Such inputs may contribute to provide neurons in the posterior part with functionally unique response characteristics, such as described later. They may also contribute to a blurring of the distinction between the primary somatosensory cortex and the posterior parietal association cortices.

The presence of the hierarchical organization along the rostrocaudal axis over the postcentral gyrus necessarily requires integration mechanisms at a cortical locus. The classical columnar hypothesis emphasized the similarity of neuronal receptive field characteristics in a neuronal array perpendicular to the cortical surface (Mountcastle, 1984). However, receptive field characteristics vary among neurons along a neuronal array perpendicular to the cortical surface. We discussed this subject elsewhere (Iwamura et al., 1985a,b). Figure 5 shows an example of such penetration. In this array neuronal receptive fields were on the ulnar side of the volar skin. There are four or five variations in the receptive field size, position and most adequate stimulus. The smallest one was limited to nail beds. The largest ones covered all others. All but one neuron were activated by skin stimuli moving in a radial-to-ulnar direction. There is a neuron which was activated by the self-initiated hand movement only.

5. Functional surfaces and bone spaces

Figure 6 is a scatter map showing the distribution of large receptive fields found in the caudal part of the gyrus such as described in the previous section (Figure 5). We classified them into 8 types and

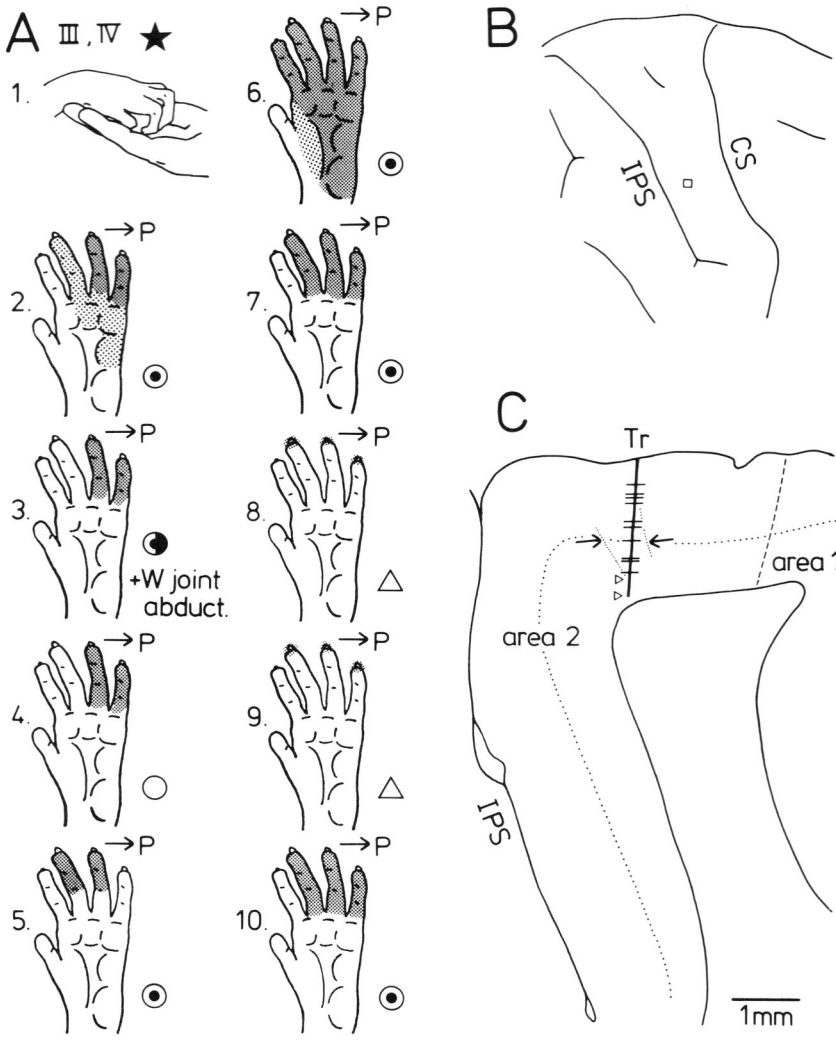

Figure 5. A:Receptive fields and the most adequate stimuli of neurons recorded along a vertical penetration in area 2. All but one neuron responded to a moving stimulus across the skin or mails, and all of them had a directional selectivity from the radial to ulnar side. Neuron 3 responded also to manipulation of the wrist joint. Neuron 1 had no receptive field to passive stimulation but was activated weakly when the animal scratched the experimenters palm. B and C, sites of penetration and recording. In C the width of the gray matter the electrode traversed (indicated by two arrows) is 720 u (adapted from Iwamura et al 1985b).

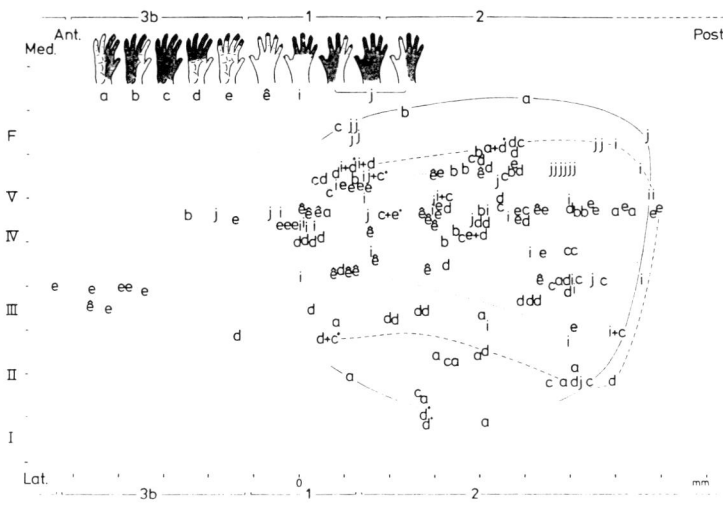

Figure 6. Functional surfaces and their distribution in the postcentral gyrus. a-j are functional surfaces so defined. In the map each surface was plotted by letter and the aproximate extent of distribution is shown(Adapted from Iwamura et al 1985c).

designated them as functional surfaces (Iwamura et al., 1985c,d). We consider functional surfaces as the skin surfaces which come into contact with objects in particular actions or postures. The idea originally came from the observation of receptive fields of neurons in the cat somatosensory cortex (Iwamura and Tanaka, 1978a,b). We noticed that their characteristic shapes represent the surfaces of contact between the skin and the ground when a cat is crouching, or contacts between the different skin areas of a cat in grooming (Figure 7). We realized that the same is true in monkey somatosensory cortex. Some of large surfaces were the surfaces of contact between the hand and an object in the hand. It is possible that the neurons with large receptive fields, and other specific types of neurons, emerged as the results of integration within the local vertical neuronal arrays. We postulated that these functional surfaces are the results of integration of information originated in areas 3a and 3b through cortico-cortical connections (Iwamura et al 1985d). According to this scheme area 3b is composed of several functional subdivisions each of which represents a functional surface and single neurons represent a part of the functional surface.

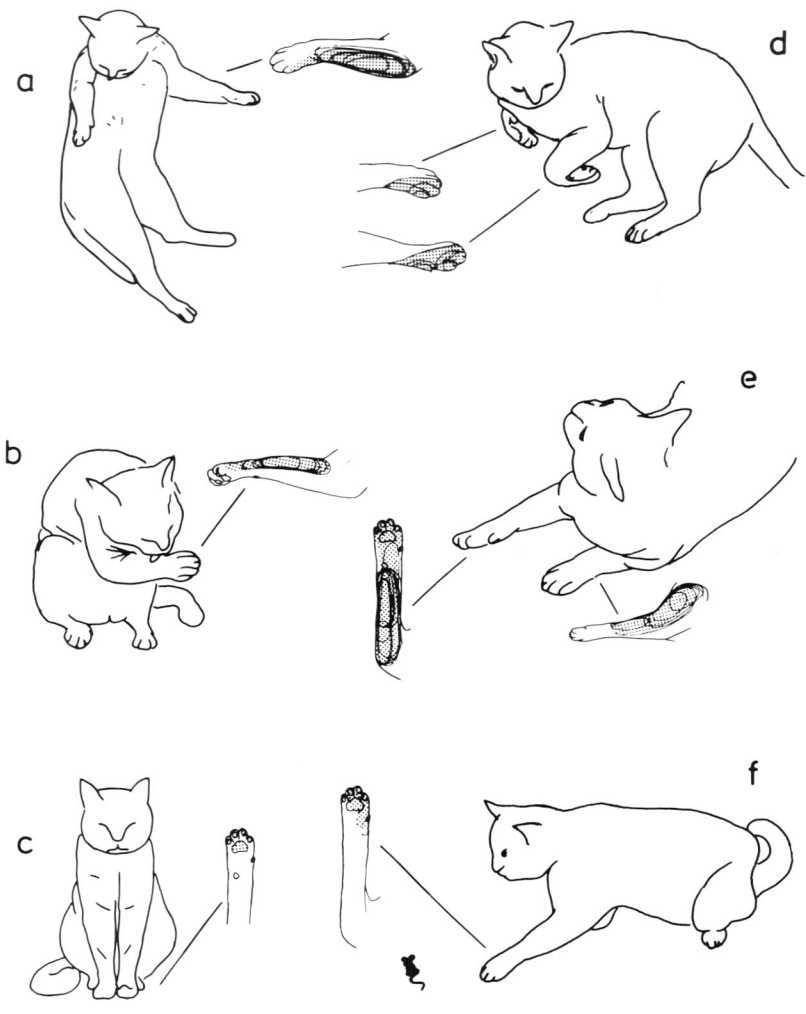

Figure 7. Relationship between large receptive fields found in the cat somatosensory cortex and possible postures by which these skin areas are stimulated.

Distribution of deep neurons with multi-loci type receptive fields coincided with these functional surfaces, that is, those which were activated by radial digit joints were found in the lateral part, those activated by ulnar digits and those activated by four digits were found in the medial part. I regard these deep neurons as representing three dimensional bone spaces made by digits, and they are functional when each of them are combined with corresponding skin surfaces to hold an object or to make contact with an object in a particular hand posture.

6. Neurons driven by monkey's self-initiated hand and arm actions

Figure 5 showed a neuron which was activated only when the animal used its hand in a self-initiated action. These types of neurons were among those neurons which did not respond or only poorly responded to passive stimulation (Iwamura and Tanaka, 1978, Iwamura et al., 1985b, Iwamura, 1993). In testing response characteristics of these neurons, we offered the animal sitting on the primate chair a piece of food and let them retrieve it from the experimenter's hand. Many neurons fired very briskly in this way. A group of neurons were selective to the precision grip which requires the flexion of radial digit joints and the contact of radial digit surfaces to the object, while other groups of neurons were driven by active grasping of an object with the ulnar side of the hand. Still another group of neurons were active when the monkey scratched-up food pieces from a flat surface with the digit tips. To induce all the variation of hand actions one after another, the cooperation of monkeys was necessaryand it was obtained in a brief period of training before and during experiments.

A total of 149 neurons were driven only or more briskly by at least one of the self-initiated actions illustrated in Figure 8. All but 11 neurons were found in area 2 and a further caudal part of the gyrus. Each neuron responded preferentially to either one of the actions. In these neurons the peak firing rate was more than 80 spikes/sec during one type of the action while it was less than 50% of the maximal value by other types of actions. With these criteria, we found a total of 46 neurons being selective to either one of the three hand actions as described above.Reaching out an arm toward an object was also one of the effective means of driving those neurons which failed to respond to passive stimulation. A total of 21 neurons were selectively activated when the monkey stretched its arm and digits to obtain bait. In some of them, over-stretching of digits was necessary. In 6 neurons, high rates

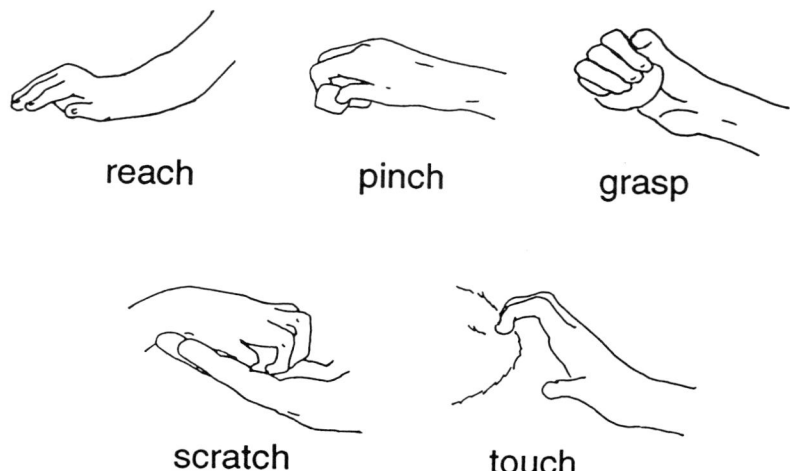

reach   pinch   grasp

scratch   touch

Figure 8. Five types of self-initiated hand actions by which neurons were activated.

of background activity (10-20 Hz) existed. The excitation during reaching was followed by the inhibition of the background activity as soon as the monkey got bait in the hand. The background activity was inhibited also by passive contact of an object to any part of the volar skin in one neuron but not in 5 other neurons.

Neurons driven by self-initiated hand actions were found intermingled with other neurons. In the crown of the cortex where they were perpendicular to the cortical surface, they were found in clusters of either skin, deep submodality or mixed types (Iwamura et al., 1985b). The radial grasp neurons were found in the lateral half of the finger region. The ulnar grasp neurons, the scratch-up neurons and touch neurons were found in the medial to middle part of the finger region, intermingled with each other. The distributions of the self-initiated hand movement neurons coincided with those of neurons representing several different skin surfaces described as functional surfaces (Figure 6) (Iwamura et al., 1985c, d). The distribution of the radial grasp neurons roughly overlapped with that of surface **a**, the ulnar grasp neurons with surface **b**, scratch-up or touch neurons with surface **e** respectively. We postulate that the neurons activated by self-initiated actions emerge as the results of integration of mainly skin and additional deep inputs in the caudal part of the gyrus. Similarly the distribution of reach neurons roughly coincided with that of the proximal arm deep submodality neurons shown in Figure 3.

The self-initiated hand action neurons described above fired only at or after the movement of the arm or digits started, or after the digits contacted objects. The present sample does not include those which might be related to the motor command (Mountcastle et al., 1975), the direct motor control (Fromm and Evarts, 1982), the motor set (Nelson, 1988), or the collorary of the movement (Soso and Fetz, 1980, Fromm, 1983). Rather their activity must be the results of peripheral feedback (Bioulac and Lamarre, 1979; Wannier et al., 1986, 1991; Inase et al., 1989), and thus they may subserve the awareness of manipulative movement or manipulated objects.

7. Neurons selectively activated by the contact of the hand with objects

As described in the previous sections, many neurons in the caudal part of the postcentral gyrus failed to respond to simple skin or joint stimulation. Specific types of stimulation were effective for activating some of these neurons: rubbing of the skin in certain directions (Iwamura et al., 1985a; Costanzo and Gardner, 1980; Warren et al., 1986), static application of an edge in certain orientations, the contact of flat surfaces (Iwamura and Tanaka, 1978a; Iwamura et al., 1985c), and active grasping of objects (Iwamura and Tanaka, 1978a). Koch and Fuster (1989) confirmed the presence of neurons in area 2 and 5 that are concerned with the discrimination of a cube from a ball and described further that there are neurons concerned with short term haptic memory (Fuster, 1994).

Figure 9 shows another example of a neuron concerned with the detection of complex features of objects. It was activated when the monkey's right fingers touched the vertically oriented surface of an acrylic plate. Neuronal response was evoked when the monkey's hand was passively contacted with flat surfaces of a perpendicular plastic plate (0.5 mm thick) to which only the digit skin was contacted (A), or

passive

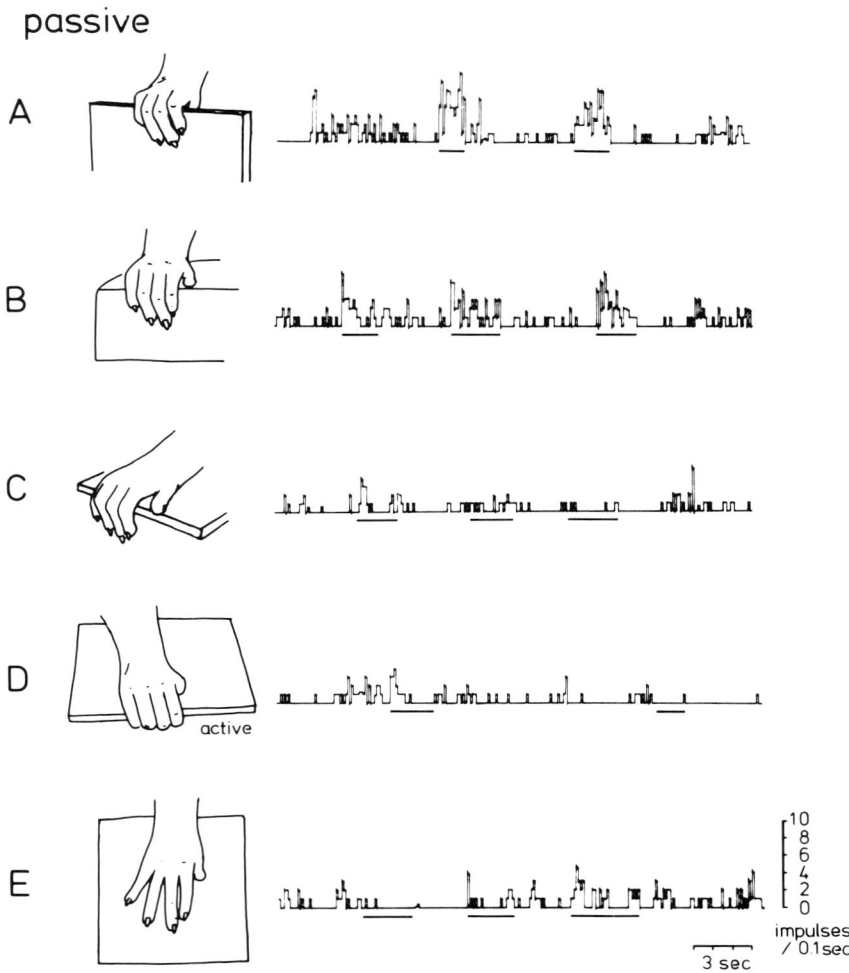

Figure 9. Peristimulus histograms of a neuron recorded in the anterior bank of IPS. Response was evoked when the monkey's hand was passively contacted with flat surfaces. A:: a perpendicular plastic plate (0.5mm thick), only the digit skin was contacted. B: a wooden block, digits were bent,the whole palmar and digit skin was contacted. C: a horizontal plastic plate, only the palmar skin was contacted. D: the monkey grasped the plastic plate. E: a horizontal plastic plate, the whole palmar and digit skin was contacted with fingers stretched. No response to punctate skin stimulus nor to joint manipulation (Iwamura, Tanaka and Hikosaka, unpublished).

with a vertical surface of a wooden block with digits bent, and the whole palmar and digit skin was contacted (B). Contact with a horizontal plastic plate (only the palmar skin was contacted), grasping the plastic plate (D), or contact with a horizontal plastic plate with the whole palmar and digit skin (E) was not effective at all.

8. Neurons with selectivity to material composition

We found neurons which were possibly concerned with material composition, that is, softness or firmness of objects. We applied common objects to the monkey's hand or brought the monkey's hand to the objects (Iwamura et al., 1995). The first example is shown in Figure 10. This neuron was one of 12 neurons found along a perpendicular penetration in area 2. The receptive field of this neuron was on the whole palmar skin when tested with punctate skin stimuli, but the response was weak. In contrast, the passive contact of a bottle brush, a piece of rabbit fur, or the experimenter's beard was effective. This neuron was never activated when the monkey's right hand was touched with the left hand dorsum, knees, or other parts of the monkey's own body. Rubbing with a coarse surface such as sandpaper or carpet material was not effective. Pressing the palm skin gently to smooth and hard surfaces such as an acrylic plate induced a weak inhibitory response. The stronger inhibition was induced when the palmar skin was kneaded by the experimenter's fingers, indicating that activation of subcutaneous or more deeply located receptors may be responsible for the inhibition. Similar results were obtained when the monkey was actively exploring objects. Active grasping of a bottle brush or fur gave brisk responses while grasping of a small piece of hard-baked biscuit brought about an inhibition of the background activity. We found a neuron which showed nearly the opposite selectivity for material, activated by the contact of firm material and inhibited by soft material.

9. Neurons selective to stability or mobility

Figure 11 shows the responses of another pair of neurons. Neuron "a" (upper records), fired most briskly when the monkey's contralateral hand, either the palmar or dorsal side, was brought in contact with any part of the monkey chair including its parts invisible to the monkey.

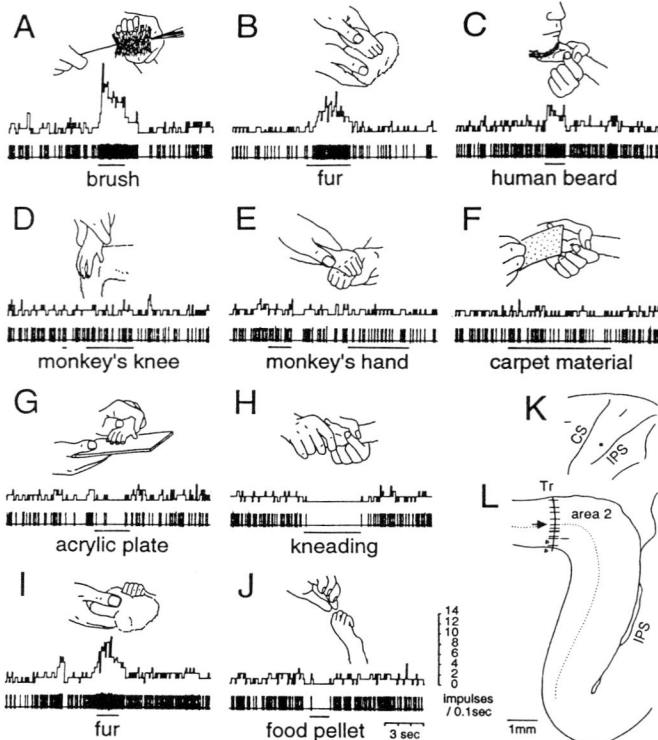

Figure 10. An area 2 neuron selective to soft and shaggy objects. The palmar skin was contacted with various objects. In each of A-J, the upper trace indicates the peristimulus time histogram and the lower trace indicates record of single neuronal activity. A: a bottle brush, B: a piece of rabbit fur, C: the beard of one of the experimenters, D: the monkey's own hair at the knee, E: the dorsal surfaces of the monkey's opposite hand. F: rubbing with a piece of tightly woven coarse carpet material. G: pressing the palm skin with an acrylic plate. H: an inhibitory response evoked by kneading of the palm. I and J: responses evoked when the monkey grasped actively the rabbit fur (I) and a small piece of hard-baked food pellet (J). Bars indicate the period of contact judged from video records. K: the site of electrode penetration on the dorsal surface of the postcentral gyrus. CS: central sulcus, IPS: intraparietal sulcus. L: a sagittal section of the postcentral gyrus indicating the electrode track (Tr) and the recording sites of this (an arrow) and other neurons (short bars). A dotted line is drawn through layer IV. (Adapted from Iwamura et al 1995)

The possibility that this neuron was activated by visual stimuli was thus excluded. It was activated also when the monkey's hand was contacted with another heavy wooden table placed in front of the monkey. The response was weaker when a hard acrylic plate, the same material as the table of the monkey chair, was held by an experimenter (it was thus unstable) and was contacted to the monkey's hand. No response was evoked by contact with the monkey's own belly. The activity of this neuron was inhibited when the monkey actively reached and picked up a slice of an orange from the experimenter's hand. The inhibition lasted as long as the monkey kept the orange piece in the same hand, and returned to the original high level when the monkey rested its empty contralateral hand on the monkey chair again.

The activity of the partner neuron "b" was inhibited by the contact of the hand with the monkey chair or a stabilized table, but the neuron fired most briskly when the monkey reached, grabbed and held food in the hand. Thus we concluded that the neuron "a" was activated by the contact of the hand with stabilized objects while the neuron "b" was activated by the contact of the hand with mobile objects.

10. Neurons activated by the contact of the hand with objects other than the monkey's own body

Neurons described in the above two sections failed to fire by contact of the hand with the monkey's own body. The object to which each neuron was selective varied, but the unresponsiveness to the monkey's own body was common among them. It would require a mechanism to cancel excitatory signals from the hand by those from other body parts which are stimulated at the same time. We found six additional neurons that shared an intriguing common character: unresponsiveness to touching the animals own body. One of these neurons was most peculiar. Active or passive contact of the hand to any part of the monkey chair, the monkey's own body, or food offered by the experimenter did not evoke any response (Figure 12). It evoked brisk and reproducible responses irrespective of their material or shape. When tested out of sight, the contact of the hand with an orange piece activated this neuron. However once the monkey started to grope it (thus the monkey recognized what it was), the neuron became silent.

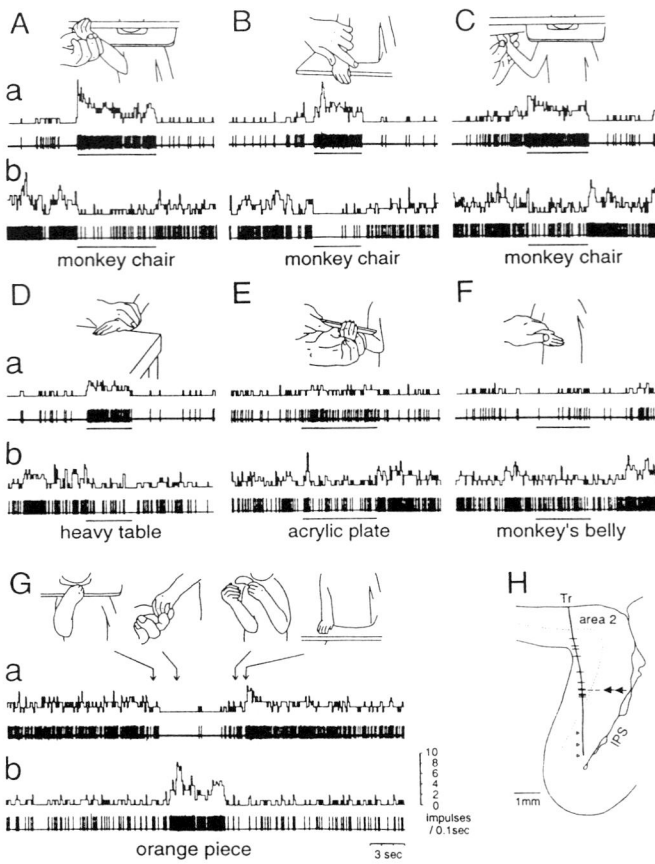

Figure 11. Two neurons "a" and "b", selective to either immobile or mobile objects respectively. They were recorded at the same time in the anterior bank of the intraparietal sulcus, and the responses of them were nearly opposite to each other when the hand was contacted with various objects. A-C: various parts of the monkey chair, D: another heavy table in front of the monkey, E: an acrylic plate held by an experimenter, F: the monkey's own belly, G: an inhibitory or excitatory response evoked by active grabbing of a slice of orange from the human hand. Note the difference in the level of the spontaneous discharges in the pre-reaching period when the monkey rested his right hand on a part of the monkey chair. H: a sagittal section of the postcentral gyrus indicating the electrode track and the recording sites of these and other neurons. Abbreviations and symbols are the same as in Figure 9. (Adapted from Iwamura et al 1995)

We interpreted that this neuron may serve to differentiate tactually unfamiliar objects from familiar ones, such as the monkey's own body, the monkey's primate chair and food.

The unresponsiveness to the own body and the monkey chair was observed also in the polysensory region of the temporal cortex(Mistlin and Perrett 1990). They found that a group of the temporal cortex neurons which responded only to unexpected touch, did not respond to the own body or primate chair as highly familiar objects and thus the contact was expected.

Each of the neurons described above was selective to one feature of an object, the object itself or the objects which constitute the immediate environment around the monkey in the laboratory. What is represented in these neuronal activity may be best explained if I refer to the concept of affordance by Gibson (1979). It is information the environment affords animals, features meaningful for animals.

11. Site for haptic discrimination of objects

All of the neurons described above were found in the caudal part of the postcentral gyrus, near to or in the anterior bank of the intraparietal sulcus. Earlier clinical observations in human and cortical ablation studies in animals have indicated that the tactile discrimination of common objects, shapes, or texture is impaired after lesions in the postcentral gyrus, rather than the parietal association cortices(Corkin et al., 1970; Moffett et al., 1967; Semmes, 1965, Semmes and Turner, 1977). Moffett et al. (1967) pointed out that in tactile discrimination tasks there exists a functional focus along the posterior margin of area 2 and the anterior extremity of the intraparietal sulcus; a small removal from this focal area gives rise to more severe impairment than a removal of comparable size from elsewhere within the posterior parietal region. The sites of recording of the present neurons fell in the same cortical region. This region thus may be critical for discrimination of not only shape and texture but also material of objects by touch and groping.

The strong selectivity seen in the present neurons might be accounted for if sensory inputs of multiple sources converge on them.

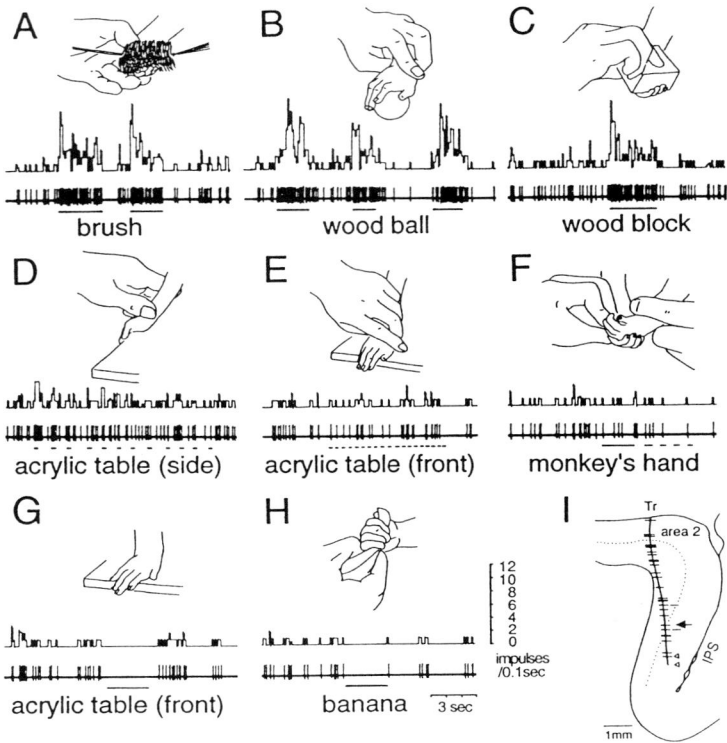

Figure 12. Peristimulus histograms of a neuron in the anterior bank of the IPS. This neuron was activated when the monkey's hand was passively contacted with objects such as a bottle brush (A), wooden ball (B), a wooden rectangular block (C) etc. It was never activated however by the contact of the hand to a part of the primate chair on which the monkey was sitting (D ,E), or to its own hand (F). It was not activated either when the monkey actively felt the table surface (G) or groped a cut piece of an orange (H) (Iwamura ,Tanaka and Hikosaka,unpublished).

Indeed, neurons with a variety of receptive field characteristics, cutaneous, deep, and the proximal arm origins, and those responding to visual stimulation are intermingled in this region of the cortex (Iwamura et al., 1993). The neurons described here might even receive non-somatosensory signals as well to acquire such complex response characteristics.

12. Functional differentiation within the postcentral digit region revealed in deficits in manipulative behaviors after muscimol injection

As shown in the previous sections, the distribution of functional surfaces, that of neurons activated by the self-initiated hand movements and that of neurons selectively activated by the contact of certain objects, all suggest the possibility that functional differentiation exists within the finger region the lateral region for fine digit use and the medial region for coarse and more powerful use of the hand. To prove this we investigated the effects of muscimol injection on manipulative behavior in various sites of the finger region in conscious monkeys. Muscimol is a GABA agonist. GABA has been presumed to be an inhibitory neurotransmitter in the cerebral cortex, and inactivation of local neurons by muscimol may be used as a substitute for localized surgical lesion. Within 24 hours after the injection, disturbances disappear completely, and thus the injection experiment is repeatable (Hikosaka et al., 1985; Iwamura and Tanaka, 1991).

Monkeys were trained to retrieve diced apple pieces from several containers of different shapes, both with and without blindfolding. We used several containers but here I describe only the results obtained using a transparent plastic dish (diameter 8 cm). The monkey adopted the most efficient ways of retrieving an apple piece from the container through practice.

Figure 13 shows sites of recording and muscimol injection in the postcentral gyrus of one hemisphere. Figs. 13B and C show examples of receptive fields of neurons recorded at each of these tracks prior to the injection. Based on the receptive field size we classified these tracks into two groups, anterolateral and posteromedial, as indicated. Among the anterolateral tracks (Figure 13B), receptive fields of neurons tended to be small, single-digit type on the thumb, the second or the third digit, and either skin or deep, while among the posteromedial tracks (Figure 13C), multi-digit type receptive fields were found more frequently. Moreover, there were neurons selective to moving stimuli (tracks 8, 10, 12, P: preferred direction), or sensitive to the presence of an edge (tracks 9, 12) in specific orientations. These results are consistent with our previous findings in mapping studies of the SI cortex in chronic awake monkeys (Iwamura et al., 1983a, b; 1985a, b).

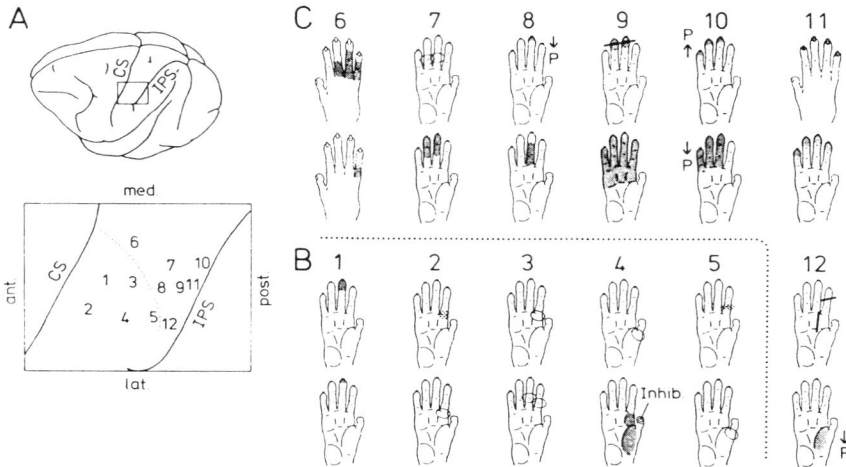

Figure 13. A: Sites of recording and injection (numbered) on the right hemisphere of a monkey. The dotted line shows the border between the anterolateral and posteromedial groups. B and C: Specimens of neuronal receptive fields recorded along tracks (numbered) in the anterolateral (B) and the posteromedial regions (C). Shaded areas: receptive fields for neurons responding to skin (dark) or deep (light) stimulation. Open circles: those for neurons responding to joint manipulation. (Adapted from Iwamura and Tanaka 1991)

Deficits in apple retrieval behavior started immediately after the injection into the anterolateral region of the SI cortex. For example, at track 14, use of digits to retrieve an apple piece from a dish became uncertain even under vision. The affected digit (the second digit) corresponded to the receptive field position of local neurons. It was quickly overcome, but the way of retrieving was changed. These observations were interpreted as indicating that a sensory cue to guide the use of the second digit in the original mode was disturbed by the muscimol injection, and that the animal changed the strategy to use other available sensory cues. Within 30-40 min. after muscimol injection to the anterolateral part, the monkey's hand became much more clumsy, slow, and disorganized. The monkey made mistakes or spent more time in retrieving an apple piece, repeating the same procedures several times or adopting odd strategies to retrieve an apple piece. Thus, the time to retrieve was prolonged up to four times that of the control (0.5-1.5 sec when eyes were open). Nevertheless, the monkey usually managed to retrieve an apple piece after all and the rate of success remained high as long as the eyes were open (94.9 + 4.5%, N=5). When blindfolded the monkey was reluctant to try any of these tests. No signs of paresis nor of weakness in the muscles were observed. Thus, the symptoms observed after anterolateral injection could be attributed to the loss of information essential to guide fine digit movements (Johansson et al., 1987).

Muscimol was injected into the posteromedial part of the digit region where the receptive fields of neurons are of the multi-finger type. Even 30-40 minutes after the injection, it did not disturb very much the retrieval of an apple piece from the dish when performed under visual control. When blindfolded, however, the animal showed several signs of perceptual difficulty: it frequently ignored a piece of apple coming into contact with the fingers, it lost orientation (Figure 14), and it brought an empty hand to the mouth. Thus the animal failed to get a reward in the majority of these trials, the rate of success being below 43.5% (26.5+11.7%, N=7) for posteromedial injection sites when blindfolded, contrasted to the 100% success rate when the eyes were open. Moreover, even when the monkey succeeded to retrieve an apple piece in the blindfolded condition, the time for retrieval was prolonged as the animal spent time in searching and identifying the apple piece. The monkey showed signs of perceptual deficits, difficulty in identifying apple pieces or the container. These deficits were improved temporarily after the test was done under vision, or after the hand of the unaffected side was used. Thus, the deficit was deemed to be in the comprehension of objects

Organization of SI has been described in terms of somatotopic or cytoarchitectonic subdivisions. The present results suggest that the SI cortex functions beyond these subdivisions when the monkey uses the hand to manipulate real objects. The anterolateral part provides information regarding the much finer features of objects which can be

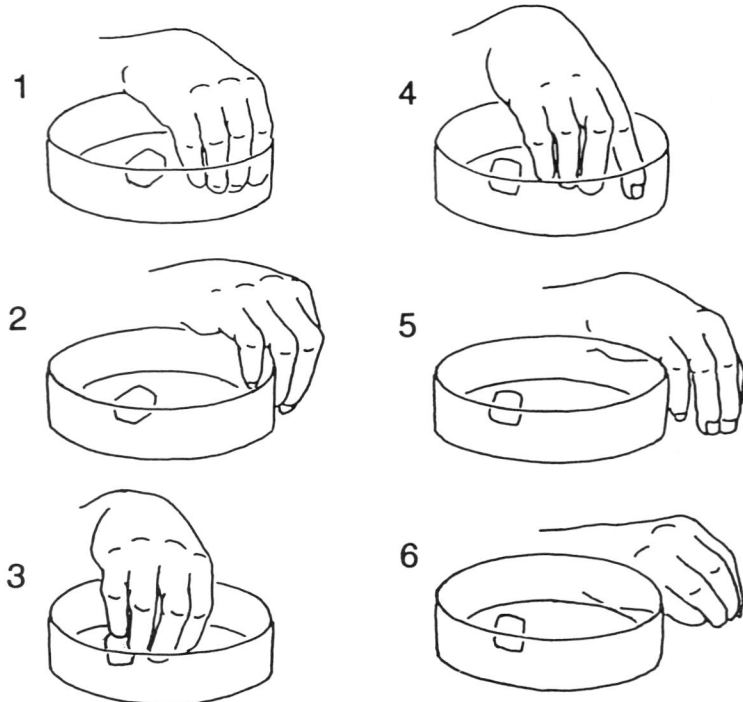

Figure 14. Difficulty in detecting an apple piece in the blindfolded condition, observed at track 10, 138 min after the injection. The monkey's eyes were closed. (Adapted from Iwamura and Tanaka 1991).

perceived at very localized finger tips, and are necessary for the control of fine and skilled finger movements. On the other hand the posteromedial part provides information for comprehension of objects to be manipulated, the global features of objects (size, shape, texture etc.) perceived at relatively large hand or finger skin area. The postcentral region, its caudal part in particular, has strong and direct cortico-cortical connections with the precentral motor regions (Jones, 1986; Zarzecki, 1986). The SI cortex thus provides the motor system with the information necessary for selection of the most adequate strategy to manipulate a particular object.

It is possible to say that the resulting deficits were more severe when muscimol was injected to the anterolateral part. This is consistent with the results of Randolf and Semmes (1974), who showed that an extensive lesion in area 3 disturbed all of the discrimination tasks they used, while lesions in other areas only disturbed some of them. We have proposed a serial and hierarchical model of information processing within the postcentral gyrus. The present finding that injection of muscimol into the anterolateral part of the finger region induced more severe deficits seems to be consistent with such a model.

13. Limb-Kinetic Apraxia

It has been known that damage to the somatosensory pathway causes significant deficits in motor behavior. After dorsal rhizotomy or dorsal column section, animals lose control of distal limb movements (Mott and Sherrington, 1895; Wall, 1970; Taub, 1976; Leonard et al., 1988). Patients with localized damage in the postcentral gyrus lose skill in everyday hand function. This symptom has been known as limb-kinetic apraxia (Liepmann, 1920). These patients have no motor weakness. They have higher order sensory deficits such as elevation of the two point discrimination threshold or astereognosis but no persistent elementary sensory loss (Kawamura et al., 1986). The present observation in monkeys may be compatible with these clinical symptoms.

14. Attention to objects alters responsiveness of SI neurons

Our perceptual system can select particular sets of stimuli while ignoring irrelevant ones, depending on current interest. Detection of a sensory stimulus is facilitated when attention is directed towards the stimulus source. Neurological studies in monkeys have shown enhanced sensitivities or selectivity of neurons in somatosensory cortex to the stimulation to which monkeys were directing attention (Hyvarinen et al., 1980; Poranen and Hyvarinen, 1984; Hsiao et al., 1993). Thus the responsiveness of the postcentral neurons to the peripheral stimulation is under the influence of top down control. This possibility was suggested by our earlier observation that many neurons in the caudal part of the gyrus fired only when the animal moved its hand actively and intentionally (Iwamura et al 1985b).

Iriki et al (1995, 1996) recently found a correlation between neuronal activity and pupil size. The latter was considered as an indicator of the intensity and time course of attention. Pupillometrics has been used as an objective indicator of mental activity. The pupil was measured together with neuronal activity while monkeys were presented with somatosensory stimulation. Monkeys were trained to be passive while an instruction LED was red for 2.5 sec, and green for the next 2.5 sec. During this period the receptive field of the recorded neurons was periodically stimulated with a probe. Reward was delivered simultaneously with receptive field stimulation only when the LED turned green. In this paradigm, red light is a warning signal and green light is a reward signal, and the monkey would become attentive to detect the skin stimulation delivered during the interval between these signals. Either the facilitation or inhibition of ongoing neuronal activity was observed depending on individual neurons during both the warning signal and the reward signal. The onset and extent of these changes correlated well with changes in pupil area. The neurons with the facilitatory or inhibitory effect were found all over the postcentral gyrus. The attention-related effects were seen only in those neurons whose receptive fields were on the glabrous skin. Thus the effect is selective to certain neurons, and not a general elevation of arousal level.

15. Bilateral and ipsilateral representation of fingers in the postcentral gyrus

It has been generally believed that the representation of the body in the postcentral somatosensory cortex is contralateral. At variance with this concept is the presence of neurons with bilateral receptive fields. They have been found in the postcentral regions for representation of face, oral cavity (Schwartz and Fredrickson, 1971; Dreyer et al., 1974; Ogawa et al., 1989) or trunk (Conti et al., 1986; Manzoni et al., 1989). The receptive fields of these neurons were across the midline of the body where two halves of the body meet, and the ipsilateral properties of these neurons were attributed to the activity through the callosal connection.

The presence of bilateral or ipsilateral receptive field neurons in the postcentral hand region had been noticed by occasional and anecdotal observations (Iwamura et al., 1993): in the anterior bank of

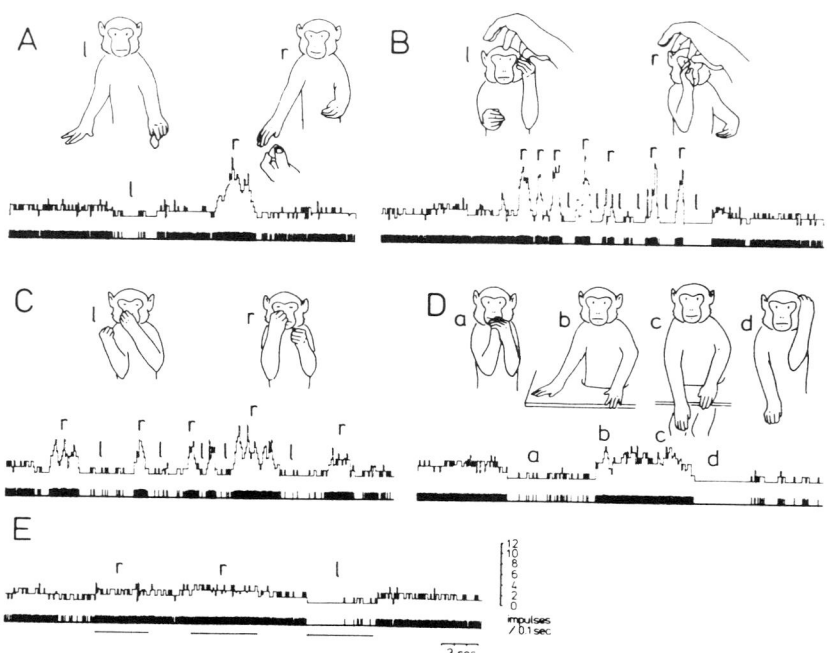

Figure 15. A neuron in the anterior bank of the IPS activated or inhibited by the use of, or passive stimulation of either the right or left hand respectively. A, B: reaching and grasping bait by either hand. C: grooming its own nose by either hand. Soft banana meat was put on the nose. D,a: grooming the nails of the right hand which was held by the left hand, b: feeling the surface of the table or its own knee by the right hand, d: grooming its own head by the left hand. E: either the right (r) or left (l) hand was held by an experimenter (Iwamura, Tanaka and Hikosaka, unpublished).

the IPS, we encountered neurons responding to stimulation of psilateral hands. In one of them, retrieving bait with the contralateral (left) hand was inhibitory to the background activity while it was excitatory with the ipsilateral hand (Figure 15A). The responses were obtained irrespective of the type of hand actions: not only grasping or holding something but also palpating table surface, or grooming various parts of its own body were all effective (Figure 15B, C). Passive manipulation of either hand by an experimenter evoked a response in the same sense, although the response was much weaker (Figure 15D).

Thus it appears that this neuron is signaling which side of the hand is currently in attention or in consciousness.

We recently re-explored systematically the anterior bank of the IPS and found a substantial number of neurons with bilateral or ipsilateral receptive fields clustered in the caudalmost part (areas 2 and 5) of the postcentral finger region (Iwamura et al., 1994). In one hemisphere, the entire postcentral finger region was explored in a total of 50 penetrations and the responses of 863 units were studied in detail. Many neurons along penetrations through the upper bank of the intraparietal sulcus had bilateral receptive fields (Figure 16). In figure 16, nine neurons had bilateral receptive fields which covered either multiple finger tips, radial fingers, the dorsum of the hand or the entire volar surface. The multiple finger tip type constituted the majority in this and many other penetrations (31 out of 72 neurons with bilateral hand receptive fields recorded in this hemisphere) and, in general, the receptive field size and orientation as plotted by stimulation of one hand was symmetrical to that studied through the other hand. In many cases, the spatial configuration of the receptive fields in the two hands was the same, but the intensity of the response was not necessarily the same. We found that among 105 bilateral receptive field neurons in this hemisphere, 61 responded better to the contralateral side (58.1 %), 9 preferred the ipsilateral side (8.6 %), and 35 responded about equally to either side (33.3 %); their response intensity was estimated by the maximal instantaneous firing rate in the peristimulus histogram.

Bilateral receptive fields were found in the upper part of the bank (area 2), but the majority were found in the middle part of the bank, across the border separating area 2 from areas 5 and 7. They were found in both supra- and infragranular layers. Among 105 bilateral receptive fields, 72 had receptive fields restricted to the hand. Bilateral receptive field of neurons in the medial region included, in addition to the multiple finger tip type, those which extended to the forearm from the hand or those which were restricted to the axial trunk on the body midline. Bilateral neurons with receptive field confined within the forearm or upper-arm were rare.

These results were confirmed in an additional three monkeys that served as controls in a study whose purpose was to learn whether

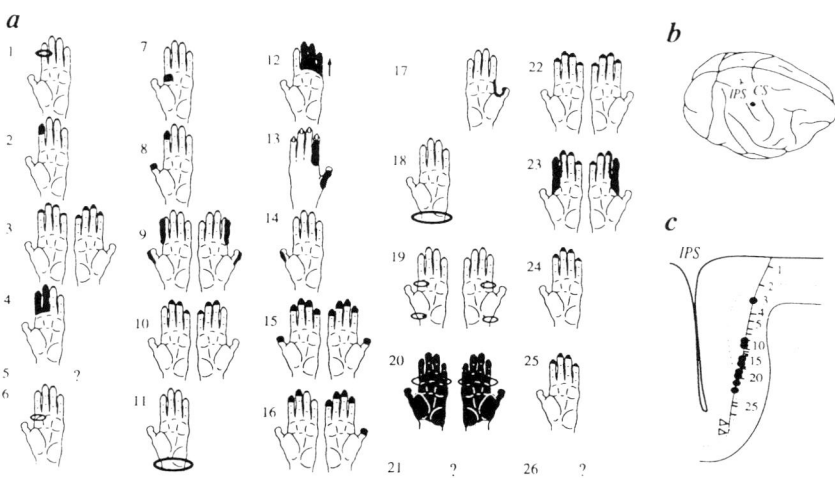

Figure 16. a: Receptive fields (numbered 1-26) of neurons recorded along an electrode penetration in the upper bank of the intraparietal sulcus. The extents of skin receptive fields are shaded dark. Effective positions of joint manipulation neurons are encircled. Neurons having no identifiable receptive fields are indicated by a question mark. The arrow at neuron 12 indicates the preferred direction of moving stimulus across the receptive field. A total of 9 neurons had bilateral receptive fields along this penetration. Their recording sites are indicated by filled circles in c. Note the symmetry in the ipsilateral and contralateral receptive field configuration in most of these pairs. b: A dot on the postcentral gyrus indicates the entry of the electrode penetration in the same animal from which receptive fields were determined in a. CS: central sulcus. IPS: intraparietal sulcus. c: A section perpendicular to CS that includes the electrode track shown in a. Sites of neuronal recordings are indicated by either horizontal bars (for unilateral receptive fields) or filled circles (for bilateral receptive fields). Numbers represent the location of neurons corresponding to receptive fields depicted in a. Open triangles at the end of the electrode track indicate sites of lesioning used as landmarks for later reconstruction of recording sites. A dotted line indicates layer IV. (Adapted from Iwamura et al 1994)

ipsilateral responses are transmitted through the contralateral hemisphere. In one monkey, the finger region of the postcentral gyrus was ablated completely; in the other two, multiple injections of ibotenic acid were made into several sites of bilateral finger representation in the postcentral gyrus, after identifying the sites by multi-unit recording. After either procedure, no bilateral receptive fields were found in the postcentral gyrus of the opposite side. The bilateral receptive fields disappeared after the destruction of the opposite hemisphere indicating their dependence on the callosal connections. Previous anatomical studies indicated that this cortical region is endowed with rich callosal connection. This is contrasted with the practical absence of callosal fibers in the more rostral part of the finger region (Jones and Powell, 1969; Pandya and Vigonolo, 1969; Karol and Pandya,1971; Jones et al., ,1979; Jones and Hendry, 1980; Killackey et al., 1983; Shanks et al., 1985; Caminiti and Sbriccoli, 1985).

16. The bilateral neurons in the postcentral gyrus and hierarchical processing of information in the somatosensory cortices

We reported that the bilateral receptive fields of neurons found in the upper bank of the IPS were far more complex than those of cells found in the more rostral part of the gyrus and thus were considered to be at the higher level of the hierarchical chain (Hyvarinen and Poranen, 1978; Iwamura et al., 1983a,b; 1985a,b; 1993; 1994). We interpret this to mean that the bilaterality conferred by the callosal connections is postponed in the finger region until the very end of the hierarchical processing.

The region of bilateral hand representation in the upper bank of the intraparietal sulcus would be the station for integrating information necessary for the cooperative-operative actions of the two hands, which includes the unitary perception of objects explored by the two hands in unison. It would also be the station for information transfer across the midline for tactile discrimination. Such transfer is lost after callosal sections (Ebner and Myers, 1962; Hunter et al., 1975). It has been thought that SII is the most likely candidate for the function (Manzoni et al., 1989; Berlucchi, 1990; Ridley and Ettlinger, 1978). The present results indicate that the intraparietal zone may be another candidate for that function. The intraparietal zone receives intrinsic cortico-cortical projections from the more rostral region (Seltzer and Pandya, 1986) and projects to SII and its surrounding region (Pandya and Seltzer, 1982). The hierarchical relationship between SI and the second somatosensory area (SII) has been proposed, that is, the latter has been placed at the higher station on the basis of anatomical connections between them (Pons et al., 1992), but this hypothesis has not yet been tested at the single neuronal level.

17. Bilateral receptive field neurons in SII

It had been known that neurons with bilateral receptive fields on hands or arms exist in the second somatosensory cortex (SII) and neighboring areas (Robinson and Burton, 1980a,b), and in the parietal association cortices (Duffy and Burchfiel, 1971; Sakata et al., 1973; Mountcastle et al., 1975; Burbaud et al., 1991). All these cortical regions do have rich callosal connections. Thus it is possible that the caudal postcentral region where we found bilateral hand neurons is a part of these caudal somatosensory association cortices.

To discuss the implication of our findings of bilateral receptive field neurons in the postcentral gyrus in regard to the information processing in the somatosensory cortical system, it seems important to compare the receptive field properties of neurons between two cortical regions. Previous experiments in SII and neighboring areas have been done in anaesthetized conditions (Robinson and Burton, 1980a,b). We are currently exploring the same cortical region within the lateral sulcus including SII in awake monkeys to clarify the difference between the postcentral bilateral region and the SII region (Iwamura et al., 1995, 1996).

18. Conclusions

I have reviewed the results of our studies on the functional organization of the postcentral hand and digit region in awake monkeys. Our study has started to test a hypothesis that the information processing proceeds in a hierarchical manner in this cortex. We found a systematic increase in the receptive field size and complexity of response characteristics toward the caudalmost part of the postcentral gyrus. It is the convergence of information from multiple sources both distal and proximal arm, cutaneous and deep. Thus conditions required for haptic or dynamic touching are provided in this part of the cortex. In fact there emerged neurons activated only by self-paced hand movements. There emerged also a set of feature detection neurons which showed selectivity to specific features of tactile objects that fit in with the concept of "affordance" proposed by Gibson (1966,1979), that is, information the environment affords animals. Seeking feature detection neurons in the anterior bank of the intraparietal sulcus will be promising if the experimental paradigm is selected along this line. Study of effects of muscimol injection into various parts of the digit region revealed functional differentiation within the digit region in regard to the performance of the hand for manipulation of objects. We found also that responsiveness of postcentral neurons is under the control of attentional influences. Elucidation of attentional process will be one of the important subjects in future study. Finally we found bilateral receptive field neurons in the caudalmost part of the digit region. They are considered as the highest in the hierarchical process in this gyrus. How these neurons are related to the use of bilateral hand or to the information transfer from one hand to the other is the subject of future study. We currently are examining how these neurons are different from SII neurons with bilateral receptive field. Clarification of the differences may lead to a better understanding of SII cortex. I emphasize that single neuronal studies in awake animals is one of the strongest methods to gain insight into the central mechanisms of tactile perception.

References

Ageranioti-Belanger,S.A. and Chapman, C.E.(1992) Discharge properties of neurones in the hand area of primary somatosensory cortex in monkeys in relation to the performance of an active tactile discrimination task. II. Area 2 as compared to areas 3b and 1. Experimental Brain Research 91,207-228.

Berlucchi, G. (1990) Commissurotomy studies in animals. in Boler, F. and Grafman, J.(eds) Handbook of Neuropsychology 4, Elsevier, Amsterdam, pp9-47.

Bioulac, B. and Lamarre, Y.(1979) Activity of postcentral cortical neurons of the monkey during movements of lower jaw.Brain Research 172,427-437.

Bonin, G. von and Bailey, P. (1947) The neocortex of Macaca mulatta. University of Illinois Press, Urbana

Brodmann, K. (1909) Vergleichende Lokalisationslehre der Grosshirnrinde in ihren Prinzipien dargestellt auf Grund des Zellenbaues. Barth, Leipzig

Burbaud, P., Doegle, C., Gross, C. and Bioulac, B. (1991) A quantitative study of neuronal discharge in areas 5,2, and 4 of the monkey during fast arm movements. Journal of Neurophysiology 66, 429-443.

Caminiti, R. and Sbriccoli, A. (1985) The callosal system of the superior parietal lobule in the monkey. Journal of Comparative Neurology 237, 85-99.

Campbell, F. (1905) Histological studies on the localization of cerebral function. Cambridge University Press

Conti, F.,Fabri, M. and Manzoni, T. (1986) Bilateral receptive fields and callosal connectivity of the body midline representation in the first somatosensory area of primates. Somatosensory Research 3, 273-289.

Costanzo, R.M. and Gardner, E.P.(1980) A quantitative analysis of responses of direction-sensitive neurons in somatosensory cortex of awake monkeys. Journal of Neurophysiology 43,1319-1341.

Corkin, S., Milner, B. and Rasmussen, T. (1970) Somatosensory thresholds, Archieves of Neurology 23, 41-58.

Darian-Smith ,I., Goodwin, A., Sugitani, M. and Heywood, J. (1984) The tangible features of textured surfaces: their representation in the monkey's somatosensory cortex. In Edelman GM, Gall WE, Cowan W.M. (eds) Dynamic aspects of neocortical function. John Wiley, New York, pp 475-500

Dreyer, D.A., Loe, P.R., Metz, C.B. and Whitsel, B.L. (1975) Representation of head and face in postcentral gyrus of the macaque. Journal of Neurophysiology 38,714-733.

Duffy, F.H. and Burchfiel, J.L. (1971) Somatosensory system: organizational hierarchy from single units in monkey area 5. Science 172, 273-275.

Ebner, F.F. and Myers, R.E. (1962). Corpus callosum and the interhemispheric transmission of tactual learning. Journal of Neurophysiology 25, 380-391.

Economo ,C. von, Koskinas, G. (1925) Die Cytoarchitektonik der Hirnrinde des erwachsenen Menschen. Springer, Berlin

Fromm, C. (1983) Contrasting properties of pyramidal tract neurons located in the precentral or postcentral areas and of corticorubral neurons in the behaving monkey. In (1970) J.E.Desmedt(ed),Motor control mechanisms in health and disease. Raven Press, New York, pp 329-345.

Fromm, C. and Evarts, E.V. (1982) Pyramidal tract neurons in somatosensory cortex: central and peripheral inputs during voluntary movement. Brain Research 238,186-191.

Fuster, J. M. (1995) Memory in the cerebral cortex. MIT Press, Cambridge, Massachusetts, pp 358.

Gardner, E.P. (1988) Somatosensory cortical mechanisms of feature detection in tactile and kinesthetic discrimination. Canadian Journal of Physiology and Pharmacology 66, 439-454.

Gibson, J. J.(1962) Observation on active touch, Psychological Reviews 69,477-491.

Gibson, J. J.(1966) The senses considered as perceptual systems. Houghton Mifflin, Boston, 335p.

Gibson, J. J.(1979) The ecological approach to visual perception. Houghton Mifflin, Boston, 332p.

Hikosaka, O., Tanaka, M., Sakamoto, M. and Iwamura, Y. (1985) Deficits in manipulative behaviors induced by local injection of muscimol in the first somatosensory cortex of the conscious monkey. Brain Research 325, 375-380.

Hikosaka, O. and Wurtz, R.A.H. (1985). Modification of saccade eye movements by GABA-related substances. I. Effect of muscimol and bicuculline in monkey superior colliculus. Journal of Neurophysiology 53, 266-291.

Hsiao, S.S. O'Shaughnessy, D.M. and Johnson, K.O.(1993) Effects of selective attention on spatial form processing in monkey primary and secondary somatosensory cortex. Journal of Neurophysiology 70, 444-447.

Hubel, D.H. and Wiesel, T.N. (1959) Receptive fields of single neurones in the cat's striate cortex. Journal of Physiology (London) 148, 574-591.

Hunter, M., Ettlinger, G. and Maccabe, J.J. (1975) Intermanual transfer in the monkey as a function of amount of callosal sparing. Brain Research 93, 223-240.

Hyvarinen, J. and Poranen, A. (1978) Receptive field integration and submodality convergence in the hand area of the postcentral gyrus of the alert monkey. Journal of Physiology (London) 257, 199-227.

Hyvarinen, J., Poranen, A. and Yokinen, Y. (1980) Influence of attentive behavior on neuronal response to vibration in primary somatosensory cortex of the monkey. Journal of Neurophysiology 43, 870-882.

Inase, M., Mushiake,H., Shima,K., Aya,K. and Tanji,J. (1989) Activity of digital area neurons of the primary somatosensory cortex in relation to sensorially triggered and self-initiated digital movements of monkeys. Neuroscience Research 7,219-234.

Iriki, A., Tanaka, M. and Iwamura, Y. (1995) Attention-induced changes in neuronal activities of the somatosensory cortex. Journal of Physiological Society of Japan 57(Supplement), 175-181.

Iriki, A., Tanaka, M. and Iwamura, Y. (1996) Attention-induced neuronal activities in the monkey somatosensory cortex revealed by pupillometrics. Neuroscience Research, submitted.

Iwamura, Y.(1993) Dynamic and hierarchical processing in the monkey somatosensory cortex, Biomedical Research 14, supplement. 10, 107-111.

Iwamura, Y., Iriki, A. and Tanaka, M. (1994) Bilateral hand representation in the postcentral somatosensory cortex. Nature 369, 554-556.

Iwamura, Y., Iriki, A., Tanaka, M., Taoka, M. and Toda, T. (1995) Bilateral hands are represented in the postcentral somatosensory cortex of the monkey: a new midline fusion theory. Abstracts of 4th IBRO World Congress of Neuroscience.

Iwamura, Y., Iriki, A., Tanaka, M., Taoka, M. and Toda, T. (1996) Bilateral receptive field neurons in the postcentral gyrus: two hands meet at the midline. In Ono, T. et al (eds), Perception, Memory, and Emotion :Frontier in Neuroscience, Oxford University Press.

Iwamura, Y. and Tanaka, M. (1978a) Postcentral neurons in hand region of area 2: their possible role in the form discrimination of tactile objects. Brain Research 150, 662-666.

Iwamura, Y. and Tanaka, M. (1978b) Functional organization of receptive fields in the cat somatosensory cortex. I: Integration within the coronal region. Brain Research 151, 49-60.

Iwamura, Y. and Tanaka, M. (1978c) Functional organization of receptive fields in the cat somatosensory cortex. II. Second representation of the forepaw in the ansate region. Brain Research 151, 61-72.

Iwamura, Y. and Tanaka, M. (1991) Organization of the first somatosensory cortex for manipulation of objects: an analysis ofbehavioral changes induced by muscimol injection into identified cortical loci of awake monkeys. In O. Franzen and J. Westman (eds) Information processing in the somatosensory system. Wenner-Gren International symposium series, 57. Stockton, New York, pp 371-380

Iwamura, Y., Tanaka, M. and Hikosaka, O. (1980) Overlapping representation of fingers in the somatosensory cortex (area 2) of the conscious monkey. Brain Research 197, 516-520.

Iwamura, Y., Tanaka, M., Hikosaka, O. and Sakamoto, M. (1995) Postcentral neurons of alert monkeys activated by the contact of the hand with objects other than the monkey's own body. Neuroscience Letters 186, 127-130.

Iwamura, Y., Tanaka, M., Sakamoto, M. and Hikosaka, O.(1983a) Functional subdivisions representing different finger regions in area 3 of the first somatosensory cortex of the conscious monkey. Experimental Brain Research 51, 315-326.

Iwamura, Y., Tanaka, M., Sakamoto, M. and Hikosaka, O. (1983b) Converging patterns of finger representation and complex response properties of neurons in area 1 of the first somatosensory cortex of the conscious monkey. Experimental Brain Research 51, 327-337.

Iwamura, Y., Tanaka, M., Sakamoto, M. and Hikosaka, O. (1985a) Diversity in receptive field properties of vertical neuronal arrays in the crown of the postcentral gyrus of the conscious monkey. Experimental Brain Research 58, 400-411.

Iwamura, Y., Tanaka, M., Sakamoto, M. and Hikosaka, O. (1985b) Vertical neuronal arrays in the postcentral gyrus signaling active touch: a receptive field study in the conscious monkey. Experimental Brain Research 58, 412-420.

Iwamura, Y., Tanaka, M., Sakamoto, M. and Hikosaka, O. (1985c) Functional surface integration, submodality convergence, and tactile feature detection in area 2 of the monkey somatosensory cortex. In Goodwin, A.W. and Darian-Smith, I. (eds) Hand function and the neocortex, Experimental Brain Research Supplement 10. Springer, Berlin, pp 44-58.

Iwamura, Y., Tanaka, M., Sakamoto ,M. and Hikosaka, O. (1985d) Comparison of the hand and finger representation in areas 3, 1, and 2 of the monkey somatosensory cortex. In: Rowe, M. and Willis, D. (eds) Development, organization, and processing in somatosensory pathways. Alan R. Liss, New York, pp 239-245.

Iwamura, Y., Tanaka, M., Sakamoto, M. and Hikosaka, O. (1993) Rostrocaudal gradients in neuronal receptive field complexity in the finger region of alert monkey's postcentral gyrus. Experimental Brain Research 92,360-368.

Jones, E.G. (1975) Lamination and differential distribution of thalamic afferents within the sensory-motor cortex of the squirrel monkey. Journal of Comparative Neurology 160, 167-204.

Jones, E.G. (1986) Connectivity of the primate sensori-motor cortex. In Jones, E.G. and Peters, A. (eds) Cerebral Cortex, Vol 5, Plenum, New York.

Jones, E.G. and Burton, H. (1976) Areal differences in the laminar distribution of thalamic afferents in cortical fields of insular, parietal and temporal regions of primates. Journal of Comparative Neurology 168, 197-247.

Jones, E.G,, Coulter, J.D., and Hendry, S.H.C. (1978) Intracortical connectivity of architectonic fields in the somatic sensory, motor and parietal cortex of monkeys. Journal of Comparative Neurology 181, 291-348.

Jones, E.G., Coulter, J.D. and Wise ,S.P. (1979) Commissural columns in the sensory-motor cortex of monkeys. Journal of Comparative Neurology 188, 113-136.

Jones, E.G. and Friedman, D.P. (1982) Projection pattern of functional components of thalamic ventrobasal complex on monkey somatosensory cortex. Journal of Neurophysiology 48, 521-44.

Jones E.G. and Hendry S.H.C. (1980) Distribution of callosal fibers around the hand representations in monkey somatic sensory cortex. Neuroscience Letters 19, 167-172.

Jones, E.G. and Porter, R. (1980) What is area 3a? Brain Research 203, 1-43.

Jones, E.G. and Powell, T.P.S.(1969) Connections of the somatic sensory cortex of the rhesus monkey.II. Contralateral cortical connections. Brain 92,717-730.

Jones, E.G. and Powell, T.P.S. (1970) Connections of the somatic sensory cortex of the rhesus monkey. III Thalamic connections. Brain 93, 37-56.

Kaas, J.H., Nelson, R.J., Sur, M., Lin, C.S. and Merzenich, M.M. (1979) Multiple representations of the body within the primary somatosensory cortex of primates. Science 204, 521-523.

Kane, S.A., Mink, J.W. and Thach, W.T. (1988) Fastigial, interposed and dentate nuclei: somatotopic organization and the movements differentially controlled by each. Abstracts of Neuroscience Society 14, part 2, 954.

Karol, E.A. and Pandya, D.N.(1971) The distribution of the corpus callosum in the rhesus monkey. Brain 94, 471-486.

Kawamura, M., Hirayama, K. and Shirota, J. (1986) Limb-kinetic apraxia produced by a localized infarction of the central region (Liepmann). Studies on two cases. Clinical Neurology 26, 20-27.

Koch, K.W. and Fuster, J.M. (1989) Unit activity in monkey parietal cortex related to haptic perception and temporary memory. Experimental Brain Research 76, 292-306.

Killackey, H.P., Gould, H.J.III, Cusick, C.G., Pons, T.P. and Kaas, J.H.(1983) The relation of corpus callosum connections to architectonic fields and body surface maps in sensorimotor cortex of new and old world monkeys. Journal of Comparative Neurology 219, 384-419.

Kunzle, H. (1978) Cortico-cortical efferents of primary motor and somatosensory regions of the cerebral cortex in macaca fascicularis. Neuroscience 3, 25-39.

Leonard, C.M., Vierck, C.J. and Cooper, B.Y. (1988) Field observations of hand movements after dorsal column section. Abstracts of Society of Neuroscience 14, part 2, 954.

Liepmann,H (1920) Apraxie.Ergebnis der gesamte Medizin 1,516-543.

Manzoni, T., Barbaresi, F., Conti, P. and Fabri, M.(1989) The callosal connections of the primary somatosensory cortex and the neural bases of midline fusion. Experimental Brain Research 76, 251-266.

McKenna, T.M., Whitsel, B.L.and Dreyer,D.A. (1982) Anterior parietal cortical topographic organization in macaque monkey: a reevaluation. Journal of Neurophysiology 48,289-317.

Mistlin, A. J. and Perrett, D.I.(1990) Visual and somatosensory processing in the macaque temporal cortex:the role of 'expectation'. Experimental Brain Research 82, 437-450.

Moffett, A., Ettlinger, G., Morton, H. B. and Piercy, M. F.(1967) Tactile discrimination performance in the monkey: the effect of ablation of various subdivisions of posterior parietal cortex. Cortex 3, 59-96.

Mott, F.W. and Sherrington, C.S. (1895) Experiments upon the influence of sensory nerves upon movements and nutrition of the limbs. Proceedings of Royal Society B 57, 481-488.

Mountcastle, V.B. (1984) Central nervous mechanisms in mechanoreceptive sensibility. In Darian-Smith,I.(ed), Handbook of Physiology, Section 1, The nervous system, Vol.III. Sensory processes, Part 2, Bethesda, MD., American Physiological Society, pp.789-878.

Mountcastle, V.B., Lynch,J.C., Georgopoulos, A., Sakata, H. and Acuna, C. (1975) Posterior parietal association cortex of the monkey: command functions for operations within extrapersonal space. Journal of Neurophysiology 38, 871-908.

Nelson, R.J. (1988) Set related and premovement related activity of primary somatosensory cortical neurons depends upon stimulus modality and subsequent movement. Brain Research Bulletin 21,411-424.

Ogawa, H., Ito, S. and Nomura, T. (1989) Oral cavity representation at the frontal operculum of macaque monkeys. Neuroscience Research 6, 283-298.

Pandya, D.N. and Seltzer, B.(1982) Intrinsic connections and architectonics of posterior parietal cortex in the rhesus monkey. Journal of Comparative Neurology 204, 196-210.

Pandya, D.N. and Vignolo, L.A. (1969) Interhemispheric projections of the parietal lobe in the rhesus monkey. Brain Research 15, 49-65.

Pons, T.P., Garraghty, P.E. and Mishkin, M. (1992) Serial and parallel processing of tactual information in somatosensory cortex of rhesus monkeys. Journal of Neurophysiology 68,518-527.

Pons, T.P., Garraghty, P.E., Cusick, C.G. and Kaas, J.H.(1985) The somatotopic organization of area 2 in macaque monkeys. Journal of Comparative Neurology 241, 445-466.

Pons, T.P. and Kaas, J.H. (1985) Connections of area 2 of somatosensory cortex with the anterior pulvinar and subdivisions of the ventroposterior complex in macaque monkeys. Journal of Comparative Neurology 240, 16-36.

Poranen, A. and Hyvarinen, J. (1982) Effects of attention on multiunit responses to vibration in the somatosensory regions of the monkey brain. Electroencephalography and clinical Neurophysiology 53, 525-537.

Powell, T.P.S and Mountcastle, V.B. (1959a) The cytoarchitecture of the postcentral gyrus of the monkey macaca mulatta. Bulletin of Johns Hopkins Hospital 105, 108-131

Powell, T.P.S. and Mountcastle, V.B. (1959b) Some aspects of the functional organization of the postcentral gyrus of the monkey: a correlation of findings obtained in a single unit analysis with cytoarchitecture. Bulletin of Johns Hopkins Hospital 105, 133-162

Randolf, M. and Semmes, J. (1974) Behavioral consequences of selective subtotal ablations in the postcentral gyrus of Macaca mulatta. Brain Research 70, 55-70.

Ridley, R.M. and Ettlinger, G.(1978) Further evidence of impaired tactile learning after removals of the second somatic sensory projection cortex (SII) in the monkey. Experimental Brain Research 31, 475-488.

Robinson, C.J. and Burton, H. (1980a) Somatotopic organization in the second somatosensory area of M.fascicularis. Journal of Comparative Neurology 192, 043-067.

Robinson, C.J. and Burton, H. (1980b) Organization of somatosensory receptive fields in cortical areas 7b, retroinsula, postauditory and granular insula of M.fascicularis. Journal of Comparative Neurology 192, 069-092

Sakata, H. (1971) Somatic sensory responses of neurons in the parietal association area (area 5) of monkeys. In Kornhuber H.H. (ed), The somatosensory system, Georg Thieme, Stuttgart, pp250-261.

Sakata, H., Takaoka, Y., Kawarasaki A. and Shibutani, H. (1973) Somatosensory properties of neurons in the superior parietal cortex(area 5) of the rhesus monkey. Brain Research 64, 85-102.

Schwartz, D.W.F. and Frederickson, J.M. (1971) Tactile direction sensitivity of area 2 oral neurons in the rhesus monkey cortex. Brain Research 27,397-401.

Seltzer, B. and Pandya, D.N. (1986) Posterior parietal projections to intraparietal sulcus of the rhesus monkey. Experimental Brain Research 62, 459-469.

Semmes, J. (1965) A non-tactual factor in astereognosis. Neuropsychologia 3, 295-315.

Semmes, J. and Turner, B. (1977) Effects of cortical lesion on somatosensory tasks, Journal of Investigative Dermatology 69, 181-189.

Shanks, M.F., Pearson, R.C.A. and Powell, T.P.S.(1985) The callosal connections of the primary somatic sensory cortex in the monkey. Brain Research Reviews 9, 43-65.

Shanks, M.F. and Powell, T.P.S. (1981) An electron microscopic study of the termination of thalamocortical fibres in areas 3b, 1 and 2 of the somatic sensory cortex in the monkey. Brain Research 218, 35-47.

Soso, M.J. and Fetz, E.E. (1980) Responses of identified cells in postcentral cortex of awake monkeys during comparable active and passive joint movements. Journal of Neurophysiology 43,1090-1110.

Taub, E. (1976) Movement in nonhuman primates deprived of somatosensory feedback. Exercise and Sport Science Review 4, 335-374.

Vogt, B.A. and Pandya, D.N. (1978) Corticocortical connections of somatic sensory cortex (areas 3,1, and 2) in the rhesus monkey. Journal of Comparative Neurology 177, 179-192.

Vogt, C. and Vogt, O.(1919) Ergebnisse unserer Hirnforschung. Vierte mitteilung: die physiologische bedeutung der architektonischen rindenreizungen. Journal of Psychology and Neuorlogy (Leipzig) 25, 279-461

Wall, P.D. (1970) The sensory and motor role of impulses traveling in the dorsal columns towards cerebral cortex. Brain 93, 505-524.

Wannier, T.M.J., Maier, M.A. and Hepp-Reymond, M. C. (1991) Contrasting properties of monkey somatosensory and motor cortex neurons activated during the control of force in precision grip. Journal of Neurophysiology, 65,572-589.

Warren, S., Heikki, A., Hamalainen, A. and Gardner, E.P. (1986) Objective classification of motion- and direction-sensitive neurons in primary somatosensory cortex of awake monkeys, Journal of Neurophysiology 56, 598-622.

Zarzecki, P. (1986)   Functions of corticocortical neurons of somatosensory, motor and parietal cortex.   In   Jones,E.G. and Peters, A.(eds),Cerebral Cortex, Vol.5, Plenum, New York.

Neural Aspects of Tactile Sensation
J.W. Morley (Editor)
© 1998 Elsevier Science B.V. All rights reserved

Processing of Somesthetic Stimuli in Primate Sensory-motor Cortex

R. Romo, A. Zainos, H. Merchant, A. Hernández and W. García.

In the last fifty years, neurobiologists have revealed the peripheral and brain structures associated with somatosensory perception. This knowledge goes from the description of the mechanoreceptive organs, peripheral fibers, and brain centers which process sensory stimuli in the somatic modality. With this knowledge, sensory physiologists have begun to address the question of how sensory stimuli are represented in the peripheral and central nervous system (CNS) of mammals. It is implicit in this idea, that understanding the representation of an external stimulus in the CNS, can be a powerful tool for revealing perception, memory, learning, and purposive motor behavior (Figure 1). Indeed, it has been demonstrated that both the temporal (Talbot et al., 1968) and spatial (Phillips et al., 1988) properties of the tactile stimuli are encoded in the evoked activity of cutaneous afferents. This observation suggests that the discharges associated with the stimulus are transmitting information to the CNS, and that this could be considered as the primary material for sensation and perception in highly evolved brains. The next level of inquiry on the representation of tactile stimuli is the primary somatic sensory (SI) cortex (Mountcastle et al., 1969). This is due to the fact that the processing of the tactile stimuli in SI cortex, is then widely distributed to other cortical and subcortical structures (Mountcastle, 1978). Therefore, it is very likely that in this wide internal dynamic representation of the stimulus, the neural signal(s) responsible for sensation and perception, might be generated.

In this chapter, we deal with three aspects of the somatosensory perception: (1) the question of how SI cortex represents moving tactile stimuli and, (2) how this representation in SI cortex is relevant for the perception of the stimulus itself. Also, we would like to address the question (3) of how motor areas of the frontal lobe participate in the output of the sensory perceptual process through voluntary motor reactions. Accordingly, SI cortex could be considered as the starting point of the cortical information processing for sensory perception,

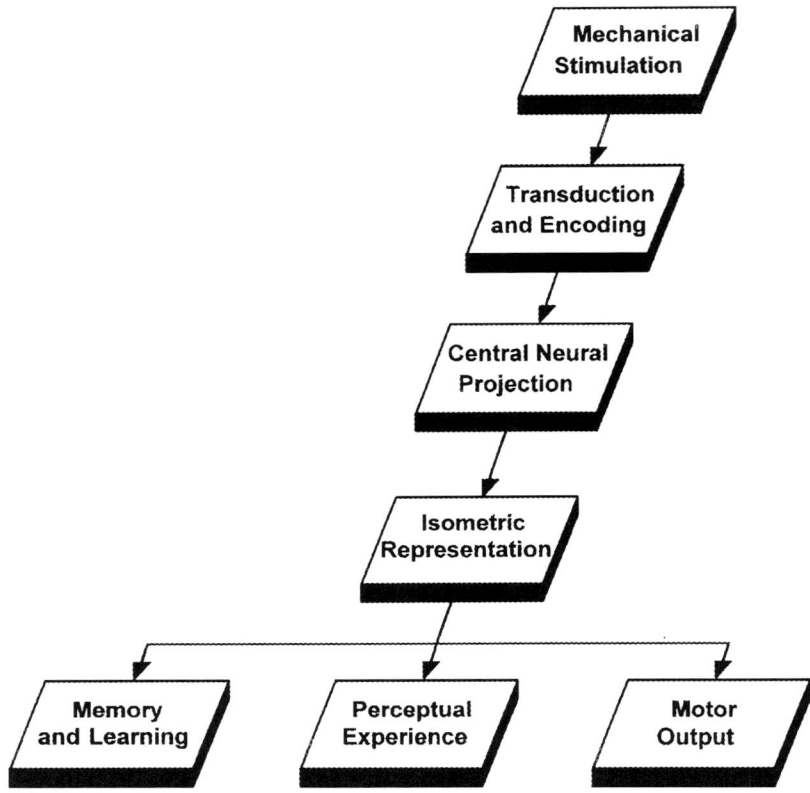

Figure 1. Schematic outline of the sensory processing mechanisms triggered by a somesthetic stimulus (modified from an original diagram of V.B. Mountcastle published in Romo et al., 1993a).

and motor areas of the frontal lobe, as the output cortical node, which in their activity not only encode the parameters of the motor act, but also the indication of the sensory perception of the stimulus as well (Figure 2).

This knowledge comes from experiments in which the neuronal activity of SI cortex (Darian-Smith et al., 1984b; Hsiao et al., 1993; Mountcastle et al., 1990; Sinclair and Burton, 1991) and cortical motor areas are studied as monkeys perform in a sensory somesthetic task (Mountcastle et al., 1992; Romo et al., 1993c). This strategy seems appropriate for investigating how sensory and motor areas of the brain interact to construct the sensori-motor behavior during perception.

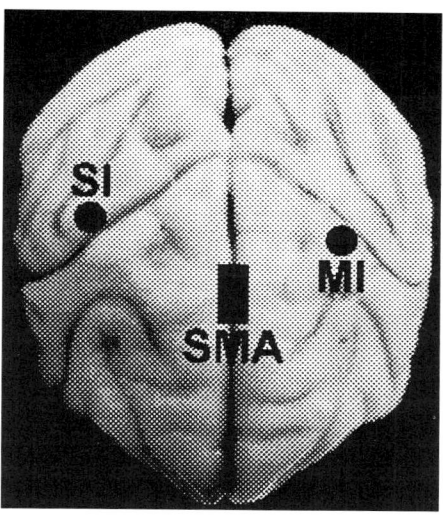

Figure 2. Picture of the monkey brain (Rhesus monkey). SI, primary somatic sensory cortex; MI, primary motor cortex; SMA, supplementary motor area. The evoked neuronal activity of SI cortex contralateral to the stimulated hand was studied in both naïve monkeys and in monkeys trained in a somesthetic task. The neuronal activity of MI and SMA was studied as animals performed a sensory somesthetic task.

Before we address the above questions, we would like to present a brief description of the mechanoreceptors and primary afferents which encode cutaneous information in their activity. We also give a brief description of the connectivity between SI cortex and motor areas of the frontal lobe.

1. Cutaneous afferents, SI cortex, and motor areas of the frontal lobe

Three separate and clearly identified primary afferents innervate three separate skin receptors of the primate hand (Darian-Smith, 1984a). These primary afferents are classified according to their temporal adaptation to a light, steady mechanical stimulus applied to their receptive field and the receptor organ with which they are linked (Talbot et al., 1968). These fiber types include rapidly adapting Pacinian (PC), rapidly adapting Meissner (RA), and slowly adapting Merkel (SA-I) afferents. In addition, in the human hand a fourth type (SA-II) has been described (Johansson and Vallbo, 1979), although it has not been found in the primate hand. These peripheral channels maintain their segregation at the levels of the dorsal column nuclei, ventrobasal complex of the thalamus, and in SI cortex.

SI cortex of primates is divided into four areas, each of which contains a somatotopic representation of the body form (Kass et al., 1979; Iwamura et al., 1980; Nelson et al. 1980). They are areas 3a, 3b, 1 and 2. The areas of interest for the study of tactile representation are 3b, 1 and 2, since neurons here replicate the dynamic properties of cutaneous afferent fibers: RAs, SAs and PCs (Powell and Mountcastle, 1959; Mountcastle et al., 1969). These three subsets of neocortical neurons are also referred to as RAs, SAs and PCs and are arranged in columns (Mountcastle, 1957; Powell and Mountcastle, 1959). It has been observed that in area 3b there is a preponderance of SA columns and in areas 1 and 2, RA columns (Powell and Mountcastle, 1959; Sur et al., 1984). It seems that PC columns are more often encountered in area 2, but they are also found in areas 3b and 1 (Powell and Mountcastle, 1959; Mountcastle et al., 1969). Area 3a is more related to the processing of sensory inputs coming from muscle afferents (Phillips et al., 1971).

It may appear obvious that subjects indicate sensory perception through voluntary motor acts. However, there are few studies which have addressed the roles of motor areas of the frontal lobe in sensory perception (Mountcastle et al., 1992; Romo et al., 1993). Indeed, a large number of anatomical studies have indicated that somatic sensory areas of the anterior and posterior parietal lobe maintain direct connections with motor areas of the frontal lobe (for reviews: Jones, 1984; Kass, 1993; Mountcastle, 1984; Mountcastle et al., 1992). For example, apart from the well known projections from SI cortex to areas 5, 7b and second somatic sensory cortex (Jones and Powell, 1969; Jones et al., 1978; Pearson et al., 1985; Pons and Kass, 1986; Shanks et al., 1985), SI cortex projects directly to primary motor (MI) cortex, premotor (PM) and supplementary motor area (SMA) of the frontal lobe (Darian-Smith et al., 1993; Jones and Powell, 1969; Jones et al., 1978; Luppino et al., 1993; Pandya and Kuypers, 1969; Tokuno and Tanji, 1993; Wiesendanger et al., 1985). Also, posterior parietal areas are directly connected with MI, PM and SMA (Cavada and Goldman-Rakic, 1989; Luppino et al., 1993; Petrides and Pandya, 1984; Tokuno and Tanji, 1993). Both anterior and posterior parietal somatic areas also send projections to subcortical structures (Flaherty and Graybiel, 1991; Galea et al., 1994; Jones et al., 1977; Kemp and Powell, 1970).

These observations have paved the way for the study of the representation of tactile stimuli at the peripheral, cortical, and subcortical structures which receive a direct input from SI cortex and motor areas of the frontal lobe. The question, therefore, is: how are the spatiotemporal properties of the tactile stimuli dynamically represented in the activity of the peripheral fibers and in the neuronal ensembles of CNS? A number of studies have addressed this question and found that the parameters of the stimulus can be represented in the activity of the cutaneous afferents and in SI cortex (Phillips et al. 1988; Talbot et al. 1968; Mountcastle et al. 1969; Ruiz et al. 1995). Also, that the sensation elicited by the tactile stimulus puts in motion one type of primary afferent and a set of cortical neurons of SI cortex which share the same functional properties of the peripheral fibers (Phillips et al. 1988; Mountcastle et al. 1969; Talbot et al. 1968); therefore, at the peripheral level there is a transmitting channel which predominates over the others. This segregation is still maintained in SI cortex. However, the most difficult question to be solved by sensory physiologists is to reveal the neuronal signals in somatic sensory cortices which determine the perception of the somesthetic stimuli, and how these neural signals are transmitted to motor areas of the frontal lobe to indicate perception through a voluntary motor reaction.

2. Coding of moving tactile stimuli in primate SI cortex

A number of studies have investigated the processing of moving tactile stimuli in SI cortex. They have shown that neurons of SI cortex possessing cutaneous receptive fields respond vigorously to the speed and are sensitive to the direction of mechanical stimuli (Constanzo and Gardner, 1980; Essick and Whitsel, 1985; Gardner and Constanzo, 1980; Hyvarinen and Poranen, 1978; Warren et al., 1986; Whitsel et al., 1972). According to these studies, neurons of area 3b are sensitive to motion (Warren et al., 1986), and neurons in area 2, or in the transition between area 1 and 2, are also sensitive to the direction of the moving tactile stimuli (Hyvarinen et al., 1978; Warren et al., 1986). However, these studies have not revealed how the moving tactile stimuli are encoded in the neuronal activity evoked in SI cortex.

We have addressed this question with the use of a vector-averaging model (Georgopoulos et al., 1988, 1993) in a population of neurons of SI cortex (Figure 3) whose cutaneous receptive fields were scanned in different directions, but with a fixed traverse distance

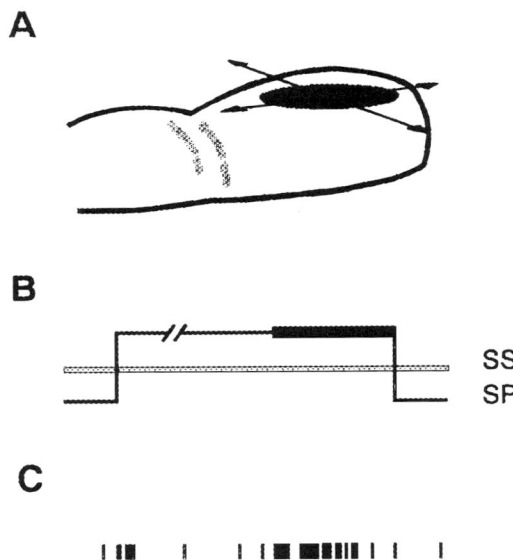

Figure 3. A. Diagram of a cutaneous receptive field of a neuron recorded in the somatic sensory (SI) cortex that was scanned in 4 different directions as indicated by the arrows. B: protocol followed to test the responses of the neuron to the moving tactile stimulus. SS, skin surface; SP, stimulus probe. Stimulation consisted of indentation of the skin and, after a variable delay (broken line) period between 2-4 s, movement of a round metal probe (2 mm) across the receptive field (bold line). The probe was lifted off from the skin at the completion of the traverse distance. C: vertical tics indicate the responses of the neuron to indentation and to the scanning of the receptive field (reproduced from Ruiz et al., 1995).

and force (Figure 4). We assumed that the responses of this neuronal ensemble are weighted and pooled to obtain an accurate estimate of the direction of the tactile stimulus. According to this model, each neuron "votes" for its preferred direction with a weight proportional to its

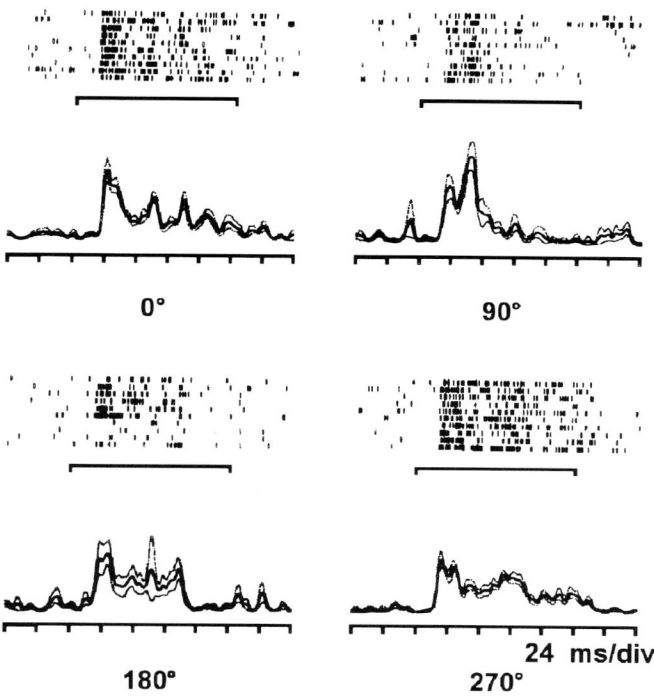

Figure 4. Responses of an SI cortical neuron in area 1 whose cutaneous receptive field was scanned in 4 different directions. The receptive field of this neuron was located in the distal pad of digit 3, and it was classified as rapidly adapting. Horizontal lines represent the duration of the moving stimuli. Neuronal impulses are represented as vertical tics. Each line represents 1 trial; trials of the 4 directions were presented randomly. Below the 10 repetitions are the mean spike-density functions (continuous lines). The dotted lines superimposed about the mean, represent 1 SE of the mean. Stimulus parameters: traverse distance, 6 mm; command static force, 20 g; speed, 50 mm/s (reproduced from Ruiz et al., 1995).

response intensity (Georgopoulos et al., 1988, 1993). To test the vector-averaging model in this population of neurons of SI cortex, we

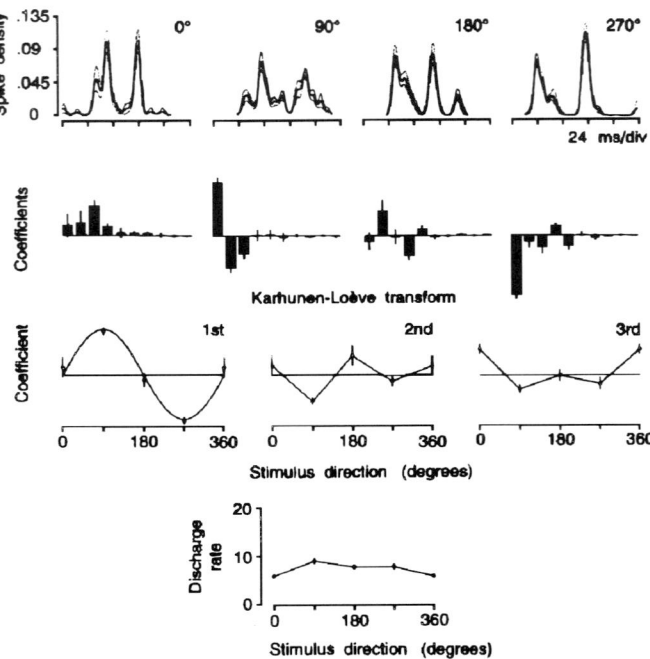

Figure 5. Variation of the activity of an SI cortical neuron as a function of the scanning of the receptive field in 4 different directions. Top: mean spike-density functions (± SE) associated with the 4 different directions of the stimulus. Second row: mean (± SE) values of the 1st 10 coefficients of the Karhunen-Loève (KL) transform (line spectra). Horizontal line indicates a value of 0. Third row: variations of the 1st, 2nd, and 3rd coefficients of the KL transform as a function of the direction of the stimulus. Only the 1st coefficient varied orderly as a function of the stimulus direction. Bottom: mean discharge rates calculated from the 4 directions. The receptive field was scanned at a speed of 50 mm/s; traverse distance, 6 mm; and with static force of 20 g (reproduced from Ruiz et al., 1995).

calculated the mean impulse rate and the coefficients of the Karhunen-Loève (KL) transform (Richmond et al., 1987) during the stimulus-evoked responses (Figure 5). The purpose was to determine whether the mean impulse rate or the temporal covariance (represented by the coefficients of the KL transform) fitted a weighting function model (Ruiz et al., 1995). The results indicate that the first coefficient of the KL transform fitted the proposed model, in one-third of the neurons studied. This was not the case when the mean impulse rates were used

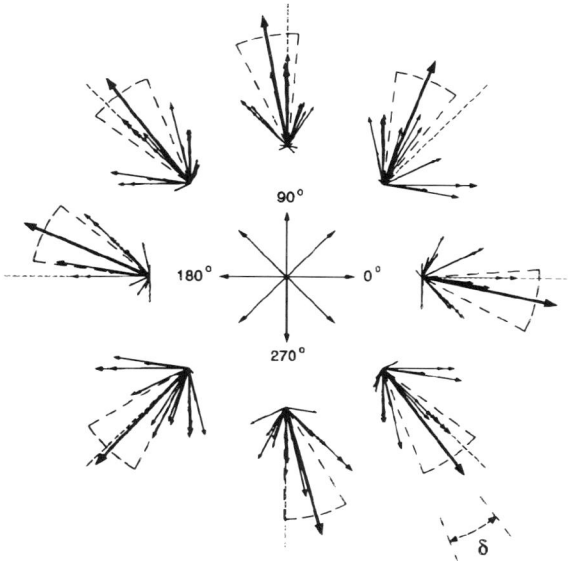

Figure 6. Neuronal population response to the direction of the moving tactile stimulus. Lines with arrows represent the vectorial contributions of each neuron when the direction of the stimuli moved at 23 mm/s across the cutaneous receptive fields. Heavy lines and arrows represent the direction of the neuronal population vector. Broken lines represent the 95% confidence sectors (2δ), calculated for each direction. Stimulus direction generally falls within each confidence sector (reproduced from Ruiz et al., 1995).

as a measure of the neuronal activity associated with the direction of the moving tactile stimuli (Figure 5). The vector-averaging approach showed how an ensemble of neurons of SI cortex represents the direction of stimuli (Figure 6). This representation is in the form of a neuronal population vector, whose magnitude is modulated by the stimulus speed (Figure 7).

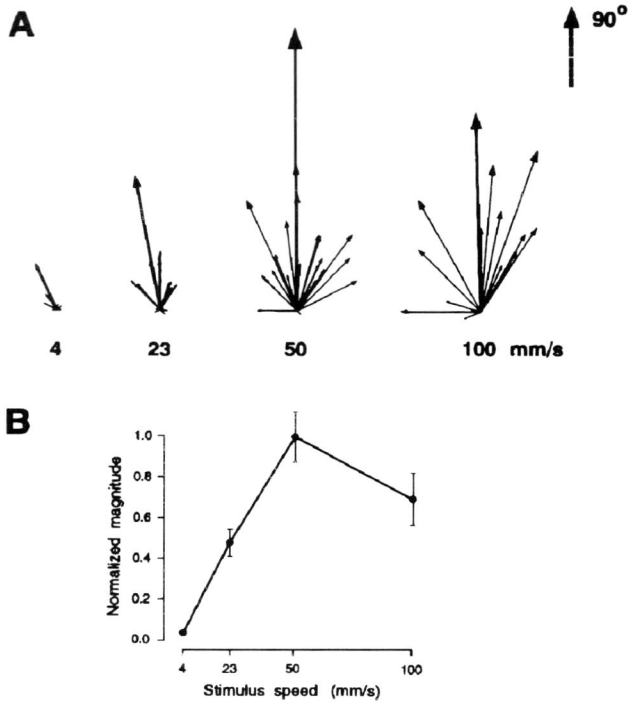

Figure 7. Modulation of the magnitude of the neuronal population vector as a function of the stimulus speed (A and B). Continuous lines and arrows indicate the individual neuronal vector contributions, when the direction of the tactile stimulus moved at 90° and at variable speeds (A). Heavy lines and arrows indicate the magnitudes of the neuronal population vector. B: normalized magnitude of the population vector (reproduced from Ruiz et al., 1995).

The results obtained suggest that the direction of a moving tactile stimulus is represented in the evoked neuronal activity of SI cortex in the form of a neuronal population vector, whose magnitude is modulated by the speed of the stimulus. We have suggested that this dynamic internal representation provides the initial substrate for higher order processing of moving tactile stimuli (Ruiz et al., 1995). Part of this question has been addressed by recording the evoked neuronal activity in SI cortex as monkeys perform in a sensory somesthetic task.

3. Categorical perception of somesthetic stimuli

A salient question raised by the above study is whether that type of representation of the moving tactile stimuli in SI cortex can be correlated directly with the sensory performance of the animal. We addressed this question in a sensory somesthetic categorization task.

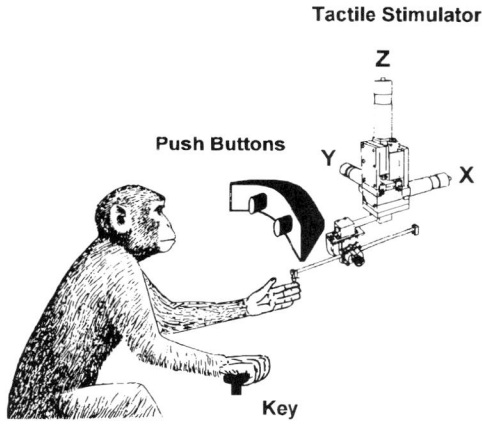

Figure 8. Drawing of a monkey working in the tactile categorization task. Descriptions of the task sequences, tactile stimulator, stimulus set, and sensory-motor performance are given in the text.

We trained monkeys in a somesthetic task in which they were required to categorize the speed of a probe (2-mm round tip) moving across the glabrous skin of one of the fingers of the left, restrained hand, and indicate the speed by interrupting with the free hand one of two target switches (Figure 8). The left arm of the animal was secured in a half cast and maintained in a palm up position (Ruiz et al., 1995). The free hand operated an immovable key (elbow joint at about 90°) and two target switches (the centers located at 70 and 90 mm to the right midsagittal plane) placed at reaching distance (250 mm from the animal's shoulder and eye level). The stimuli consisted of a set of 10 speeds from 12 to 30 mm/s, in a fixed traverse distance (6, 8 or 10 mm), direction and force (20 g) in which half of them were considered as low (12, 14, 16, 18 and 20 mm/s) and the rest as high (22, 24, 26, 28 and 30 mm/s). Stimuli were presented by a tactile stimulator built in our laboratory for studying motion processing in the somatosensory system of primates (Romo et al., 1993b).

The trained monkey began a trial when he detected a step indentation of the skin by placing his free hand into an immovable key in a period which did not exceed 1 s. He maintained this position through a variable delay period (1.5-4.5 s, beginning with detection of the indentation of the skin) until the probe moved at any of the 10 speeds. The subject indicated the detection of the end of the scanning by removing his hand from the key within 600 ms, and indicated whether the speed was low or high by projecting his free hand to one of the two target switches within 1 s (medial switch was used to indicate low speeds and lateral one for high speeds). The animal was rewarded for correct categorization of the speeds by a drop of water. The tactile stimuli were neither visible nor audible in any part of the task. The number of correct and wrong categorizations in a run [which consisted of 10 trials per class (speeds) presented randomly] were used to construct psychometric functions. These psychometric functions were plotted as the percentage of judgments of the speeds as > 20 or < 22 mm/s. Motor performace was evaluated by measuring the reaction (RT) and movement (MT) times in single and averaged trials.

In the top of Figure 9 (A-D) are the psychometric curves of four monkeys, represented in the form of logistic functions fitted to the data points (not shown). They are plotted as a percentage of judgments of the speeds as > 20 mm/s. The data were obtained from one run performed by each animal (10 trials per class). In the middle of Figure 9 (A-D) are the mean values of the RTs, and in the bottom, are the MTs during the categorization of the differents stimulus speeds. It can be appreciated that no significant differences between the mean (± SE) values of the RTs and MTs for low (352.5 ± 2.0 ms and 175.1 ± 2.4 ms, respectively) and high (337.9 ± 2.1 ms and 194.9 ± 3.6 ms, respectively) speeds were detected for each animal and between the different animals as well. Figure 9B shows that the performance of the categorization task, RTs and MTs were not affected when the stimuli were delivered for the first time to digit 2 or 4, compared with digit 3 in Figure 9A. It is also remarkable that the categorization of the stimulus speeds is not affected if the traverse distance is modified. Figure 9C shows the logistic curves, the RTs and MTs when the same set of the

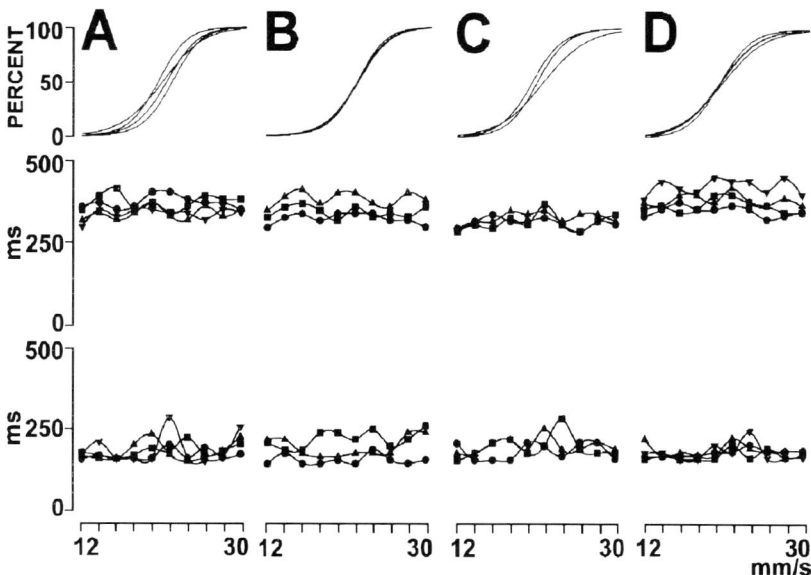

Figure 9. Logistic functions (top), mean values of the RTs (middle) and MTs (bottom), calculated during the categorization of the stimulus speeds. A: performance of 4 monkeys when the 3rd finger of the left, restrained hand, was stimulated from distal to proximal and with constant traverse distance of 6 mm and force (20g). B: performance of 1 monkey that categorized the stimulus speeds, but when the stimuli were delivered in digits 2, 3 or 4 (in separate runs). C: performance of 1 monkey when the traverse distance varied from 6, 8 or 10 mm. D: performance of 1 monkey when the direction of the stimuli was from distal to proximal and opposite, medial to lateral and opposite (unpublished results from Romo et al.)

stimuli was delivered to digit 3, but with traverse distances of 6, 8, or 10 mm/s. Finally, the direction of the stimulus speed did not affect the categorization task. Figure 9D shows the logistic curves when the digit 3 was scanned in four different directions (distal to proximal and opposite, medial to lateral and opposite). Mean values of the RTs and MTs were not affected by the directions of the scanning and remained similar to situations A-C of Figure 9.

According to these results, animals indeed categorized the stimulus speeds. This is supported by the fact that the sensory and motor performance was invariant through the delivery of the stimuli in different fingers, traverse distances and directions. Also, animals had a clear tendency to generalize the categorization, since performance was similar when the stimuli were presented for the first time in digit 2 or in digit 4, or when the speeds were delivered in new traverse distances, or directions. It is interesting that in this task, animals categorized the stimulus speeds on the basis of one single stimulus. To perform this task, it is very likely that animals had to conform a "mnemonic template" of the edges of the stimulus set during the training period (the lowest and the highest speed). This "mnemonic template" must read and classify the evoked neuronal activity elicited by the stimulus to create a decision process. By contrast, in a sensory discrimination task, animals use two stimuli, separated by a fixed interval of time, in which the second stimulus is compared against the first one to create a decision process during sensory discrimination (Mountcastle et al., 1990). Therefore, this task is neither a simple sensory detection, nor a discrimination task. Instead, we propose this is a sensory categorization task. In fact, psychologists have defined categorical perception as the discrete perceptual responses derived from a range of continuum stimuli (Harnard, 1989). In this somesthetic categorization task, the range of the speeds is the continuum stimuli and the arm movements (binary), the discrete responses.

4. SI cortex in the categorical perception of somesthetic stimuli

During the performance of the categorization task, we studied the evoked discharges of single neurons in SI cortex. All these neurons possessed cutaneous receptive fields confined to one single digit (distal segment of digit 2, 3 or 4 of the left, restrained hand). These neurons were recorded between the cortical surface and 2000 microns below; therefore, they were located in area 1 of SI cortex. Also, these neurons were classified according to their adaptation to a light, sustained indentation of the skin in their receptive fields (RAs and SAs). Therefore, the cutaneous receptive fields of these neurons were appropriate for studying the postcentral neural events associated with the categorization of the stimulus speeds.

Figure 10A shows the responses of a RA neuron of SI cortex during the categorization of the stimulus speeds. This neuron responded with a train of impulses to the contact of the stimulus probe with the skin in the distal segment of digit 3, where the receptive field

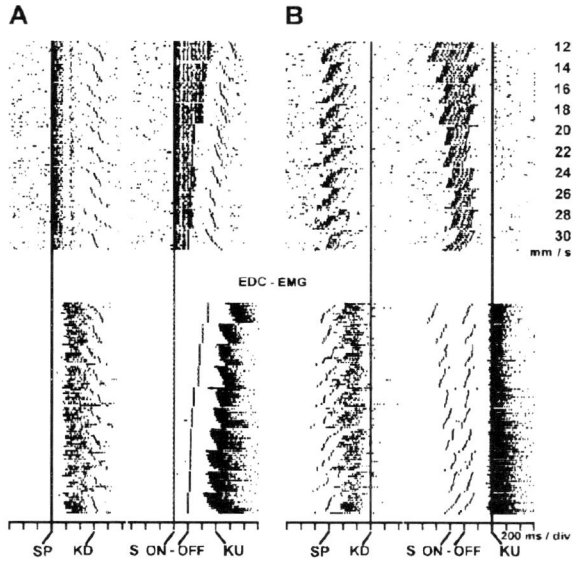

Figure 10. Responses of a neuron of SI cortex (area 1) whose cutaneous receptive field was scanned with the set of stimulus speeds that the animal categorized. A: vertical lines indicate beginning of indentation by the stimulus probe (SP), and beginning of the scanning (S-ON). Vertical lines after the beginning of the stimuli indicate the end of the scanning (OFF). Small verticals lines indicate detection of the stimulus probe (KD), and detection of the end of the stimulus (KU). In the top of the figure are the neuronal impulses and in the bottom the EMG of the extensor digitorum communis (EDC) of the responding arm. B: the same neuronal and EMG activity, but now aligned with respect to the KD and KU. Stimulus parameters: traverse distance, 6 mm; direction, distal to proximal; constant force, 20 g; speeds, 12-30 mm/s (unpublished results from Romo et al.).

is located, and responded again during the scanning. The muscle activity of the extensor digitorum communis (EDC) of the responding arm was recorded simultaneously. It is shown in Figure 10B that the neuronal response is not associated with the muscle activity, indicating that this evoked neural activity depends entirely on the tactile stimulus.

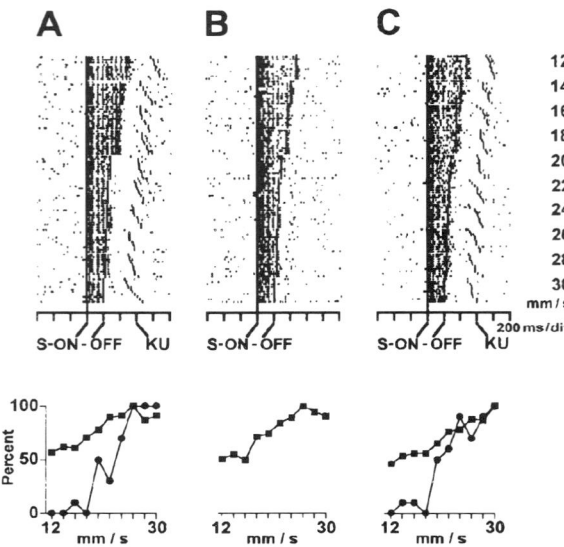

Figure 11. Responses of a SI cortical neuron during the tactile categorization task (A and C) and when the same stimuli were delivered passively (B), in the non working situation. Below the rasters are shown the relations between the percentage of trials in which the animal judged that the stimulus speed was high (filled circles in A and C) and the mean frequency rates (± SE, filled squares) as a function of the stimulus speeds. B: discharge rates during the passive delivery of the stimuli. The receptive field of this neuron was located in the distal segment of digit 3, and it was classified as rapidly adapting. Same stimulus parameters as in Figure 10 (unpublished results from Romo et al.).

A striking property of neurons of SI cortex (area 1) is that all of them, which were tested in the categorization task, responded similarly when the same set of stimuli was delivered passively (Figure 11B). Also, similar neuronal responses were observed when the animal made correct or incorrect categorizations. Finally, the response latencies relative to the beginning of the moving stimuli was determined in 45 neurons of SI cortex. These latencies ranged from 18 to 38 ms (25.8 ± 0.6 ms) and did not vary as a function of the stimulus speeds during the categorization task or in the passive mode (24.6 ± 4.0 ms). These findings suggest that neurons of SI cortex increase their impulse rates as a function of the stimulus speeds during the categorization task. However, these neuronal responses occurred also when the same set of stimuli was delivered passively, in the non working condition. The response latencies of neurons of SI cortex relative to the beginning of the moving stimuli were similar between the different classes during the categorization and non categorization task. This raises the question that the neuronal signals associated with the perception of the somesthetic stimuli may be occurring in more central areas anatomically linked to SI cortex. The somatic areas of the posterior parietal lobe are obvious candidates for investigating the neural signals responsibles for the perception of the moving tactile stimuli. In preliminary experiments, we have investigated this possibility. We have recorded neurons in areas 2 and 5 which are linked to area 1. However, we have not seen any neural signal which may determine the categorization of the moving tactile stimuli. It is very likely that other somatic areas of the parietal lobe (area 7b or SII), which receive direct projections from SI, are associated with the higher order processing of the moving tactile stimuli and that in their activity they may be coding the categorization processing.

5. SMA in the categorical perception of somesthetic stimuli

The SMA appears as an interesting cortical structure for associating parietal somatic areas with MI cortex during the categorization task. Indeed, anatomical studies have demonstrated direct connections between parietal somatic areas and the SMA (Jones, 1984; Mountcastle, 1984; Mountcastle et al., 1992; Luppino et al., 1993). The output of the SMA is directed to MI cortex and to the spinal cord (Dum αvδ Strick, 1991; He et al., 1995; Hummelsheim et al., 1986; Macpherson et al., 1982). In addition, the SMA maintains an important one way connection with the brain stem and the neostriatum (McGuire et al., 1991; Wiesendanger et al., 1985). Given this prominent connection with motor brain centers, neurophysiologists have looked to the role of SMA in motor behavior. However, it might be possible that the SMA is engaged also in some aspects of the sensory perceptual process (Romo et al., 1993c). We have investigated this possibility by recording single neurons in the

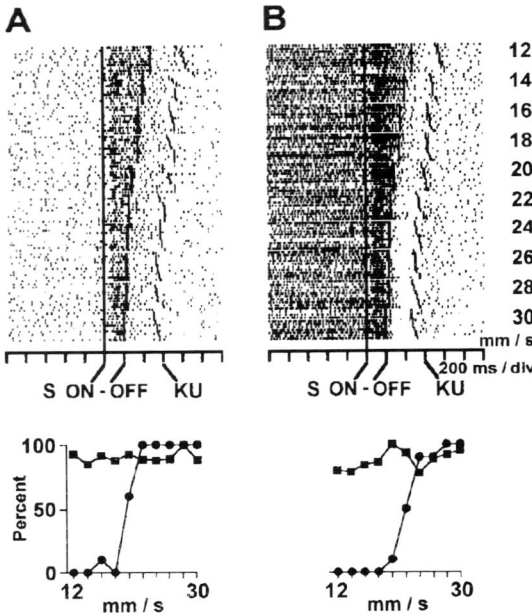

Figure 12. Sensory responses of two SMA neurons during the tactile categorization task. Neuron in B was preceded by preparatory activity. In the bottom of A and B are the relations between sensory performance and the neuronal responses during the stimulus periods. Filled circles are the mean of the percentage of stimulus speeds judged high, and filled squares the percentage of maximum neuronal response (unpublished results from Romo et al.).

SMA contralateral and ipsilateral to the stimulated hand during the execution of the categorization task

We studied a large number of SMA neurons (1836) during the categorization task in four trained monkeys. About 15% of the studied neurons in the SMA contralateral and ipsilateral to the stimulated hand showed responses confined exclusively to the stimulus period of all speeds during the categorization task (Figure 12A). We classified these discharges as "sensory responses". Many of these sensory responses were preceded by preparatory activity (Figure 12B). The response latency for the sensory neurons of the SMA contralateral (123.4 ± 4.9 ms) and ipsilateral (123.5 ± 3.4 ms) to the stimulated hand were almost identical. The response latencies and the magnitudes of the discharge rates did not vary as a function of the stimulus speeds. However, most of these sensory responses of the SMA neurons occurred during the categorization task, since they disappeared when the same set of stimuli was delivered passively. Indeed, we explored carefully the possible existence of cutaneous receptive fields for these SMA neurons which had sensory responses during the categorization task, but none of them showed a clear receptive field during the manual stimulation of the skin of the left, restrained hand.

These findings suggest that a population of neurons of the SMA contralateral and ipsilateral to the stimulated hand is linked to the somesthetic stimulus during the categorization task. These responses occur exclusively when the subjects categorize the tactile stimulus speeds. However, they code neither the stimulus speed nor the categorization process. In view of their response latency, this neuronal population responds after SI cortex. If we assume that the processing of the moving stimulus begins in SI cortex, this signal must be distributed from parietal somatic areas to the SMA.

Almost a third of the responsive neurons of the SMA contralateral and ipsilateral to the stimulated hand discharged during the moving tactile stimuli but continued discharging until the end of the arm movement (Figure 13A). We classified these discharges as "sensory-motor responses". Some of these sensory-motor responses were preceded by preparatory activity (Figure 13B). The response latency of sensory-motor neurons of the SMA contralateral (152.4 ± 2.9 ms) and ipsilateral (145.7 ± 3.5 ms) to the stimulated hand were similar and occurred 20 ms after the sensory responses. The response latency and the magnitude of the discharge rates did not vary as a function of the stimulus speeds. These sensory-motor responses occurred exclusively during the categorization task, since they did not discharge when the same set of stimuli was delivered passively. As for the sensory responses, they did not possess a clear cutaneous or deep receptive field during passive exploration in the stimulated hand or in the responding arm.

These findings suggest that sensory-motor neurons of the SMA may be well suited for translating a sensory event into a behavioral motor reaction during the categorization task. This is supported by the

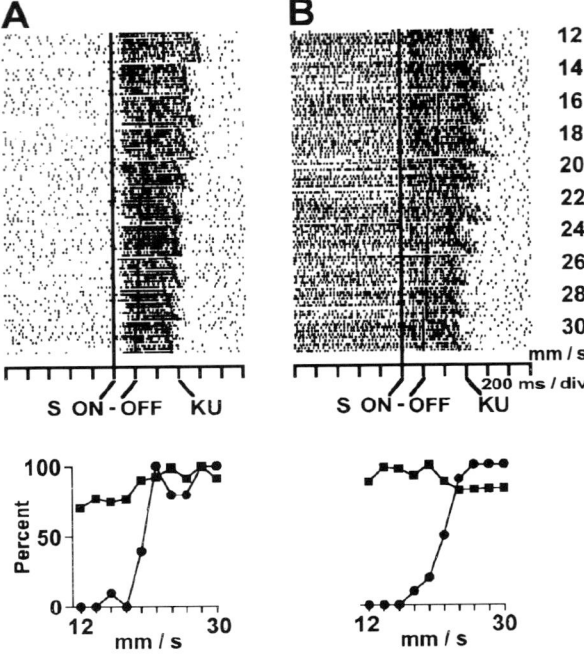

Figure 13. Sensory-motor responses of two SMA neurons during the tactile categorization task. These neurons responded during the stimulus periods and ended their discharges at the end of the reaction times. Neuron in B was preceded by preparatory activity. Below the rasters are the relations between the sensory performance and the neuronal responses during the stimulus + reaction time periods. Filled circles are the mean of the percentage of stimulus speeds judged high, and filled squares mean the percentage of maximum neuronal response (unpublished results from Zainos et al.).

fact that sensory-motor responses occurred after the sensory discharges. Therefore, these two neuronal populations (the sensory and sensory-motor) of the SMA may interact to construct a sensory-motor continuum during the categorization task. Indeed, as for the sensory responses, they occurred exclusively during the categorization task. However, sensory-motor neurons of the SMA code neither the stimulus parameters nor the categorization process.

A third class of responsive neurons of the SMA contralateral and ipsilateral to the stimulated hand discharged differentially during the categorization task (30%). We called these differential discharges "categorical responses". These categorical responses occurred mainly during the stimulus period (sensory; see Figure 14), during the stimulus-RT periods (sensory-motor; see Figure 15), and during the RT period (motor; see Figure 16). These categorical responses predicted in their activity whether the stimulus speed is low or high. A number of the categorical neurons were tested when the same set of stimuli was delivered passively. None of them responded in this condition.

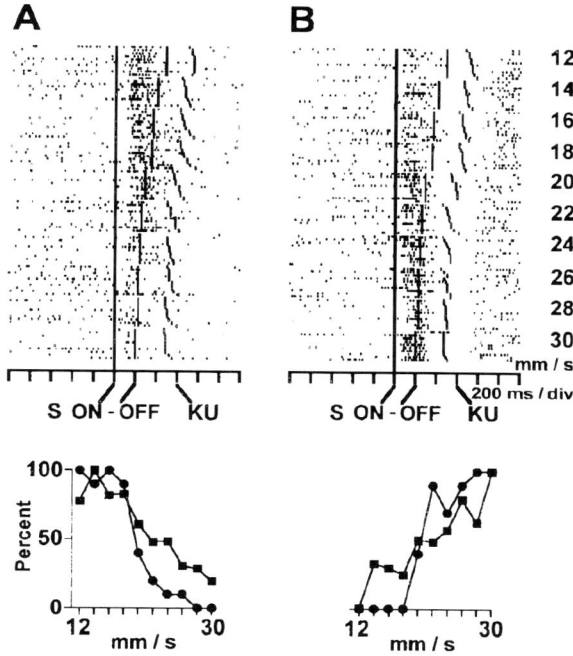

Figure 14. Differential responses of two SMA neurons during the tactile categorization task. These two neurons responded selectively for low (A) or high (B) stimulus speeds. Below the rasters are the relations between the sensory performance and the neuronal responses during the stimulus periods. Filled circles are the mean of the percentage of stimulus speeds judged low (A) or high (B), and filled squares means percentage of maximum neuronal response (unpublished results from Romo et al.).

The response latency of sensory categorical neurons of the SMA contralateral (176.9 ± 10 ms) and ipsilateral (151.0 ± 9 ms) to the stimulated hand discharged before the sensory-motor categorical neurons (196.4 ± 6.8 ms and 198.0 ± 11.2 ms forSMA contralateral and ipsilateral to the stimuli, respectively). However, these two populations of categorical neurons responded after the sensory and sensory-motor neurons described above.

These differential responses, however, may be reflecting in their activity the intention to press, or the direction of the arm

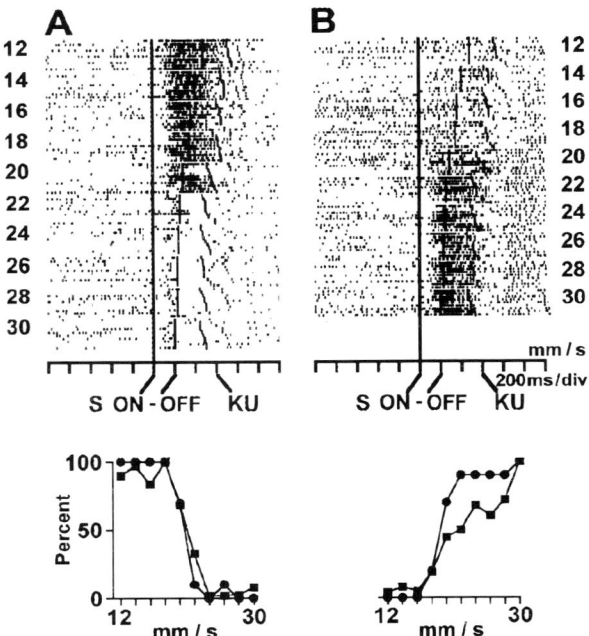

Figure 15. Differential responses of two SMA neurons during the tactile categorization task. These two neurons responded selectively during the stimulus + reaction time period for low (A) or high (B) speeds. Below the rasters are the relations between the sensory performance and the neuronal responses during the stimulus + reaction time periods. Filled circles indicate the percentage of stimulus speeds judged low (A) or high (B), and filled squares indicate the percentage of maximum neuronal response (unpublished results from Romo et al.).

movement to one of the two target switches. To rule out this possibility, a number of the categorical neurons were tested in a light instruction task. In this situation, each trial began as in the somesthetic task, but one of the two target switches was illuminated beginning with the skin indentation, continued during the delay period and turned off when the probe was lifted off from the skin (stimulus-triggers). This condition instructed the animal which target interrupt switch was required to be pressed for reward. Only 15% of the categorical neurons tested in the

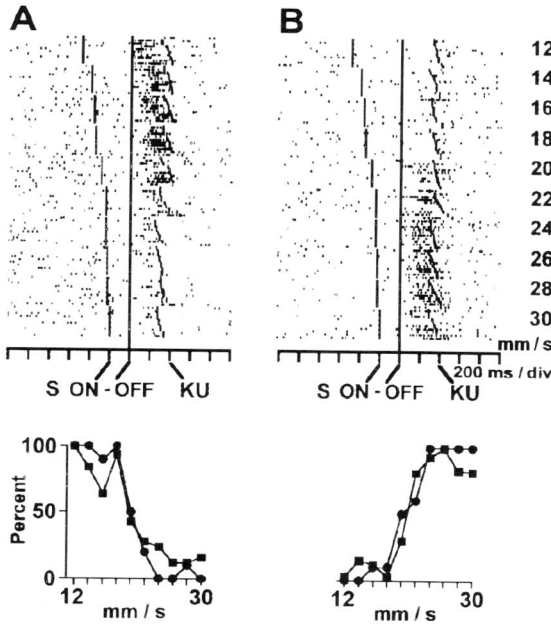

Figure 16. Differential responses of two SMA neurons during the tactile categorization task. These two neurons responded selectively during the reaction time period for low (A) or high (B) speeds. Below the rasters are the relations between the sensory performance and the neuronal responses. Filled circles indicate the percentage of stimulus speeds judged as low (A) or high (B), and filled squares indicate the percentage of maximum neuronal response (unpublished results from Romo et al.).

light instruction task maintained their differential responses. This suggests that, indeed, most of the differential responses are coding the categorization of the moving tactile stimulus.

These results suggest that there are neurons in the SMA contralateral and ipsilateral to the stimulated hand that reflect in their activity the sensory decision process during the execution of the categorization task. These categorical responses are entirely dependent on the categorization process, since they did not discharge when the same set of stimuli was delivered passively. Also, most of these categorical neurons are not associated with the intention to press, or with the trajectory of the arm movements to the target switches. The fact that categorical responses occurred during the stimulus, stimulus-RT, and during the arm movement, suggests that they may participate in the construction of the sensory decision process. Therefore, it appears that there two neuronal continuums in the SMA: one for the sensory decision process and, a second, for translating the sensory stimulus into a voluntary motor act in this learned somesthetic task.

6. Error signals in the SI cortex and SMA during the categorization task

We determined the responses of neurons of SI cortex and SMA during error trials. The results indicate that there are no differences between the neuronal discharges of SI cortex and SMA (in the populations of neurons with sensory and sensory-motor responses) during correct and incorrect categorizations of the stimulus speeds. However, this was not the case in the categorical neurons of the SMA. Therefore, the analysis of the pattern of discharges of categorical neurons of the SMA during the error trials gives additional information for interpreting these responses. The first interpretation could be in terms of the direction of the trajectory of the arm movements towards the target switches and, the second, in terms of the categorization process itself. According to the first strategy, we found four classes of neural signals in the categorical neurons of the SMA (Figure 17).

The first type of discharge during the error trials could be interpreted as if these neurons were coding the intention, or the direction of the arm movement to one target switch (medial target

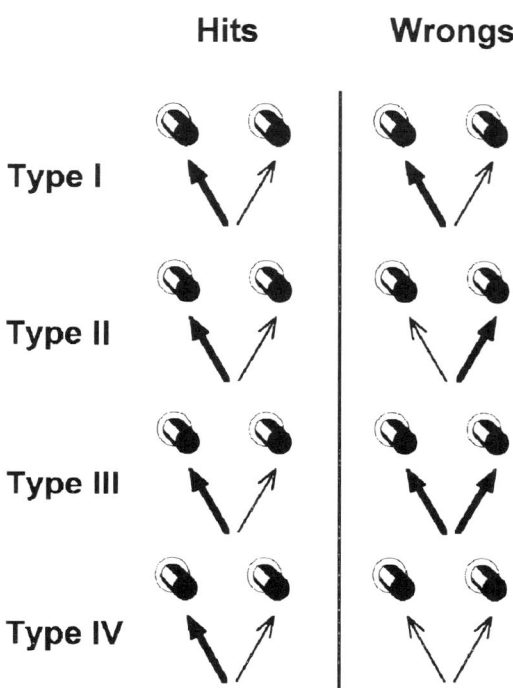

Figure 17. Diagram illustrating the four types of discharges of categorical neurons during the error trials. Heavy arrows mean preferential discharge of the categorical neurons when the animal indicated correctly whether the stimulus was low (left). To the right is indicated the types of errors of the categorical neurons shown in the left. The opposite occurred for those neurons which were selective for high stimulus speeds (not shown in the diagram). Unpublished results from Romo et al.

for low and lateral for high stimulus speeds), instead of reflecting in their activity whether the speed of the stimulus is low or high. Therefore, neurons of this type would discharge every time the animal projects its hand to only one target switch, no matter if the categorization was correct or incorrect, but not to the other target when the categorization was correct or incorrect (Figure 17). Mountcastle and colleagues (1992) have observed a similar type of response for a population of neurons of MI cortex in a sensory discrimination task. However, they interpreted these responses during the error trials as if the "error was made in the discrimination process itself; followed by a correctly matched arm projection, and thus to the wrong target". This may be true also for the first type of categorical neurons of the SMA. Indeed, in the light instruction task, in which animals performed the same movements as in the categorization task, most of these categorical neurons failed to discharge. Only few of them kept their selective responses observed in the categorization task. We conclude that these types of neurons of the SMA reflect in their activity the categorization of the stimulus speed. However, in agreement with previous studies, some of them may be also associated with the coding of the intention, or the arm trajectory to the target switch (Alexander ανδ Crutcher 1990; Matsuzaka et al., 1992).

Contrary to the first type, the second type of neural discharge during the error trials consisted in that this class was not associated with the direction of the arm trajectory to the target switch. For example, if these neurons were selective for low speeds during the correct trials, now they will discharge when the low speeds were classified as high, but not to the opposite target (Figure 17). Therefore, these categorical responses reflect in their activity that the stimulus was processed correctly, but during the execution the animal made an error in the selection of the corresponding target. Thus, the error would be in the motor-output itself. Mountcastle and colleagues (1992) have seen that some neurons of MI cortex, in a sensory discrimination task, display this type of discharge during the error trials. Some of these neurons were tested in the light instruction task, again, most of these neurons lost their differential responses seen during the categorization task.

The third type of neural discharge during the error trials consisted in that the categorical neurons lost their differential responses during the incorrect categorizations. The magnitude of their discharge rates was high during the error trials (Figure 17). Therefore, this discharge should not be interpreted in terms of the coding of the direction of the arm trajectory to one of the two target switches. Instead, this might indicate that the error itself was reflected in the neuronal discharge, but very likely indicating the difficulties for a correct sensory decision. Some of these categorical neurons were studied in the light instruction task, to see whether they were associated with the coding of the direction of the arm movement. Most of these neurons did not display differential responses in this task. Therefore, most of these categorical responses are not associated with the direction of the arm movements per se, although a few of them still had their selective responses observed in the categorization task.

The fourth type of categorical neurons did not discharge during the error trials. This type of neuron cannot be associated with any aspect of the motor output (Figure 17). Therefore, the responses of these neurons are exclusively related with the correct categorization of the stimulus speed.

In conclusion, these results reveal the existence of different classes of categorical neurons in the SMA. These categorical neurons might be working together to construct a sensory decision process during the performance of this learned somesthetic task. We have found no comparable observations to those we have made in the SMA, except one made in MI cortex in monkeys performing in a sensory somesthetic discrimination task (Mountcastle et al., 1992).

7. Conclusions

The experimental findings obtained on the representation of moving tactile stimuli at the level of SI cortex have thrown new light on this important subject (Ruiz et al., 1995). They suggest that the physical properties of the mechanical stimuli are represented isometrically in SI cortex in the form of a neuronal population vector. This neuronal population vector is modulated as a function of the stimulus speeds. Thus, these results are in agreement with previous neurophysiological studies made in SI cortex showing that both the temporal and spatial properties of the tactile stimulus can be observed in the evoked neuronal activity (Mountcastle et al., 1990; Phillips et al., 1988). This must suggest that a function of SI cortex is the coding of the parameters of the tactile stimuli. Also, that SI cortex provides the initial substrate to other cortical and subcortical structures for somesthetic perception. In fact, modules of SI cortex are well suited for distributing this internal dynamic representation of the tactile stimulus to other brain structures. Indeed, anatomical studies have demonstrated that SI cortex is densely connected with the somatic areas of the parietal lobe, and through them, with the precentral gyrus and frontal motor areas as well (Jones, 1984; Mountcastle, 1984; Mountcastle et al., 1992). Therefore, the SI cortex should be considered as only one of several cortical areas that are thought to play roles in the cortical mechanisms in somesthetic perception.

The role of SI cortex in somesthetic perceptions has been explored in different sensory somesthetic tasks (Darian-Smith, et al., 1984b; Hsiao et al., 1993; Mountcastle et al., 1990; Sinclair and Burton, 1991). In general, most of these studies agree that in the evoked neuronal activity of SI cortex there is a sort of representation of the tactile stimuli during the sensory performance. However, these types of representations of the tactile stimuli occur also when the same tactile stimuli, which were used to trigger sensory performance, are delivered passively, in trained, or naive animals (Mountcastle et al., 1990; Phillips et al., 1988; Ruiz et al., 1985). These observations indicate that SI cortex represents in the evoked neuronal activity the physical properties of the somesthetic stimuli, although it has been difficult to relate this neural representation to the perception of the stimulus. Therefore, SI cortex is important for representing the parameters of the tactile stimuli and for distributing these representations to more central structures, where the neuronal perceptual process, very likely occurs.

Sensory physiologists evaluate sensory performance in terms of the relevant motor responses. Indeed, in monkeys trained in the somesthetic categorization task, we recorded neurons in the SMA that respond to the stimulus speeds, but only when the animals perform the categorization task (Romo et al., 1993c). Interestingly, many of these neurons reflect in their activity the sensory decision process. The same observation was made by Mountcastle and colleagues (1992) in the MI cortex in a sensory somesthetic discrimination task. Although the outputs of the somesthetic perception are observed in the motor areas of the frontal lobe, these neuronal events are quite different from those observed in SI cortex (Mountcastle et al., 1992; and results shown above). This raises the question that the perception of the somesthetic stimuli may have several different representations in parietal areas, frontal motor areas, and in those subcortical structures (Romo et al., 1995) associated with the performance of this learned somesthetic task.

Acknowledgments

The research of R. Romo was supported in part by an International Research Scholars Award from the Howard Hughes Medical Institute, DGAPA-UNAM (Proyecto IN203994) and CONACyT (Proyecto 400346-5-3421 N9309).

References

Alexander, G. E. and Crutcher, M.D. (1990) Preparation for movement: neural representations of intended direction in three motor areas of the monkey. Journal of Neurophysiology, 64, 133-150.

Cavada, C. and Goldman-Rakic, P. (1989) Posterior parietal cortex in rhesus monkey: II. Evidence for segregated corticortical networks linking sensory and limbic areas with the frontal lobe. Journal of Comparative Neurology, 287, 422-445.

Constanzo, R.M. and Gardner, E.P. (1980) A quantitative analysis of responses of direction-sensitive neurons in the somatosensory cortex of awake monkeys. Journal of Neurophysiology, 43, 1319-1341.

Darian-Smith, C., Darian-Smith, I., Burman, K. and Ratclife, N. (1993) Ipsilateral cortical projections to areas 3a, 3b, and 4 in the macaque monkey. Journal of Comparative Neurology, 335, 200-213.

Darian-Smith, I. (1984a) The sense of touch: performance and peripheral neural processes. In Brookhart, J.M. and Mountcastle, V.B. (eds) The Nervous System, Sec 2. Darian-Smith (ed) Vol 2, The American Physiological Society, Wash. D.C., pp739-788.

Darian-Smith, I. Goodwin, A., Sugitani, M. and Hewood, J. (1984b) The tangible features of textured surfaces: Their representation in the monkey's somatosensory cortex. In: Edelman, G.M., Gall, W.E. and Cowan, W.M. (eds) Dynamic aspects of neocortical function, Wiley, N.Y., pp 475-500.

Dum, R. P. and Strick, P.L. (1991) The origin of corticospinal projections from premotor areas in the frontal lobe. Journal of Neuroscience, 11, 667-689.

Essik, G.K. and Whitsel, B.L. (1985) Assessment of the capacity of humans subjects and S-I neurons to distinguish opposing directions of stimulus motion across the skin. Brain Research Reviews, 10, 187-212.

Galea, M.P. and Darian-Smith, I. (1994) Multiple corticospinal neuron populations in the macaque monkey are specified by their unique cortical origins, spinal terminations and connections. Cerebral Cortex, 4, 166-194.

Gardner, E.P. and Constanzo, R.M. (1980) Neuronal mechanisms underlying direction sensitivity of somatosensory cortical neurons in awake monkeys. Journal of Neurophysiology, 43, 1342-1354.

Georgopoulos, A.P., Kettner, R.E. and Schwartz, A.B. (1988) Primate motor cortex and free arm movements to visual targets in three-dimensional space. II. Coding of the direction of movement by a neuronal population. Journal of Neuroscience, 8, 2928-2937.

Georgopoulos, A.P., Taira, M. and Lukashin, A. (1993) Cognitive Neurophysiology of the motor cortex. Science, 260, 47-52.

Iwamura, Y., Tanaka, M. and Hikosaka, O. (1980) Overlapping representation of fingers in the somatosensory cortex (area 2) of conscious monkey. Brain Research, 197, 516-520.

Harnard, S. (1989) Categorical perception. Cambridge University Press.

He, S.Q., Dum, R.P. and Strick, P.L. (1995) Topographic organization of corticospinal projections from the frontal lobe: motor areas of the medial surface of the hemisphere. Journal of Neuroscience, 15, 3284-3306.

Hsiao, S.S., O'Shauhnessy, D.M. and Johnson, K.O. (1993) Effects of selective attention on spatial processing in monkey primary and secondary somatosensory cortex. Journal of Neurophysiology, 70, 444-447.

Hummelsheim, H., Wiesendanger, M., Bianchetti, M., Wiesendanger, R. and Mcpherson, J. (1986) Further investigations of the efferent linkage of the supplementary motor area (SMA) with the spinal cord in the monkey. Experimental Brain Research, 65, 75-82

Hyvarinen, J. and Poranen, A. (1978) Movement-sensitive and direction-and orientation-selective cutaneous receptive fields in the hand area of the post-central gyrus in monkeys. Journal of Phsyiology (Lond.), 283, 523-537.

Johansson, R.S. and Vallbo, A. (1979) Tactile sensitivity in the human hand: relative and absolute densities of four types of mechanoreceptive units in glabrous skin. Journal of Physiology (Lond.), 286, 283-300.

Jones, E.G. (1984) Connectivity of primate sensory-motor cortex. In Jones, E.G. and Peters, A. (eds) Cerebral Cortex: Vol 5, Sensory-motor areas and aspects of cortical connectivity. Plenum, N.Y., pp113-175.

Jones, E.G., Coulter, J.D., Burton, H. and Porter, R. (1977) Cells of origin and terminal distribution of corticostriatal fibers arising in the sensory-motor cortex of monkeys. Journal of Comparative Neurology, 173, 53-80.

Jones, E.G., Coulter, J.D. and Hendry, S.H.C. (1978) Intracortical connectivity of architectonic fields in the somatic sensory, motor and parietal cortex of monkeys. Journal of Comparative Neurology, 181, 291-348.

Jones, E.G. and Powell, T.P.S. (1969) Connexions of the somatic sensory cortex of the rhesus monkey. I. Ipsilateral cortical connexions. Brain, 92, 477-502.

Kass, J.H. (1993) Parallel and serial processing in the somatosensory system. In Rudomin, P., Arbib, M.A., Cervantes-Perez, F. and Romo, R. (eds) Neuroscience: from neural networks to artificial intelligence. Springer-Verlag, Heidelberg, pp136-153.

Kass, J.H., Nelson, R.J., Sur, M., Lin, C.S. and Merzenich, M.M. (1979) Multiple representations of the body within the primary somatosensory cortex of primates. Science, 204, 521-523.

Kemp. J.M. and Powell, T.P.S. (1970) The cortico-striate projection in the monkey. Brain, 93, 525-546.

Luppino, G., Mattelli, M., Camarda, R.M. and Rizzolatti, G. J. (1993) Cortico-cortical connections of area F3 (SMA-proper) and area F6 (pre-SMA) in the macaque monkey. Journal of Comparative Neurology, 338, 114-140.

Matsuzaka, Y., Aizawa, H. and Tanji, J. (1992) A motor area rostral to the supplementary motor area (presupplementary motor area) in the monkey: neuronal activity during a learned motor task. Journal of Neurophysiology, 68, 653-663.

McGuire, P.K., Bates, J.F. and Goldman-Rakic, P.S. (1991) Interhemispheric integration: II. Symmetry and convergence of the corticostriatal projections of the left and right supplementary motor area (SMA) in the rhesus monkey. Cerebral Cortex, 1, 408-417.

Mcpherson, J.M., Wiesendanger, M., Marangoz, C. and Miles, T.S. (1982) Corticospinal neurons of the supplementary motor area of monkeys. Experimental Brain Research, 48, 81-88.

Mountcastle, V.B. (1957) Modality and topographic properties of single neurons of cat's somatic sensory cortex. Journal of Neurophysiology, 20, 408-434.

Mountcastle, V.B. (1978) An organizing principle for cerebral function: the unit module and the distributed system. In Edelman, G.M. and Mountcastle, V.B. (eds) The Mindful Brain, MIT Press, Cambridge, pp7-50.

Mountcastle, V.B. (1984) Central nervous mechanisms in mechanoreceptive sensibility. In Brookhart, J.M. and Mountcastle, V.B. (eds), The Nervous System, Sec 2. Darian-Smith, I. (ed), Vol 2, The American Physiological Society, Was., D.C., pp789-878.

Mountcastle, V.B., Atluri, P.P. and Romo, R. (1992) Selective output-discriminative signals in the motor cortex of waking monkeys. Cerebral Cortex, 2, 277-294.

Mountcastle, V.B., Steinmetz, M.A. and Romo, R. (1990) Frequency discrimination in the sense of flutter: psychophysical measurements correlated with postcentral events in behaving monkeys. Journal of Neuroscience, 10, 3032-3044.

Mountcastle, V.B., Talbot, W.H., Sakata, H. and Hyvarinen, J. (1969) Cortical neuronal mechanisms in flutter-vibration studied in unanesthetized monkeys. Journal of Neurophysiology, 32, 453-484.

Nelson, R.J., Sur, M., Felleman, D.J. and Kass, J.H. (1980) Representations of the body surface in the postcentral parietal cortex of Macaca fascicularis. Journal of Comparative Neurology, 192, 611-643.

Pandya, D.N. and Kuypers, H.G.J. (1969) Cortico-cortical connections in the Rhesus monkey. Brain Research, 13, 13-36.

Pandya, D.N. and Vignolo, L.A. (1971) Efferent cortico-cortical projections of the precentral, premotor and arcuate areas in the rhesus monkey. Brain Research, 26, 217-233.

Pearson, R.C.A. and Powell, T.P.S. (1985) The projection of the primary somatic sensory cortex upon area 5 in the monkey. Brain Research Reviews, 9, 89-107.

Petrides, M. and Pandya, D.N. Projections to the frontal cortex from the posterior parietal region in the rhesus monkey. Journal of Comparative Neurology, 228, 105-116.

Phillips, J.R., Johnson, K.O. and Hsiao, S.S. (1988) Spatial pattern representation and transformation in monkey somatosensory cortex. Proceedings of the National Academy of Sciences (USA), 85, 1317-3121.

Phillips, C.G., Powell, T.P.S. and Wiesendanger, M. (1971) Projection from low-threshold muscle afferents of hand forearm to area 3a of baboon's cortex. Journal of Physiology (Lond.), 217, 419-446.

Pons, T.P., and Kass, J.H. (1986) Corticortical connections of area 2 of somatosensory cortex in macaque monkeys: a correlative and electrophysiological study. Journal of Comparative Neurology, 248, 313-335.

Pons, T.P., Garraghty, P.E., Friedman, D.P. and Miskin, M. (1987) Physiological evidence for serial processing in somatosensory cortex. Science, 237, 417-420.

Powell, T.P.S. and Mountcastle, V.B. (1959) Some aspects of the functional organization of the cortex of the postcentral gyrus of the monkey: a correlation of findings obtained in a single unit analysis with cytoarchitecture. Bulletin of the Johns Hopkins Hospital, 105, 133-162.

Richmond, B.J., Optican, L.M., Podell, M. and Spitzer, H. (1987) Temporal encoding of two-dimensional patterns by single units in primate inferior temporal cortex. I. Response characteristics. Journal of Neurophysiology, 57, 132-146.

Romo, R., Ruiz, S. and Crespo, P. (1993a) Cortical representationof touch. In Rudomin, P., Arbib, M.A., Cervantes-Perez, F. and Romo, R. (eds) Neuroscience: from neural networks to artificial intelligence. Springer-Verlag, Heidelberg, pp154-170.

Romo, R., Ruiz, S., Crespo, P. and Hsiao, S.S. (1993b) A tactile stimulator for studying motion processing in the somatic sensory system of primates. Journal of Neuroscience Methods, 46, 139-146.

Romo, R., Ruiz, S., Crespo, P., Zainos, A. and Merchant, H. (1993c). Representation of tactile signals in primate supplementary motor area. Journal of Neurophysiology, 70, 2690-2694.

Romo, R., Merchant, H., Ruiz, S., Crespo, P. and Zainos, A. (1995). Neuronal activity of primate putamen during categorical perception of somaesthetic stimuli. NeuroReport, 6, 1013-1017.

Ruiz, S., Crespo, P. and Romo, R. (1995) Representation of moving tactile stimuli in the somatic sensory cortex of awake monkeys. Journal of Neurophysiology, 73, 525-537.

Shanks, M.F., Pearson, R.C.A. and Powell, T.P.S. (1985) The ipsilateral cortico-cortical connexions between the cytoarchitectonic subdivisions of the primary somatic sensory cortex in the monkey. Brain Research Reviews, 9, 67-88.

Sinclair, R.J. and Burton, H. (1991) Tactile discrimination of gratings: psychophysical and neural correlates in human and monkey. Somatosensory Motor Research, 8, 241-248.

Sur, M., Wall, J.T. and Kass, J.H. (1984) Modular distribution of neurons with slowly adapting and rapidly adapting responses in area 3b of somatosensory cortex in monkeys. Journal of Neurophysiology, 51, 724-744.

Talbot, W.H., Darian-Smith, I., Kornhuber, H.H. and Mountcastle, V.B. (1968) The sense of flutter-vibration: comparison of the human capacity with response patterns of mechanoreceptive afferents from the monkey hand. Journal of Neurophysiology, 31, 301-334.

Tokuno, H. and Tanji, J. (1993) Input organization of distal and proximal forelimb areas in the monkey primary motor cortex: retrograde double labeling study. Journal of Comparative Neurology, 333, 199-209.

Warren, S., Hamalainen, H.A. and Gardner, E.P. (1986) Objective classification of motion-and direction-sensitive neurons in primate somatosensory cortex of awake monkeys. Journal of Neurophysiology, 56, 598-622.

Whitsel, B.L., Roppolo, J.R. and Werner, G. (1972) Cortical information processing of stimulus motion on primate skin. Journal of Neurophysiology, 35, 691-717.

Wiesendanger, M., Hummelsheim, H. and Bianchetti, M. (1985) Sensory input to the motor fields of the agranular frontal cortex: a comparison of the precentral, supplementary motor and premotor cortex. Behavioral Brain Research, 18, 89-94.

Neural Aspects of Tactile Sensation
J.W. Morley (Editor)

Constancy in the Somatosensory System: Central Neural Mechanisms Underlying the Appreciation of Texture During Active Touch.

C. E. Chapman

The hand has been the object of numerous studies both within the fields of motor control and sensory neurophysiology, as well as in sensory psychophysics. Interest has been focused particularly upon studies in higher primates, where the hand shows marked specializations for its dual sensory and motor functions, exploration and manipulation of the environment. The specialization for sensory functions is reflected by the high receptor density, especially on the finger pads, the increased area devoted to the hand at all levels of the central neural pathways involved in somaesthesia, and the resultant increased perceptual acuity, especially as regards touch (Mountcastle, 1984). The specialization for motor functions is reflected in the large number of muscles devoted to controlling digit and hand movement, the existence of direct projections from the motor cortex to the motoneurons controlling the distal muscles, the increased representation of the hand region in the central structures involved in producing voluntary movement (especially primary motor cortex), and the precision with which digit movements can be generated and controlled (Phillips, 1986).

The motor and sensory functions of the hand are highly interdependent, and either may predominate depending upon the task at hand. For example, if one reaches into a pocket to retrieve a paper clip from amongst a mixture of objects (change, keys, etc.), then the exploratory movements are secondary to the goal of sampling and evaluating the sensory inputs generated during the search. Following identification, however, the motor functions of the hand may take precedence as the paper clip is first retrieved and then applied to secure several sheets of paper. In the latter situation, sensory feedback may be secondary to the execution of the motor act.

This example highlights one other important observation, namely that movement between the skin and an object is an important component of touch. Thus, humans can detect the presence of micro-features, 1 to 3 μm in height, on an otherwise smooth surface when there is relative movement between the skin and the surface (Johansson and LaMotte, 1983). In the absence of lateral movement, however, the same subjects were unable to detect the presence of features up to 2 or 3 times higher. In a similar vein, Morley et al. (1983) showed that discrimination of small changes in surface texture, provided by modifying the spatial period of gratings consisting of alternating ridges and grooves, is significantly better when there is lateral movement between the skin and the surface. A change in spacing of 52 μm could be discriminated (standard surface, 1 mm spatial period) with lateral movement as compared to only 103 μm under static conditions (i.e. no lateral movement). As reviewed by Johnson (1983), several factors contribute to this: (i) dynamic stimuli recruit all of the cutaneous mechanoreceptors involved in discriminative touch (rapidly and slowly adapting afferents), while static stimuli only recruit the slowly adapting afferents; and (ii) dynamic stimuli elicit a more intense discharge in all mechanoreceptive afferents, regardless of their adaptation rate, effectively increasing the size of the signal-to-noise ratio. In addition, central cortical neurons in primary somatosensory cortex (SI) are overwhelming rapidly adapting in nature so that, outside of the middle cortical layers of area 3b, relatively few slowly adapting cutaneous units are encountered in SI (Paul et al., 1972; Sur et al., 1984).

While movement is important for tactile perception, this introduces complications in the interpretation of the signals generated during touch. In the first instance, movement itself is associated with a powerful inhibition, or gating, of the transmission of somatosensory information from the periphery up to SI cortex (reviewed in Chapman, 1994), which diminishes the perception of tactile stimuli (Chapman et al., 1987; Milne et al., 1988; Feine et al., 1990; Post et al., 1994). In the second instance, the afferent signals from the peripheral mechanoreceptors involved in discriminative touch vary not only as a function of the physical characteristics of the scanned surface or object but also as a function of the physical parameters of the stimulus, specifically speed and force. Altogether, the sensory feedback generated during tactile exploratory movements needs to be interpreted in the light of the conditions under which it is generated.

The need for interpretation of these signals is particularly apparent when one considers the phenomenon of *perceptual constancy*. In the visual system, perceptual constancy is well-known and underlies, for example, the ability to recognize a particular object viewed under different conditions. For example, a bunch of grapes can be identified as such whether they are viewed at 1 metre or at 10 metres, even though the image on the retina is very different. Likewise changes in ambient illumination do not interfere with the ability to recognize an object. These observations imply that cognitive processes take into account the viewing conditions when interpreting the images.

For discriminative touch, the case for perceptual constancy is less clear, at least as regards the perception of texture. As pointed out above, the peripheral neural representation of surface texture is a function of the physical characteristics of the surface, and this neural image is modified by changes in contact force or the speed of exploration. The consequence of this is that the subjective impression of roughness systematically changes as a function of contact force (Lederman, 1974), i.e. perceptual inconstancy. On the other hand, there does appear to be perceptual constancy across different speeds since Lederman (1974) also reported that subjective roughness is not influenced by a five-fold change in speed. One challenge within neuroscience is to attempt to explain how perceptual constancy can be achieved. This chapter will concentrate upon one movement-related parameter, speed, and discuss how an invariant central representation of surface texture might be computed within SI cortex.

1. Peripheral representation of texture

Over the past 15 years, a number of laboratories have provided a detailed description of the discharge properties of peripheral mechanoreceptive afferents involved in signalling the physical characteristics of textured surfaces. This approach has been facilitated by employing patterned surfaces, consisting of either gratings or embossed dots, whose physical dimensions can be systematically varied by changing the size or spacing of the individual elements in the tactual array. In the pioneering work of Darian-Smith and collaborators (Darian-Smith and Oke, 1980; Darian-Smith et al., 1980), the textured surfaces were mounted upon a rotating drum and then scanned over the receptive field of an identified mechanoreceptive afferent recorded in a passive, anaesthetized monkey. Most experiments have concentrated upon studies of the glabrous skin of the monkey finger pads, this being a region where tactile sensibilities are highly developed. It should be stressed, however, that qualitatively similar results have also been obtained from recordings of primary afferents in alert humans using the microneurographic technique (Phillips et al., 1990; Phillips et al., 1992).

Darian-Smith and Oke (1980) recorded from the three types of mechanoreceptive afferents innervating the glabrous skin of the monkey digits that are considered to play a role in discriminative touch - PC, Pacinian afferents; RA, rapidly adapting afferents; SA, slowly adapting afferents. They found that all three types of mechanoreceptive afferents were activated as gratings were scanned over their receptive field. Their discharge reflected the physical characteristics of the scanned surface, that is the distance between the individual elements in the tactual array (spatial period measured from the onset of one ridge to the next, range tested 0.54 to 1.025 µm), and also the speed of presentation of the surfaces (range 25 to 160 mm/s) and the contact force (either 20 or 60 g). The authors argued that the temporal frequency of the moving surface (speed/spatial period) is represented in the discharge of the mechanoreceptive afferents and that the resolving capacity of each type of afferent is specialized. Thus, SA afferents resolve low temporal frequencies (20 - 60 Hz), PC afferents resolve high frequencies (100-300 Hz), and RA afferents resolve intermediate frequencies. Altogether, the primary afferent responses to texture were, however, ambiguous in that the signal covaried with spatial period, speed and force. This observation indicates that knowledge of the spatial dimensions of textured surfaces such as gratings appears to require independent sources of information about the parameters of movement, including the rate of movement of the stimulus over the skin and the applied force.

Subsequent work has, however, suggested that the mean discharge rate of SA afferents may provide an unambiguous signal of surface texture independent of the speed of presentation of the surface. Lamb (1983b) investigated the ability of mechanoreceptive afferents to signal information about the physical characteristics of finely textured, embossed dot surfaces (spatial periods of 2 mm or less). His results indicated that the mean discharge of SA afferents distinguished between textures with spatial periods of 1 and 2 mm (although not the smaller changes in spatial period of the order of 2 to 3% which could be discriminated by humans [Lamb, 1983a]) but was invariant over a wide range of velocities (40 to 220 mm/s tested). The mean response rate of RA and PC afferents, in contrast, covaried with speed. Their discharge also reflected slight changes in spatial period, respectively, for the large (0.67 mm diameter, 2 mm spatial period) and the small dot surfaces (0.33 mm diameter, 1 mm spatial period), suggesting that these afferents are responsible for the human discriminative ability (Lamb, 1983a).

The observation of speed invariance for SA afferents was subsequently confirmed by Goodwin and Morley (Goodwin and Morley, 1987a,b; Morley and Goodwin, 1987) who reinvestigated the peripheral discharge characteristics of mechanoreceptive afferents, this time using sinusoidal movements of the textured surfaces (periodic gratings) under the digit tips. The advantage of this approach was that the to-and-fro movements more closely resemble the types of movements employed by humans during tactile exploration. Using a wide range of spatial periods (0.75 to 3 mm) and speeds (mean speeds of 9 to 160 mm/s; peak speeds, 20 to 480 mm/s), they reported that the mean discharge rate of SA afferents generally signalled spatial period independently of speed. In contrast, the discharge of RA afferents covaried with both speed and spatial period, while PC afferents were only sensitive to speed. Systematic increases in force (0.05 to 0.3 N), on the other hand, increased the responses of all three types of mechanoreceptive afferents to the textured surfaces, but in this case the SA and PC afferents showed the greatest sensitivity, while the RA afferents were insensitive to contact force at least under certain conditions (coarser grating) (Goodwin and Morley, 1987b).

More recently, Goodwin et al. (1989) developed a model to describe how groove width and peak temporal frequency (peak speed/spatial period) affect the afferent responses of all three types of mechanoreceptive afferents. They concluded that the model fit well with their experimental data (Goodwin and Morley, 1987; Sathian et al., 1989). In particular, both RA and SA afferents increased their responses with increasing spatial period; both also decreased their discharge with increasing temporal frequency but the decrease was greater for SAs than for RAs (and PCs). Their model predicted RA sensitivity to speed as well as SA insensitivity to speed. Thus, it appears that information about the texture of the surface, independent of speed, may be provided by the discharge of SA mechanoreceptive afferents.

This speed-invariance for SA afferents may, however, be restricted to certain stimulating conditions. For example, inspection of the results of Goodwin and Morley (1987a) indicates that SA afferents show speed-sensitivity when coarser textures are presented (their Fig. 4). Furthermore, the model developed by Goodwin et al. (1989) predicts the appearance of velocity sensitivity for SA afferents tested with the largest spatial period, 3 mm. Thus, it appears that SA speed-invariance may be restricted to textured surfaces with closely spaced tactual elements.

The suggestion that the relative spacing of the tactual elements is an important factor in determining speed invariance receives support from two studies. First, LaMotte and Srinivasan (1987a,b) investigated the role of mechanoreceptive afferents in coding shape by stroking a series of flat plates with a single step increase in thickness at the midpoint (variable curvature or slope; step width, 0.45 to 3.13 mm) over the receptive field of individual mechanoreceptive afferents, i.e. a single macro-feature. Both SA and RA afferents increased their discharge rate as the step became steeper, and both increased their discharge with stroke velocity (range tested: 1 to 40 mm/s). Second, Phillips et al. (1992) reported that the mean discharge rate of SAI and RA afferents, recorded from the median nerve of humans, increased with an increase in the speed with which widely spaced embossed dots (spatial period approximately 2 mm) were scanned across the receptive field (20 vs. 60 mm/s). A further increase in speed to 90 mm/s, however, had either weaker (RA) or no (SAI) effects.

Finally, velocity sensitivity for SA afferents has also been demonstrated in a situation in which the stimulus, a moving brush, is moved across the peripheral receptive field (Edin et al., 1995; Essick and Edin, 1995). They systematically varied the speed between 5 and 320 mm/s, and found that the discharge frequency of both rapidly (RA and PC) and slowly adapting afferents (SAI and SAII) increased with increasing speed. The exponent of the power function that described the relationship between discharge frequency and speed was, however, higher for rapidly adapting receptors than for slowly adapting receptors.

In summary, the general picture that emerges from the literature is that all of the cutaneous mechanoreceptors that contribute to discriminative touch are activated when complex textured surfaces are scanned over their receptive field. Moreover, their discharge can vary with scanning speed, but the presence or absence of a velocity effect may depend upon the type of mechanoreceptive afferent, the textured surface presented and the scanning speed. The influence of a second parameter, contact force, has been less extensively studied, but the existing evidence indicates that discharge frequency of all types of mechanoreceptive afferents increases with increased force. Thus, the signal that is forwarded to the CNS for evaluation is complex, frequently reflecting the scanned texture PLUS the peripheral conditions under which the scanning occurred, i.e. the speed, and also the force, of the scanning movement.

2. Perception of roughness

How then does the CNS interpret this complex signal? Psychophysical studies of the human capacity to appreciate surface textures provide some insight. Lederman and Taylor (1972) and Lederman (1974) investigated the ability of subjects to estimate the perceived roughness of textured surfaces (gratings with spatial periods ranging from 0.375 to 1.25 mm) while varying the parameters of stimulation. Their work showed that perceived roughness increased as a function of the groove width (0.125 to 1 mm); in contrast, roughness declined modestly with an increase in ridge width (same range tested). These observations have since been confirmed by Sathian et al. (1989) who, using a wider range of gratings (spatial period ranging from 0.74

to 3.09 mm), concluded that groove width was the most important factor in determining the perceived roughness of gratings.

Lederman (1974) also investigated the effects of changing force and speed upon magnitude estimates of roughness. Perceived roughness increased with an increase in force (28, 70 and 448 g). It should be pointed out, however, that the highest force employed (448 g) greatly exceeded the pressure that subjects would usually apply during active touch (mean, 70 g; range 17 to 172 g). Indeed it was close to the maximal force that subjects could exert under these experimental conditions and still be able to move smoothly over the gratings. Speed was varied over a wide range, 10 to 250 mm/s. Perceived roughness was significantly reduced with an increase in speed, but the effects were not apparent at all speeds tested. At the two lower speeds (10 and 50 mm/s), roughness estimates were almost identical. In other words, a five-fold change in speed had no effect upon the perception of roughness. It was only at the highest speed tested (250 mm/s) that roughness declined. Overall, this evidence supports the notion that, at least with lower scanning speeds, there is perceptual constancy for texture appreciation.

One possible explanation for the relatively minor effects of speed on perceived roughness is, however, that the subjects might have used information about the speed of their active movements (feedback from the moving limb or knowledge of the motor command, i.e. the efference copy) in order to interpret the peripheral signals. This prompted Lederman (1983) to compare roughness estimates obtained with active and passive touch. For active touch, subjects moved their digit over the surface at one of three different speeds (17, 67 or 207 mm/s); for passive touch, the subjects were immobile while the surfaces were passively moved under the digit at the same speeds. There was not, however, any significant difference in performance as a function of the mode of touch, and this across a range of speeds (although, once again, roughness estimates were significantly decreased at the highest speed). These observations argue strongly against the use of an efference copy to "interpret" the texture-related signal. It appears much more likely that the peripheral signals provide sufficient information to ensure perceptual constancy across a range of speeds.

3. Central neural coding of speed and texture

One of the first studies of central cortical neurons during movements mimicking those made during active touch was that by Darian-Smith and colleagues (Darian-Smith et al., 1982; Darian-Smith et al., 1985). They trained monkeys to scan their digit tips back and forth over textured surfaces (periodic gratings) within the context of a visuomotor tracking task. All units with a digital cutaneous receptive field were activated during the scanning movements, and some varied their discharge as a function of the texture of the underlying surface, with discharge frequency increasing when coarser gratings were scanned. Many of the cells were, however, only active during a part of the cycle of movement (specifically at the turning points), in contrast to the discharge patterns of cutaneous afferents in response to similar sinusoidal movements of gratings (Goodwin and Morley, 1987a,b). They concluded that SI cortical neurons do not unambiguously represent surface texture because, as in the periphery, discharge also covaries with force and velocity.

More recently, investigators have recorded the discharge of SI cortical neurons during active touch within the context of texture discrimination tasks (see Fig. 1). Data from both this laboratory (Chapman and Ageranioti-Bélanger, 1991; Ageranioti-Bélanger and Chapman, 1992) and Sinclair and Burton (1991) indicate that a substantial proportion of SI neurons receiving cutaneous input from the hand (areas 3b, 1 and 2) are activated during active touch, and that many of these signal differences in the scanned textures (respectively, smooth vs. rough and periodic gratings with varying spatial periods). Furthermore, while the discharge of many of these neurons covaried with peripheral factors such as the velocity of movement or the force exerted on the scanned surfaces, a substantial proportion of the texture-related units were insensitive to variations in, for example, the velocity of movement.

Examples of the types of responses encountered are shown in Figs. 2, 3 and 4. The experimental paradigm is illustrated in Fig. 1 (Chapman and Ageranioti-Bélanger, 1991; Ageranioti-Bélanger and Chapman, 1992). In response to a light cue (Fig. 1C), monkeys were trained to make a single scanning movement over the surface that

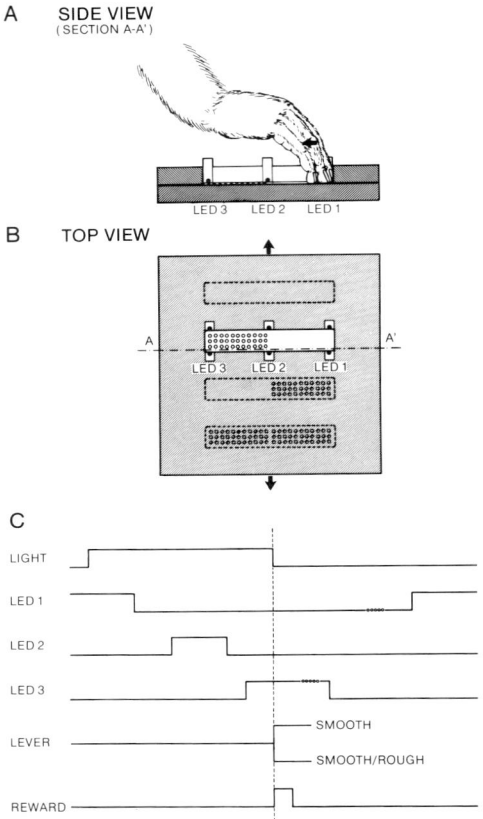

Figure 1. Active texture discrimination task. A. Side view of the behavioural apparatus (taken from level A-A' shown in B) showing the monkey's hand positioned at the start position (far end of groove) and the position of three light emitting diode (LED)/photosensor pairs that monitored digit position. The arrow indicates the direction of the scanning movement. B. Top view of the apparatus showing the groove and the four surfaces that formed the floor of the groove. In most experiments the top pair of surfaces were presented (surface changed during the intertrial interval, indicated by arrow), respectively smooth and smooth/rough. C. Time course of events in a sample trial. Note that the illumination of a light served as the GO cue; this was followed by a single scanning movement. Thereafter the monkey either pushed or pulled on a lever with the opposite hand to indicate the texture encountered over the second half of the surface (respectively, smooth or rough). Correct responses were rewarded with a drop of juice. (Reproduced with permission from Chapman and Ageranioti-Bélanger, 1991)

formed the floor of a longitudinal groove situated in front of the animal, and oriented perpendicular to the animal (Fig. 1A, direction of movement indicated by the arrow). After the scanning movement, the animal indicated the texture encountered over the second half of the surface (smooth or rough) by pushing or pulling a lever with the opposite arm. During data collection, one pair of surfaces (aluminium) was presented for each set of recordings, usually the top pair shown in Fig. 1B: one surface was smooth, while the other was initially smooth and became textured over the second half of its length (repeated arrays of 6 embossed Braille dots, 0.5 mm high, separated by 2 mm within the array and 4 mm between arrays; surface dimensions, 20 x 100 mm).

Figure 2 shows an example of an area 3b texture-related unit. During the task (active movement), this neuron showed a pronounced burst of discharge as the digits were scanned over the rough portion of the surface (compare responses to smooth/rough and smooth surfaces). Subsequently, the unit was tested for its response to texture outside the context of the task by moving the digits passively over the same surfaces (passive movement). In this case, texture-related discharge was still evident, although somewhat reduced in amplitude. The decrease in amplitude was explained by the fact that the passive movements were slightly slower than the active movements. Figure 3 plots the mean discharge rate of this unit while the digits scanned over the rough portion of the surface as a function of the speed of movement. There was a positive linear correlation between cell discharge frequency and speed (P=0.018, 18 df), i.e. this unit's discharge reflected both the texture of the scanned surface and the speed of movement. Although our sample of area 3b neurons was small (n=17), 44% of the modulated units were speed-sensitive. Such response properties suggest that area 3b neurons are at a relatively low hierarchical level, close to the peripheral texture signal that itself covaries with the parameters of movement.

In contrast, the results of our recordings in area 1 led us to suggest that this area is located at a hierarchically higher level than area 3b with respect to the processing of texture-related signals. We found that 26% (6/23) of area 1 units were only active during active touch and not during passive scanning movements (discrimination no longer required), suggesting that such units may signal peripheral events in a

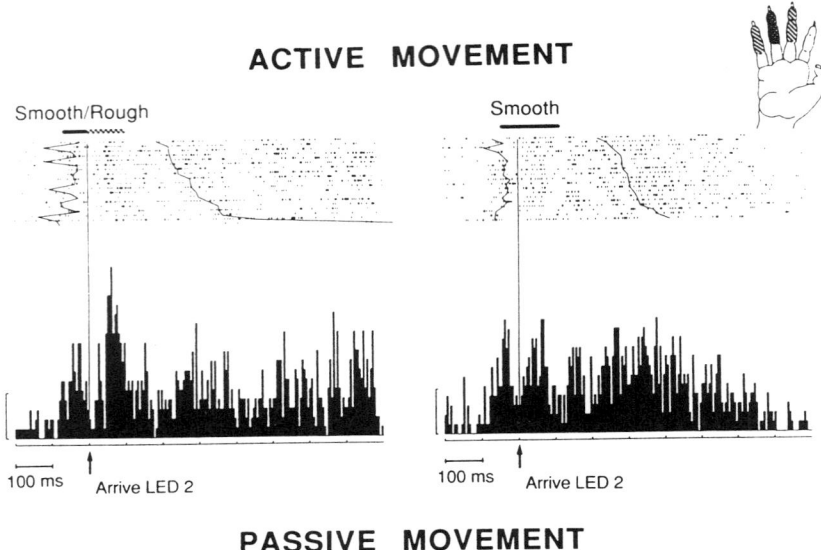

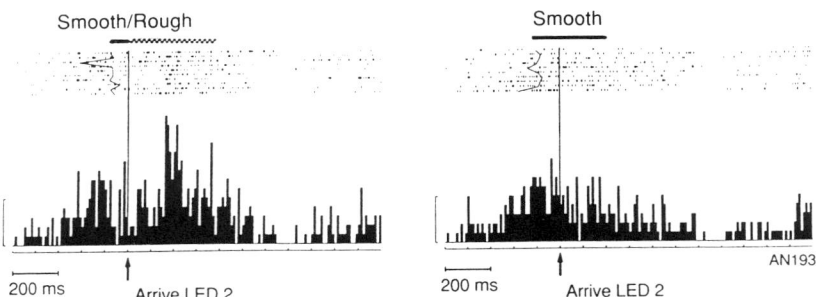

Figure 2. Area 3b unit with a clear texture-related response during performance of the active texture discrimination task (active movement) and during passive movements (no discrimination required). Receptive field: SA response to light touch (shaded area). For active touch, the trials in the rasters are rearranged in increasing order of the response time (time interval between the digits arriving at the mid-point of the surface [change in texture for smooth/rough surface] and the lever response [see the irregular line on the right side of each raster]). Trials are aligned on the time that the digits arrived at the midpoint of the surface. The first irregular line in each raster indicates the onset of movement. For passive movement, trials are displayed in order of acquisition. The schematic representation of the surface above each raster is aligned to show the average onset and end of movement. Modified from Fig. 6 in Chapman and Ageranioti-Bélanger (1991).

context-dependent manner. In contrast, area 3b, and also area 2, neurons were generally modulated in both situations (26/32). Moreover, the occasional absence of modulation with passive movements for units in areas 3b and 2 could generally be explained by peripheral factors such as the speed of movement. In addition, we found that speed sensitivity was not uniformly distributed across SI cortex. Figure 4 shows an example of a typical texture-related unit in area 1. This unit showed a weak burst of discharge when the digits

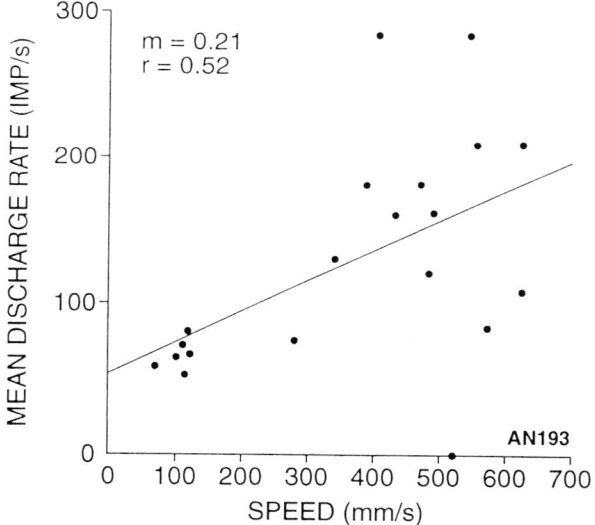

Figure 3. Relationship between discharge frequency (measured during the time that the digits were scanned over the rough portion of the smooth/rough surface) and average speed for the unit shown in Fig. 2. Data from both the active and the passive trials are pooled.

actively scanned over the rough half of the surface (top, compare smooth/rough vs smooth); in this case, a stronger texture response was evident during testing outside of the task (passive movements), possibly reflecting better contact between the small receptive field and the surface. Cell discharge frequency did not, however, significantly vary with movement speed. Figure 5 shows summary plots of the linear regression lines obtained between cell discharge frequency and

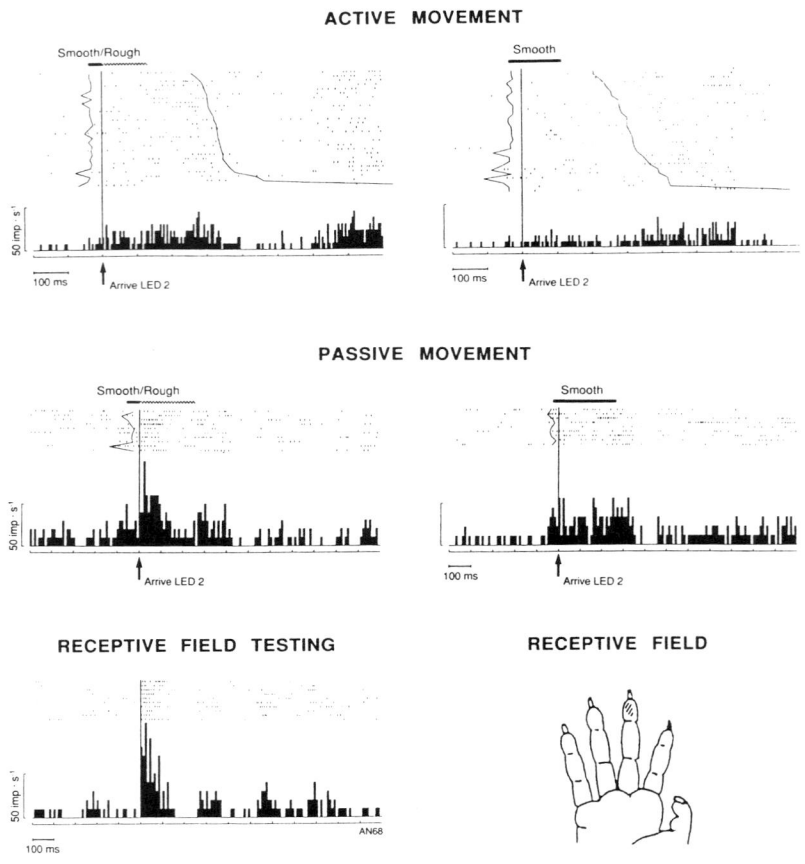

Figure 4. Area 1 texture-related unit whose discharge was unrelated to movement speed. Receptive field: weak RA response to light touch (response to light tap shown below). Data plotted as in Fig. 2. (Reproduced with permission from Chapman and Ageranioti-Bélanger, 1991).

speed for most of the texture-related cells recorded in areas 3b and 2 (A) and area 1 (B). Areas 3b and 2 were, together, characterized by a high proportion of significant relations between the texture-related response and speed (44%, 8/18). Even so, more than one-half of the texture-related units in these areas showed no change in discharge across a wide range of speeds. In area 1, on the other hand, speed

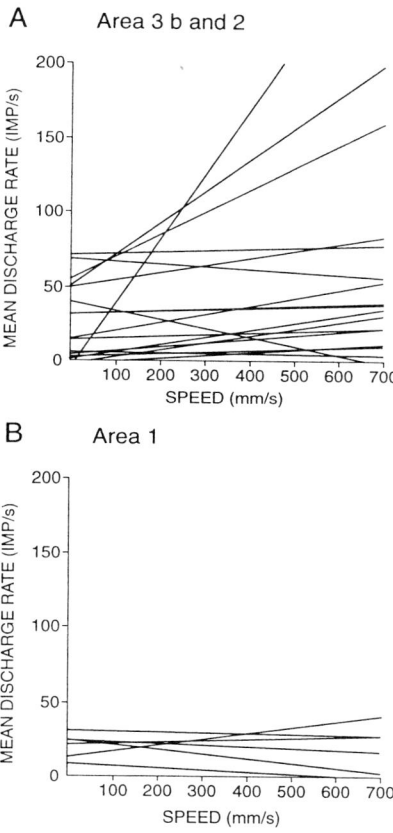

Figure 5. Plots of linear regression lines (discharge frequency versus average movement speed) calculated for 17 texture-related neurons recorded in areas 3b and 2 (A) and 6 texture-related neurons recorded in area 1 (B). For the data shown in A, 8/17 units were significantly correlated with speed (P<0.05), and their mean absolute slope was 0.13 (range -0.06 to +0.43). The mean absolute slope for the remaining 10 units was 0.03. Only 1/6 of the relations shown in B was significant (slope, -0.03), and the mean absolute slope was close to 0 (0.02).

sensitivity was infrequently encountered (1/6 texture-related units; 1/14 modulated units). In other words, the texture-related signal was invariant for speed in area 1. While our sample was small and so definitive conclusions were not possible, Sinclair and Burton (1991) also reported that 47% of their sample of texture-related cells in areas 3b and 1 were unaffected by either speed or force. This suggests that the basic observation is robust, and that such cells may contribute to perceptual constancy of texture across different conditions.

The presence of such unique response properties at the cortical level leads to the question as to how the CNS manages to extract an unambiguous texture-related signal from peripheral signals that generally confound information concerning the physical features of the surface texture with movement kinematics and kinetics.

While both we and Sinclair and Burton have argued that the extraction of an unambiguous texture-related signal must reflect the result of cortical processing, one cannot discount the possibility that the SA afferent signals are providing a "pure" texture-related signal. As reviewed above, Goodwin and Morley (1987a) have provided convincing evidence that, at least over a limited range of test conditions, the mean discharge rate of SA afferents codes spatial period independently of speed. Could this signal then explain the finding of neurons in SI that signalled texture independently of speed? Several observations argue against this interpretation. First of all, in our own studies of SI cortical discharge during active touch we encountered neurons with slowly adapting response properties that signalled differences in texture and yet their discharge also covaried with the speed of the scanning movements. An example of one such neuron was shown in Figs. 2 and 3. Sinclair and Burton (1991) likewise reported that SA units, at least in area 3b, signal both texture and speed. Thus SA cortical neurons in SI do not appear to be particularly *insensitive* to speed, and so cannot be expected to contain a "pure" texture-related signal under all conditions. Second, the surfaces employed both in our experiments and in those of Sinclair and Burton had widely spaced elements and so fell within the range where SA afferent discharge covaries with speed (spatial periods 3 mm). In this case, it becomes even more difficult to explain the presence of texture-related responses independent of speed. Overall, the available evidence suggests that the invariance of the texture-related signals encountered in SI is the result of a central computation.

A variety of signals might be employed for this feature extraction process, and these can be divided into two classes. On the one hand, the feature extraction may occur at a late stage of processing, i.e., after the exploration. Such a mechanism might be operative, for example, in a discrimination task where the discriminanda are sequentially scanned in separate passes, with the decision being taken afterwords. In this regard, we have suggested that a very precise estimate of movement duration, and so average movement speed, is provided by what we termed movement-related neurons in SI cortex (28% of our sample, 32/113). Such neurons signalled the onset and end of movement; some also signalled information about the texture scanned (7/32). In order for such a mechanism to be operational, however, the motor strategy would have to be such that movement amplitude is held constant from one scan to the next since any variation in movement amplitude would render the code inoperable. Such a constraint is met in many studies, including both our own studies of active touch and a number of psychophysical experiments in humans (e.g. Morley et al., 1983; Sathian et al., 1989), since movement amplitude was physically constrained by mechanical stops at either end of the tactile strips. It remains to be seen, however, whether free scanning movements made within the context of a texture scaling task show invariance in movement amplitude. In regard to this hypothesis it is interesting to note that a similar code was proposed by Essick, Franzen and Whitsel (1988) to explain velocity appreciation in humans. They suggested that subjects employ an estimate of the perceived duration of the stimulus, and not a computation of the ratio between the perceived distance travelled and the duration of the stimulus.

On the other hand, it may be that the feature extraction process occurs "online", i.e. during the exploration. Indeed, such a mechanism is essential to explain the results shown in Fig. 4 where the texture-related signals of a large number of neurons show essentially no variation over a wide range of speeds. A number of signals could be employed to derive a pure, on-line, texture-related signal, including efference copy or corollary discharge from the precentral motor areas. As discussed above, however, the similarity of roughness estimates with and without movement on the part of the subject argues against the use of the efference copy to interpret the peripheral feedback signals. Moreover, it is unclear if the efference copy is relayed to SI cortex since SI cortical discharge in relation to arm movements disappears following limb deafferentation (Bioulac and Lamarre,

1979). Some caution is nevertheless needed because the underlying neural mechanisms may be different for active and passive touch.

Another mechanism to abstract unambiguous information about the spatial period has been suggested by Johnson and colleagues (Connor et al., 1990; Connor and Johnson, 1992). They proposed that a peripheral neural code based upon local spatial variations in the discharge rate of a population of SA afferents innervating the skin, separated by 1-2 mm, might explain the appreciation of surface roughness. Centrally, the hypothesis depends upon demonstrating that appropriate interactions occur between inputs separated by distances approximating two afferent fibre spacings (about 2 mm). Although experimental evidence to support the specific details of the hypothesis has not yet been provided, such a code is interesting in that it is insensitive to changes in the peripheral stimulating conditions (speed, force), and so could account for the observation of SI neurons signalling texture independently of speed and force. Such a mechanism is, however, limited to relatively coarsely textured surfaces (2 mm spatial period) (Johnson and Hsiao, 1992). More finely textured surfaces that do not elicit a structured SA discharge pattern would depend upon other neural mechanisms.

Yet another potential source of information related to the temporal features of the movement might be neurons that are activated during the course of movement and signal information about speed but not texture. In our experiments (Chapman and Agerantioti-Belanger, 1991; Ageranioti-Bélanger and Chapman, 1992), such speed-sensitive neurons were particularly encountered in area 2 (25%, 9/36 units with a digital RF in contact with the surface) and area 3b (5/9, 56%). A simple subtraction process, via an inhibitory interneuron, would produce a texture-related signal that no longer varied with speed. Information about the force exerted could be subtracted in a similar fashion.

The latter mechanisms are, however, cumbersome because several steps might be required to subtract both speed and force from the signal and produce an invariant texture signal. Moreover, the speed and force sensitivities of the neurons would have to be closely matched. This might be difficult to achieve given the variability in the slopes of the curves shown in Fig. 4A; considering only the units with a significant relationship with speed, the slopes varied from -0.06 to +0.43. A simpler mechanism might take advantage of the gain control system that modulates the transmission of cutaneous inputs to SI cortex during movement. As mentioned in the Introduction, active movement is accompanied by a gating of the transmission of cutaneous signals from the periphery up to primary somatosensory cortex. Further to this, the degree of movement-related suppression of cutaneous transmission is a function of the speed of movement: the faster the movement, the stronger the suppression (Rushton et al., 1981; Chapman et al., 1988). In addition, gating of cutaneous signals also accompanies isometric contractions, with the depth of modulation increasing as a function of the rate of change of force, dF/dt (Jiang et al., 1990). This type of gain control mechanism might well assist in producing a nonvariant texture-related signal during active touch. Moreover, it has the advantage that its action would be independent of the spatial characteristics of the scanned surface. A necessary consequence of this hypothesis is, however, that the mechanisms underlying texture appreciation must differ as a function of the mode of touch. This may help to explain why we have found that the population of SI cortical neurons involved in signalling texture differences is not the same during active and passive touch (Chapman and Ageranioti, 1991): while some neurons signal texture differences during both active and passive touch, others signal the difference only during one mode of touch.

In summary, a variety of mechanisms may be involved in generating, on-line, an unambiguous central representation of texture, independent of the conditions under which the surface was explored. At present, there is insufficient evidence to decide between these different hypotheses, and in the end it may be that no single mechanism will be found sufficient to explain the experimental observations.

4. Conclusions

Given the importance of movement for discriminative touch, it is noteworthy that tactile impressions gathered during active exploratory movements are relatively constant under a wide variety of conditions. The present chapter has considered several mechanisms that might permit the CNS to interpret the complex signals generated during touch, in order to provide for perceptual constancy of texture across varying speeds. Suggestions are made of several potential sources of information that might assist the cortical regions in interpreting the signals generated during tactile exploration. It is clear, however, that further experiments are necessary in order to determine the relative importance of, for example, peripheral (non texture-related feedback) and central (movement-related gating of sensory transmission) mechanisms to the constancy of roughness estimates under different conditions.

Acknowledgments

I would like to express my thanks to the following for the excellent technical assistance provided: Marc Bourdeau, Daniel Cyr, Giovanni Filosi, Claude Gauthier and Christian Valiquette. I also thank Dr. Trevor Drew and Mr. François Tremblay for their helpful comments on the manuscript. My research is supported by the Medical Research Council of Canada and the Université de Montréal. I also gratefully acknowledge many years of salary support from the Fonds de la recherche en santé du Québec.

References

Ageranioti-Bélanger, S.A., and Chapman, C.E. (1992) Discharge properties of neurones in the hand area of primary somatosensory cortex in monkeys in relation to the performance of an active tactile discrimination task. II. Area 2 as compared to areas 3b and 1. Experimental Brain Research, 91, 207-228.

Bioulac, B. and Lamarre, Y. (1979) Activity of postcentral cortical neurons of the monkey during conditioned movements of a deafferented limb. Brain Research, 172, 427-437.

Chapman, C.E. (1994) Active versus passive touch: factors influencing the transmission of somatosensory signals to primary somatosensory cortex. Canadian Journal of Physiology and Pharmacology, 72, 558-570.

Chapman, C.E. and Ageranioti, S.A. (1991) Comparison of the discharge of primary somatosensory cortical (SI) neurones during active and passive tactile discrimination. Proceedings of the Third IBRO World Congress of Neuroscience, Montréal, Qué. p317.

Chapman, C.E. and Ageranioti-Bélanger, S.A. (1991) Discharge properties of neurones in the hand area of primary somatosensory cortex in monkeys in relation to the performance of an active tactile discrimination task. I. Areas 3b and 1. Experimental Brain Research, 87, 319-339.

Chapman, C.E., Bushnell, M.C., Miron, D., Duncan, G.H. and Lund, J.P. (1987) Sensory perception during movement in man. Experimental Brain Research, 68, 516-524.

Chapman, C.E., Jiang, W. and Lamarre, Y. (1988) Modulation of lemniscal input during conditioned arm movements in the monkey. Experimental Brain Research, 72, 316-334.

Connor, C.E., Hsiao, S.S., Phillips, J.R. and Johnson, K.O. (1990) Tactile roughness: neural codes that account for psychophysical magnitude estimates. Journal of Neuroscience, 10: 3823-3836.

Connor, C.E. and Johnson, K.O. (1992) Neural coding of tactile texture: comparison of spatial and temporal mechanisms for roughness perception. Journal of Neuroscience, 12: 3414-3426.

Darian-Smith, I., Davidson, I. and Johnson, K.O. (1980) Peripheral neural representation of spatial dimensions of a textured surface moving across the monkey's finger pad. Journal of Physiology (London), 309, 135-146.

Darian-Smith, I., Goodwin, A., Sugitani, M. and Heywood, J. (1985) Scanning a textured surface with the fingers: events in sensorimotor cortex. In: Goodwin, A. and Darian-Smith, I. (eds) Hand Function and the Neocortex, Springer-Verlag, N.Y., Experimental Brain Research Supplement, 10, 17-43.

Darian-Smith, I. and Oke, L.E. (1980) Peripheral neural representation of the spatial frequency of a grating moving across the monkey's finger pad. Journal of Physiology (London), 309, 117-133.

Darian-Smith, I., Sugitani, M., Heywood, J., Karita, K. and Goodwin, A. (1982) Touching textured surfaces: cells in somatosensory cortex respond both to finger movement and to surface features. Science, 218, 906-909.

Edin, B.B., Essick, G.K., Trulsson, M. and Olsson, K.A. (1995) Receptor encoding of moving tactile stimuli in humans. I. Temporal pattern of discharge of individual low-threshold mechanoreceptors. Journal of Neuroscience, 15, 830-847.

Essick, G.K. and Edin, B.B. (1995) Receptor encoding of moving tactile stimuli in humans. II. The mean response of individual low-threshold mechanoreceptors to motion across the receptive field. Journal of Neuroscience, 15, 848-864.

Essick, G.K., Franzen, O. and Whitsel, B.L. (1988) Discrimination and scaling of velocity of stimulus motion across the skin. Somatosensory and Motor Research, 6, 21-40.

Feine, J.S., Chapman, C.E., Lund, J.P., Duncan, G.H. and Bushnell, M.C. (1990) The perception of painful and nonpainful stimuli during voluntary motor activity in man. Somatosensory and Motor Research, 7, 113-124.

Goodwin, A.W., John, K.T., Sathian, K. and Darian-Smith, I. (1989) Spatial and temporal factors determining afferent fiber responses to a grating moving sinusoidally over the monkey's fingerpad. Journal of Neuroscience, 9, 1280-1293.

Goodwin, A.W. and Morley, J.W. (1987a) Sinusoidal movement of a grating across the monkey's fingerpad: representation of grating and movement features in afferent fiber responses. Journal of Neuroscience, 7, 2168-2180.

Goodwin, A.W. and Morley, J.W. (1987b) Sinusoidal movement of a grating across the monkey's fingerpad: effect of contact angle and force of the grating on afferent fiber responses. Journal of Neuroscience, 7, 2192-2202.

Jiang, W., Lamarre, Y. and Chapman, C.E. (1990) Modulation of cutaneous cortical evoked potentials during isometric and isotonic contractions in the monkey. Brain Research, 536, 69-78.

Johansson, R.A. and LaMotte R.H. (1983) Tactile detection thresholds for a single asperity on an otherwise smooth surface. Somatosensory Research, 1, 21-31.

Johnson, K.O. (1983) Neural mechanisms of tactual form and texture discrimination. Federation Proceedings, 41, 2542-2547.

Johnson, K.O. and Hsiao, S.S. (1992) Neural mechanisms of tactual form and texture perception. Annual Review of Neuroscience, 15: 227-250.

Lamb, G.D. (1983a) Tactile discrimination of textured surfaces: psychophysical performance measurements in humans. Journal of Physiology (London), 338, 551-565.

Lamb, G.D. (1983b) Tactile discrimination of textured surfaces: peripheral neural coding in the monkey. Journal of Physiology (London), 338, 567-587.

LaMotte, R.H. and Srinivasan, M.A. (1987a) Tactile discrimination of shape: Responses of slowly adapting mechanoreceptive afferents to a step stroked across the monkey fingerpad. Journal of Neuroscience, 7, 1655-1671.

LaMotte, R.H. and Srinivasan, M.A. (1987b) Tactile discrimination of shape: Responses of rapidly adapting mechanoreceptive afferents to a step stroked across the monkey fingerpad. Journal of Neuroscience, 7, 1672-1681.

Lederman, S.J. (1974) Tactile roughness of grooved surfaces: The touching process and effects of macro- and microsurface structure. Perception and Psychophysics, 16, 385-395.

Lederman, S.J. (1983) Tactual roughness perception: spatial and temporal determinants. Canadian Journal of Psychology, 37, 498-511.

Lederman, S.J. and Taylor, M.M. (1972) Fingertip force, surface geometry, and the perception of roughness by active touch. Perception and Psychophysics, 12, 401-408.

Milne, R.J., Aniss, A.M., Kay, N.E. and Gandevia, S.C. (1988) Reduction in perceived intensity of cutaneous stimuli during movement: a quantitative study. Experimental Brain Research, 70, 569-576.

Morley, J.W. and Goodwin, A.W. (1987) Sinusoidal movement of a grating across the monkey's fingerpad: temporal patterns of afferent fiber responses. Journal of Neuroscience, 7, 2181-2191.

Morley, J.W., Goodwin, A.W. and Darian-Smith, I. (1983) Tactile discrimination of gratings. Experimental Brain Research, 49, 291-299.

Mountcastle, V.B. (1984) Central mechanisms in mechanoreceptive sensibility. In Darian-Smith, I. (ed) Handbook of Physiology, Section 1: The Nervous System, Vol. III. Sensory Processes, Part 2, American Physiological Society, Bethesda M.D., pp789-878.

Phillips, C.G. (1986) Movements of the Hand. Liverpool University Press, Liverpool.

Phillips, J.R., Johansson, R.S. and Johnson, K.O. (1990) Representation of Braille characters in human nerve fibers. Experimental Brain Research, 81. 589-592.

Phillips, J.R., Johansson, R.S. and Johnson, K.O. (1992) Responses of human mechanoreceptive afferents to embossed dot arrays scanned across fingerpad skin. Journal of Neuroscience, 12: 827-839.

Paul, R.L., Merzenich, M.M. and Goodman, H. (1972) Representation of slowly and rapidly adapting cutaneous mechanoreceptors of the hand in Brodmann's areas 3 and 1 of *Macaca mulatta*. Brain Research, 36, 229-249.

Post, L.J., Zompa, I.C. and Chapman, C.E. (1994) Perception of vibrotactile stimuli during motor activity in human subjects. Experimental Brain Research, 100, 107-120.

Rushton, D.N., Rothwell, J.C. and Craggs, M.D. (1981) Gating of somatosensory evoked potentials during different kinds of movements in man. Brain, 104, 465-491.

Sathian, K., Goodwin, A.W., John, K.T. and Darian-Smith, I. (1989) Perceived roughness of a grating: correlation with responses of mechanoreceptive afferents innervating the monkey's fingerpad. Journal of Neuroscience, 9, 1273-1279.

Sinclair, R.J. and Burton, H. (1991) Neuronal activity in the primary somatosensory cortex in monkeys (Macaca mulatta) during active touch of textured surface gratings: responses to groove width, applied force and velocity of motion. Journal of Neurophysiology, 66, 153-169.

Sur, M., Wall, J.T. and Kaas, J.H. (1984) Modular distribution of neurons with slowly adapting and rapidly adapting responses in area 3b of somatosensory cortex in monkeys. Journal of Neurophysiology, 51, 724-744.

Neural Aspects of Tactile Sensation
J.W. Morley (Editor)
© 1998 Elsevier Science B.V. All rights reserved

Short-term Plasticity in Adult Somatosensory Cortex

M.B. Calford, J.C. Clarey, R. Tweedale

The primary sensory fields in cortex encode topographical representations of their peripheral epithelium (visual: retina; somatosensory: body surface; auditory: cochlear place or frequency) in which responsive neurons at a given locus or throughout a column have similar receptive fields. These representations provide useful model systems for the study of functional neuronal plasticity since they present an inbuilt scale upon which any changes can be measured. In the developing mammal, these representations have been shown to be capable of a remarkable degree of plasticity by virtue of the reorganization of the basic topography which follows from a changed peripheral input (Kelahan et al., 1981; Spinelli and Jensen, 1979; Waite, 1984; Wall and Cusick, 1986) or from manipulation of central pathways (Roe et al., 1990). Plasticity has also been demonstrated in the sensory cortex of adults where the removal of a restricted region of primary afferents can lead to the affected area of cortex gaining sensitivity to other regions (visual: Gilbert and Wiesel, 1992; Kaas et al., 1990; Schmid et al., 1995; 1996; auditory: Rajan et al., 1993; Robertson and Irvine, 1989; somatosensory: see below). However, such changes in the organization of adult cortex must be considered fundamentally different in nature from those in neonates where there is a critical period for plasticity (Hubel and Wiesel, 1970; Wall, 1988). In the developing nervous system there are bases for mutability of the pattern of projections through disruption of the processes of initial connections, trimming of exuberant projections (e.g. Wall and Cusick, 1986) or alterations to the normal patterns of segregation of inputs (e.g. LeVay et al., 1981). Such mechanisms of plasticity are not available in the adult brain where the connections are considered to be stable - although it does appear that some deprivation conditions may extend the developmental critical period (Daw et al., 1992; Moore, 1993; Smith et al., 1978). Nevertheless, significant alterations in the coding properties of neurons have been reported which have led some to challenge the concept of a stable adult central nervous system. It is, however, important to point out that the use of the term plasticity has itself undergone a transformation over the period of these studies. Whereas the term was once strictly used to describe cases of anatomical change, it is now widely used where there is a functional change with no apparent anatomical substrate. This functional plasticity encompasses cases where the outcome is thought to be an inherent property of the neural circuit (e.g. unmasking of previously ineffective inputs: Calford and Tweedale, 1991a; Turnbull and Rasmusson, 1990) and cases in which some change in synaptic efficacy is implicated (e.g. Glazewski et al., 1996; Kano et al., 1991).

The majority of studies demonstrating a functional plasticity in the adult brain have been of the somatosensory system. Studies of such plasticity in the dorsal horn and thalamus predate those of somatosensory cortex. Nevertheless, the present chapter will concentrate on the effects as seen in cortex and will address the other levels of the pathway only when looking at possible explanatory mechanisms - an extensive review of the lower pathway effects is recommended (Snow and Wilson, 1991).

1. Long-term representational plasticity in somatosenory cortex

Around 1980 a number of groups demonstrated that the permanent cutting of tactile afferents from a restricted body region can lead to the affected area of cortex gaining sensitivity to other body parts (Franck, 1980; Kalaska and Pomeranz, 1979; Kelahan and Doetsch, 1984; Merzenich et al., 1983a; Merzenich et al., 1983b; Rasmusson, 1982; Wall and Cusick, 1984), thus demonstrating a functional plasticity in adult cortex. Without doubt, the seminal papers in this field are those from Merzenich and colleagues who demonstrated dramatic changes in the representation of the hand in owl- and squirrel-monkey area 3b following nerve section (Merzenich et al., 1983a; Merzenich et al., 1983b) and digit amputation (Merzenich et al., 1984). These papers generated much excitement and controversy for they raised the hope that new mechanisms of neural plasticity had been unveiled or that accepted dogma of a developmental critical period for major plasticity need not always apply. The question of the mechanisms of plasticity in the adult cortex will be the major item addressed in this chapter.

However it is worth stating at the outset that alterations to the critical period concept have not been necessitated: as in the visual cortex, it is also the case in somatosensory cortex that developmental plasticity (Fox, 1992; Juliano et al., 1994; Kelahan et al., 1981; Waite, 1984; Wall and Cusick, 1986) is far more extensive than the adult counterpart. Aside from these considerations, there were two major reasons for the subsequent substantial effort that went into generating the large number of studies which followed from these first reports. Firstly, there was a need to establish a whole animal physiological model for demonstrating that principles of neuronal plasticity derived from *in vitro* studies were applicable to the intact nervous system (e.g. Kano et al., 1991; Pearson et al., 1987), and even to extend or to modify these principles (Merzenich et al., 1990; Merzenich et al., 1993). Secondly, there was the recognition that this plasticity may be relevant to the study of events that follow from many forms of neurotrauma including ischaemic stroke (e.g. Doetsch et al., 1990; Elbert et al., 1994; Jenkins and Merzenich, 1987; Johansson and Grabowski, 1994). For both of these pursuits the primary somatosensory cortex provides an excellent model not only because the detailed somatotopy supplies an inherent metric against which change can be determined, but because the peripheral components of the system are readily available for manipulation.

The fundamental datum in studies of plasticity in the adult somatosensory cortex - that of a capacity for reorganization of the somatotopic map following disruption of inputs - is well illustrated by an example from the work of Merzenich et al., (1984; Figure. 1). In that study, somatotopic maps were derived by determining the receptive fields for multineuron activation in several hundred electrode penetrations in and around the hand representation of area 3b in owl monkeys. One or 2 digits were then amputated and the same area of cortex was reexamined after 2-9 months recovery. Where single digits were removed it was found that their representation was entirely taken up by responsiveness to the adjacent digits and palmal skin. Where 2 digits were removed there was also extensive reorganization. However in these cases, there was invariably an area within the former representation that was unresponsive to mechanosensory stimulation (the issue of limits on the extent of adult cortical plasticity is examined

Owl monkey 80-39

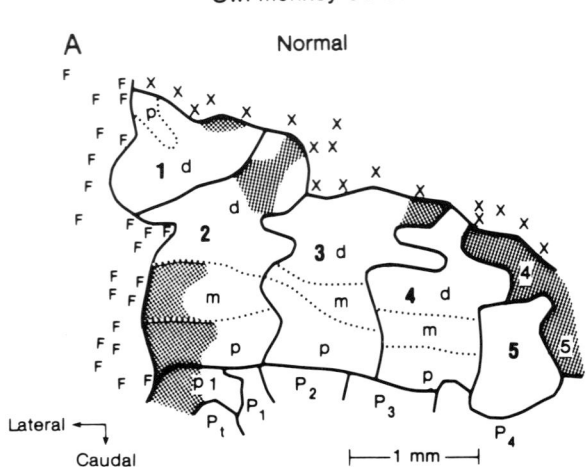

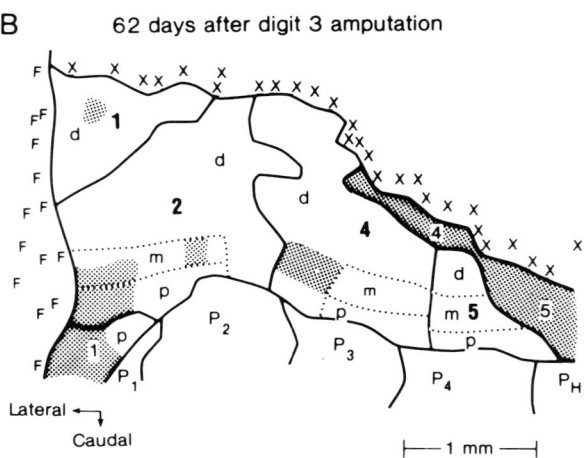

Figure 1. Example of the basic capacity for plasticity in somatosensory cortex. Somatotopic "maps" of the hand representation in primary somatosensory cortex (area 3b) of an owl monkey before (A) and 62 days after (B) amputation of digit 3. The area in which neurons previously responded to digit 3 became responsive to digits 2 and 4. Shaded areas are hairy skin- and clear areas glabrous skin-representations. X marks recording sites where neurons were driven by deep but not cutaneous stimulation and F indicates the face representation (from Merzenich et al., 1984).

below). These were dramatic demonstrations: the representations of the remaining digits expanded to around 1.8 times the original area and the linear extent of cortex with new responsiveness was over 1 mm.

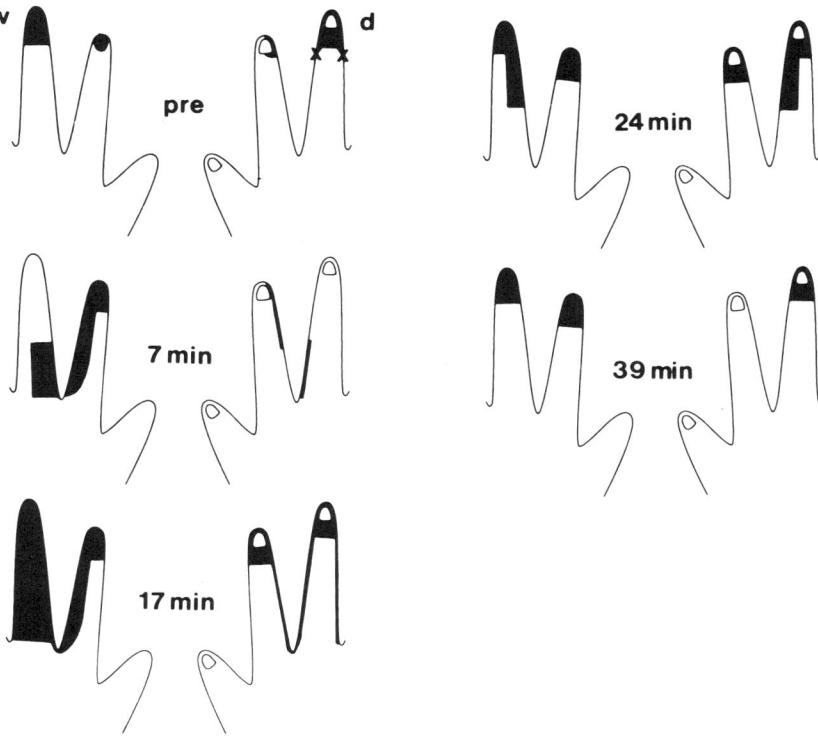

Figure 2. Sequence of changes in the receptive field of a multiunit cluster recorded at a site in area 3b of a macaque monkey. Dorsal and ventral views of part of the hand contralateral to the recording site are shown. The initial receptive field was confined to the tips of digits 2 and 3. After subcutaneous local anaesthetic injections at the sites indicated (**X**; 12 µl of 2% lignocaine at each site), the receptive field was seen to expand, such that its two aspects joined. When aesthesia returned there was a gradual shrinking of the field which by 39 minutes was close to the original (Modified from Calford and Tweedale, 1991c).

2. Time-course of representational plasticity

It was clear that any attempt to describe a mechanism to account for cortical representational plasticity would depend upon an understanding of the development, or time-course, of the new sensitivity. A number of studies were undertaken (Byrne and Calford, 1991; Calford and Tweedale, 1988; Calford and Tweedale, 1991a; Calford and Tweedale, 1991c; Cusick et al., 1990; Merzenich et al., 1983b; Rasmusson and Turnbull, 1983; Turnbull and Rasmusson, 1990; Waite, 1984; Wall and Cusick, 1984), but the results were not entirely consistent. One of the approaches to this question is to examine the immediate and short-term changes to the response profile of a single neuron (or to a group of neurons recorded at one site) consequent upon a peripheral manipulation. An example is shown in figure 2: the receptive field of a multineuron cluster in area 3b of a macaque monkey was determined before and after the application of a local anaesthetic which temporarily blocked mechanosensory afferents in a segment of the receptive field. During the period of local-anaesthesia, and for some time after the return of sensitivity, the receptive field was observed to be greatly expanded. For small denervations, this is a robust result and similar findings were reported for recordings from the digit or toe representations in flying foxes (Calford and Tweedale, 1991a) and rats (Byrne and Calford, 1991). Short-term representational plasticity has also been reported after median nerve (Merzenich et al., 1983b) or median nerve and ulnar nerve (Kolarik et al., 1994) section in New World monkeys. In these cases substantial areas remain unresponsive but there is also a dramatic expansion of the remaining radial (ulnar) nerve representation. Neural receptive fields in the expanded representations were described as having normal-variability in the Kolarik et al. study and enlarged in the Merzenich et al. study, but this property was not systematically examined in either case, where the emphasis was placed on quantifying the spatial extent of body part representations.

Implanted electrodes were used by Calford and Tweedale (1988) to follow changes in multineuron receptive field extent following amputation of the thumb at the metacarpophalangeal joint in flying foxes. It was found that the initially unmasked large receptive fields on the D1-metacarpal, wrist and adjacent wing membranes progressively shrank over the following 10 days or so (Figure. 3). Thus, in some studies, it is clear that the immediate effects of a peripheral denervation encompass the long-term cortical representational plasticity. Just as clearly this is not the case in other studies where long-term representational plasticity is far more extensive than its short-term counterpart. For example, following damage of the sciatic nerve in adult rats, the saphenous nerve representation in cortex expands progressively into the deafferented area over a period of many weeks (Cusick et al., 1990).

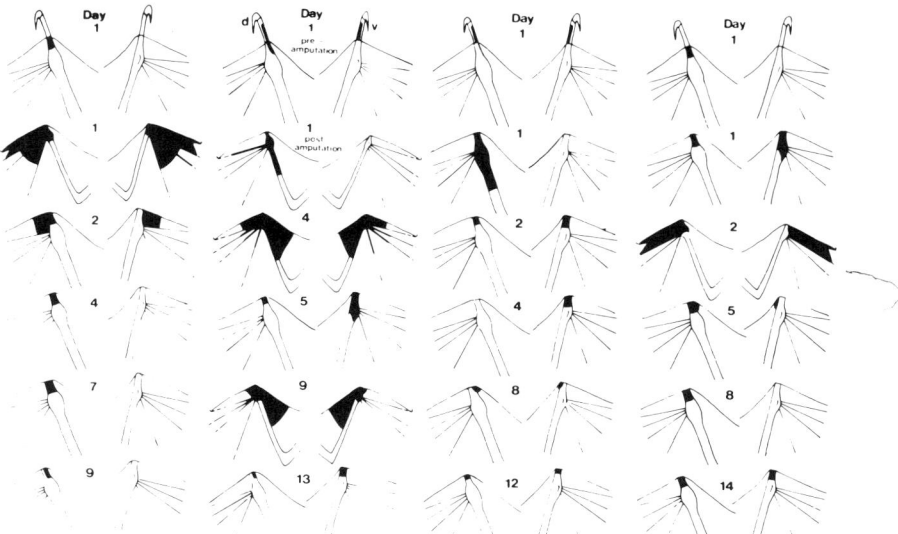

Figure 3. Receptive fields obtained from multiunit recording with implanted electrodes in four flying foxes. Each panel shows dorsal and ventral views of the wing or thumb contralateral to the recording sites. The thumb was amputated at the metacarpophalangeal joint and the receptive fields were examined immediately (Day 1) and up to two weeks later. The final receptive fields recorded were always smaller than the expanded, or new, fields unmasked shortly after the amputation - but there was fluctuation in extent in the first few days in some cases. Methods are described in Calford and Tweedale (1988) where additional examples are presented.

In some studies very limited representational plasticity has been reported following peripheral denervation (e.g Li et al., 1994; Waite, 1984), and Jain et al. (1995) reported neither short- nor long-term recovery of responsiveness in adult rat primary somatosensory cortex (S1) denervated by section of the contralateral dorsal column at midthoracic levels. Clearly the proportion of the representation, and the absolute size, of the affected area is far larger in this case than with thumb denervation in a flying fox. However, both for short- and long-term effects it appears that the size of a cortical representation is not alone sufficient to predict where there will be significant representational plasticity. For example, the cortical area affected by median nerve section is slightly larger than the representation of two digits in the squirrel monkey - yet the former undergoes considerable short-term and near complete long-term representational plasticity (Merzenich et al., 1983a,b) while the latter retains a central unresponsive region (Merzenich et al., 1984). Garraghty and Kaas and colleagues have attributed such differences to the precise nature of peripheral innervation and central representational contiguities rather than to the absolute size of the representation or to the method of denervation (Garraghty et al., 1994; Garraghty and Kaas, 1991b).

Short term effects have also been examined with digit amputation and digital nerve section in raccoons. This model is one of the most extensively studied in terms of determining the time-course of the effects of denervations. Early studies (Rasmusson, 1982; Kelahan and Doetsch, 1984) reported little short-term plasticity. Rasmusson (1982) reports that 92% of electrode penetrations failed to find units with excitatory responses to somatosensory stimuli; however around 50% of units sampled during recording sessions following digit amputation or ventral digit denervation had unusual inhibitory receptive fields (Rasmusson and Turnbull, 1983; Turnbull and Rasmusson, 1991). Kelahan and Doetsch (1984) report that within one day after digit 3 amputation 49% of penetrations had responsive units, but the stimuli required to elicit responses were unusual. Despite the considerations of the previous paragraph, the pertinent difference between these experiments and those that show short-term effects may, indeed, be the scale of the denervation in cortical terms. Digit 5, which was amputated in most of Rasmusson's studies, is represented by a surface area of approximately 20 mm^2 compared to approximately 2 mm^2 for digit 1 in the little red flying fox (Calford and Tweedale, 1991a). Later studies in which the digital nerves were separately cut did show some short-term representational plasticity (Turnbull and Rasmusson, 1990).

If one ventral nerve and/or both dorsal nerves were cut then an increased proportion of neurons with large receptive fields and a decrease in the proportion of low-threshold responses occurred. If all four digital nerves were transected then the unusual inhibitory responses (with a stimulus offset rebound discharge) were unmasked. Two factors may contribute to the difference: individual digital nerves have a smaller representation but there is also considerable scope for overlap in the representation of digital nerve territories both within a digit representation in cortex and in terms of their peripheral extent (i.e. overlapping coverage of the digit). Thus the smaller denervations may have reduced an inhibitory effect from one digital nerve allowing other inputs to that cortical area (ultimately stemming from the other digital nerves) to be unmasked; larger denervations have a similar effect but leave no inputs capable of generating an excitatory response. This explanation requires that weaker inputs from skin areas adjacent to the affected digit are capable of driving an inhibitory, but not an excitatory, response immediately after denervation. This geometric relationship, where the extensive extent of inhibition is greater than the extent of excitatory input, is consistent with a lateral inhibitory role in the formation of normal receptive fields, as discussed below.

3. Human studies

No study has looked with imaging techniques at any correlate of the peripheral-denervation-induced immediate unmasking in somatosensory cortex in humans, but there are both psychophysical and imaging studies of medium- and long-term manipulations to the human somatosensory system that have parallels to the animal work. However, before describing these studies it is worth considering some very common and anecdotal observations that most of us have made when a small body area is locally anaesthetized. For example, with dental anaesthesia the perceived body image is not that of the affected areas missing from the image but of an enlarged representation. This phenomenon has been studied by Gandevia and colleagues (Glasby and Gandevia, 1995; and personal communication) and shown to be quantitatively demonstrable with psychophysical magnitude-matching techniques. The link between enlarged neural receptive fields (and representation of some body parts) and the perception of larger body parts is not clear, and the implications for mechanisms of consciousness and perception raised by this comparison go well beyond the subject matter of this chapter. It may be that the perception of a larger body area is a direct consequence of increased activation of the affected area of the representation (see Figure. 9, for an example of increased responsiveness with unmasking) due to the disinhibitory effects of the partial loss of afferent input. Furthermore, it was shown that other body parts can be affected: thus local anaesthesia of the lips produced a perception of increased size of the thumb (the reverse relationship was also shown). In view of the fact that the thumb and lips are adjacent in the topographic representation in primate S1, this finding suggests that short-term unmasking of enlarged receptive fields, and increased responsiveness, may occur in the human cortical representation in areas adjacent to the representation of the anaesthetized body part.

Psychophysical studies of subjects with recent (weeks) or long-term amputations of an arm have shown a remarkable result in that the amputated area can be perceived as a secondary experience when parts of the face or shoulder are stimulated (Halligan et al., 1994; Halligan et al., 1993; Ramachandran et al., 1992). Halligan et al. (1994) point out that such phenomena have been reported, but not widely publicized, for over a century. Most subjects seem unaware of the secondary sensations until tested, but others could report the appearance of and time-course of changes in the sensations. One of the most interesting features of the mapping of the phantom sensations onto the face is that it involves both ipsilateral and contralateral sites.

Two groups have used magnetic-source imaging to study plasticity in human somatosensory cortex. In one study, by Elbert et al., (1994), it was shown that the equivalent dipole source for stimulation of the chin was shifted into previous hand representation (judged by comparison with the contralateral cortex) in subjects with amputations of an arm. This shift probably reflects a larger area of significant neural activation in S1 cortex for stimulation of the chin, with an expansion into the area which would have represented the area that was amputated. It was subsequently shown that the extent of this shift correlated with the degree of phantom limb pain experienced, but not with non-painful phantom experience or with remapping of a phantom hand onto the face (Flor et al., 1995). Mogilner et al., (1993) used this same technique to examine the changes in representation of the digits of two patients undergoing surgery to correct congenital syndactyly. In one case it was shown that the equivalent dipole site separation for stimulation of D3 and D5 increased by 3 mm one week after separation of D5 from a D3-D5 syndactyly. As with some of the early changes reported in the psychophysical studies (Halligan et al., 1994; Ramachandran et al., 1992), this change was too rapid to depend upon growth or axonal sprouting in any central pathway.

4. Nature of somatosensory topographic pathways: role of inhibition

The pioneering work on the physiology of the topographic representations in somatosensory cortex of Adrian (1941) and Woolsey (1942) pointed to the precision of the somatotopy. Mountcastle's single unit physiological investigations (e.g. Mountcastle, 1957), which confirmed the columnar organization hypothesis, gave further emphasis to the concept of a precise representation. More recent physiological studies, such as those on the modular nature of the representations (Dykes and Gabor, 1981; Dykes et al., 1980; Favorov and Diamond, 1990; Favorov et al., 1987; Krubitzer and Calford, 1992) have consistently reaffirmed this dogma. In contrast there has been a dramatic change in the understanding of the anatomical basis of the somatotopic representations. Whereas it was once thought that precision in the thalamocortical projection was sufficient to account for the somatotopy, later studies clearly indicate that anatomical precision is sloppy with respect to physiological precision. Merzenich and Kaas (1980), and Dykes (1983), were among the first to point out the necessity of such a relationship to allow further specialization in encoding and feature extraction along a topographically organized pathway.

There are two aspects to the divergence in the thalamocortical projection that allow interaction across elements: spread of individual afferent terminations and accuracy of projections. The extent of arborization of the terminals of individual thalamocortical axons extend up to 2 mm, often with clear patches of high density terminals

(Garraghty et al., 1989; Landry and Deschênes, 1981). This projection is thus coarse with respect to a physiological somatotopy in which receptive fields change noticeably with electrode movements of the order of 100 μm (see Favorov and Diamond, 1990; Favorov et al., 1987, for examples of rapid changes). In addition, there is the divergence contributed by the error in topography of the projection. This has not been described in fine-detail for the thalamocortical projection to S1, but an extensive divergence is clear where multiple retrograde tracers are used (Darian-Smith and Darian-Smith, 1993;

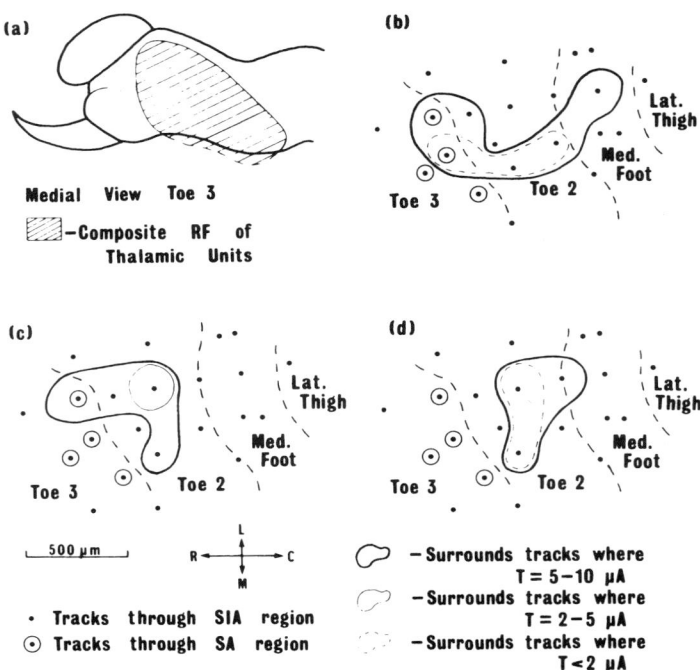

Figure 4. Physiological demonstration of the extent of topographic divergence in the somatosensory thalamocortical projection. **b**, **c**, and **d**, show antidromic stimulation thresholds at sites in cat S1 cortex against representational boundaries of the digits for 3 thalamic units. The composite receptive field (RF) of the units is shown in **a** (from Snow et al., 1988).

Darian-Smith et al., 1990), in that closely spaced (e.g. 1.5 mm) injections reveal overlapping sources (but not individual neurons) in thalamus. One study has combined both of these aspects of divergence in the thalamocortical projection by using electrical stimulation to determine the extent of spread from a given thalamic location to cortex (Snow et al., 1988). It was shown that antidromic activation of thalamic neurons could be effected from a far larger area of cortex than that sharing the same neural receptive fields (Figure. 4). The term "somatotopically inappropriate", originally coined to describe some spinal cord terminations from hair follicle afferents (Meyers and Snow, 1984), was used to describe the underlying projections which allow such activation.

In addition to divergence in the projection from a thalamic locus (or equivalently, convergence in the projection to a cortical locus) there are also somatotopically-inappropriate projections from other parts of a cortical field (intrinsic) and from other somatosensory fields (e.g. Krubitzer et al., 1993). Intrinsic connections have received more attention in studies of the visual cortex (Gilbert and Wiesel, 1979; Gilbert and Wiesel, 1989) where it has been shown that connections are most dense between modules which are linked functionally (ocular dominance, orientation coding), but not necessarily by retinotopic representation. Similar constraints on the intrinsic connectivity within somatosensory cortex have yet to be established.

A partial resolution of the disparity between anatomical and physiological precision is gained by considering inhibitory components of the receptive fields of cortical neurons. In most cases the inhibitory domains are larger than the excitatory domains (Mountcastle and Powell, 1959). Consideration of this geometry led to the concept of "afferent" or "lateral" inhibition. For most neurons lateral inhibition is not demonstrable simply by stimulating in the area surrounding its receptive field, as in anaesthetized preparations low rates of spontaneous activity generally preclude such attempts.

While inhibitory receptive fields are reported in awake animal recording, lateral inhibition is difficult to delineate because of the stochastic nature of ongoing activity. However, lateral inhibition can be shown to be a property of most neurons in S1 by using a masking paradigm. Simultaneous masking, in which a test stimulus is positioned within the receptive field while a probe is presented in surrounding regions has been used to demonstrate competitive and lateral inhibition. A preferable method, for portraying the geometry of inhibitory components of a neuron's receptive field, employs forward masking in which there is a short delay between the presentation of the masker and test stimuli (Jänig et al., 1977). Inhibition is detected by suppression of

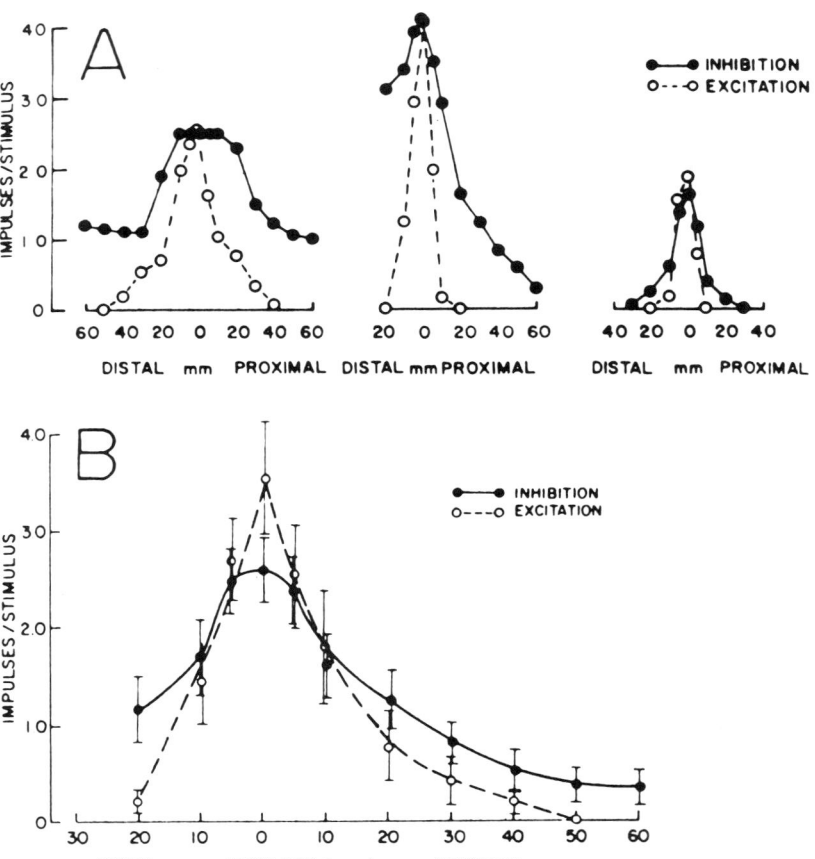

Figure 5. Examples of neural response profiles demonstrating lateral inhibition. Response measures were made using two air-puff stimuli in a delayed forward masking paradigm. Neurons were recording in cat vpl (thalamus), hairy skin forepaw representation (from Jänig et al., 1979).

the response to the test stimulus. Using this method, a variety of relationships between inhibitory and excitatory spatial domains were described in somatosensory cortex and thalamus of the cat (Jänig et al., 1979; Laskin and Spencer, 1979; Figure. 5). For many neurons the inhibitory domain extended beyond the excitatory domain and therefore demonstrated lateral inhibition. However forward masking is usually greatest for a masker stimulus presented within the excitatory domain; often maximal inhibition and excitation occur at the same point. This situation is precisely what is predicted from inhibition generated by a local interneuron network, which is responsible for sharpening physiological resolution where the anatomical precision is relatively coarse. Such alignment of the maxima of the excitatory and inhibitory domains is also found in auditory cortex (Calford and Semple, 1995) where the spatial variable is replaced by stimulus frequency. Similarly the centre-surround receptive fields of cells in the retina and the lateral geniculate nucleus (LGn) are composed of aligned and opposing Gaussian distributions (Rodieck, 1965). However, simple cells of primary visual cortex have displaced ON and OFF domains, apparently resulting from a combination of LGN inputs rather than effects of local interneuron inhibition (Hawken and Parker, 1987; Hubel and Wiesel, 1962).

Direct demonstration of the cortical locus of a functional inhibition and the inherent viability of a wider afferent input is provided by studies showing pharmacological interference of inhibition. The principal inhibitory transmitter in somatosensory cortex is undoubtedly GABA (Houser et al., 1984). However, a role for glycine-like amino acid transmitters (glycine, taurine, beta-alanine) is suggested by pharmacological effects on the majority of cutaneous-responsive neurons (Tremblay et al., 1988). A number of studies have shown that antagonism of GABAa-receptors by direct cortical iontophoretic application of GABA-site antagonists (Alloway and Burton, 1991; Alloway et al., 1989; Dykes et al., 1984; Hicks and Dykes, 1983; Lamour et al., 1988), Cl-channel antagonists (Batuev et al., 1982) or benzodiazapine-site modulators ("inverse-agonists": Oka and Hicks, 1990; Oka et al., 1986), lead to a rapid expansion of the receptive field of most S1-neurons (Figure. 6). In addition, GABAa-receptor antagonism severely reduces directional tuning of cat S1 neurons to vibrissal deflection (Batuev et al., 1989), indicating that this is a property produced by cortical inhibition. Interestingly, in both cat (Alloway et al., 1989; Dykes et al., 1984) and macaque (Alloway and Burton, 1991) the effects of functional antagonism of GABAa-receptor mediated inhibition were far more extensive for neurons with a rapidly adapting (RA; or transient) response profile than for neurons with a slowly adapting (SA; or sustained) response profile. Furthermore, Alloway and Burton (1991) report that the GABA-site competitive antagonist bicuculline had an effect of greatly expanding the receptive field for the onset transient component of SA neurons but had a less extensive effect on the spatial extent of the sustained component. A circumstantial argument can be made relating the rapid time-course of GABAa-mediated inhibition and RA responses and the inappropriateness of this form of inhibition to SA-type responses. Studies with agonists and antagonists of longer time-course GABAb-receptor mediated inhibition have, however, not related this form of inhibition to control of sustained responses (Kaneko and Hicks, 1988; Kaneko and Hicks, 1990). In some cases the GABAb-receptor agonist, baclofen, enhanced responsiveness and the antagonists, phaclofen and gamma-amino-n-valeric acid, reduced responsiveness; suggesting a presynaptic role for GABAb-receptor on GABA containing terminals (Kaneko and Hicks, 1990). Overall, however, the work with intracortical iontophoretic disruption of inhibition confirms that the spatial extent of neural receptive fields is finally a product of local inhibitory tuning from the wide array of potential inputs that are contributed by the limited precision of the anatomical inputs. Disruption of inhibition at the thalamic level also produces neural receptive field expansion (cat: Hicks et al., 1986; but not in rat: Roberts et al., 1992) indicating that the process of divergence in topographic projection and physiological sharpening of resolution probably occurs throughout the pathway.

While afferent divergence and point-to-point convergence are obvious in many topographic projections, and appear to be integral to the feature extraction processing mechanisms, some projections have more precise connections. Two systems that have been studied in detail are the barrel field of rodent somatosensory cortex and primary visual cortex. Both of these representations are distinguished by precise modular organizations that have cellular-architectural, connectional, metabolic (cytochrome oxidase distribution) and physiological manifestations (Purves et al., 1992). However, even in these areas extensive interaction between modules is demonstrable. This takes form in the so called "horizontal" connections, which preferentially connect functionally-similar modules (Gilbert and Wiesel, 1989).

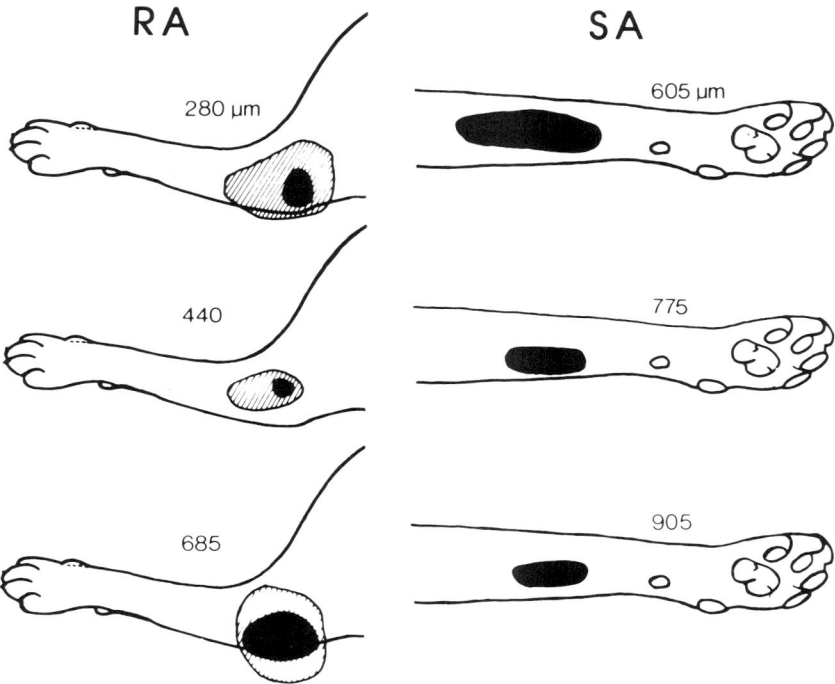

Figure 6. Direct demonsration of underlying larger RFs in cat S1 with disruption of GABAergic interneuron-mediated inhibition. Receptive fields of neurons at the indicated depths are shown in black. For rapidly adapting (RA) neurons, iontophoretic application of the GABAa-receptor antagonist, bicuculline, produced an expansion of the RF (hatched area). Such application did not affect the RFs of three slowly adapting (SA) neurons (from Hicks and Dykes,1983).

From a functional point of view these instrinsic connections may play a similar role to that of a large afferent axonal arborization, and similar local inhibitory mechanisms are thought to be necessary for local processing. Indeed, disruption of GABA-mediated inhibition in visual cortex has the dual effect of producing loss of some cortically

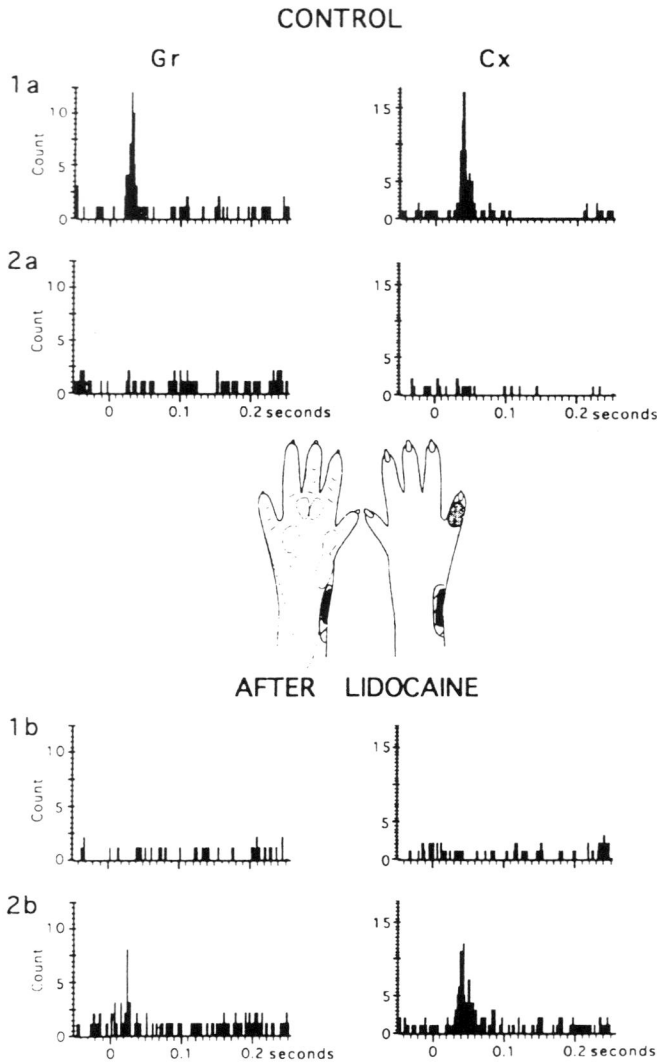

Figure 7. Parallel changes in the responses of single neurons in nucleus gracilis and cortex of adult rat after subcutaneous local anaesthesia (black shaded area) of the original receptive field area (cross hatched). Responses in 1a and 1b were obtained to 20 repetitions of an air-jet stimulus aimed at the original receptive field and responses in 2a and 2b were to stimulation of the stippled area for which responses were unmasked during the local anaesthesia (from Panestos et al., 1995).

derived properties (e.g. Morin and Molotchnikoff, 1992) and loss of resolution (Volchan and Gilbert, 1992).

5. Subcortical contributions to representational plasticity in cortex

Whereas the majority of recent studies have examined the representation at the cortical level, it is possible that some of the changes in the representation at that level could be the result of changes induced lower in the pathway. A capacity for plasticity in the central somatosensory representation was reported from studies of the representation in the **dorsal horn** of the spinal cord around 25 years ago (Basbaum and Wall, 1976; Wall and Egger, 1971). Demonstrations of changes at this level induced by various disruptions of the normal peripheral input were for many years controversial and contradictory. However, following from a review of this literature by Snow and Wilson (1991) it appears that many of the inconsistencies were due to differences in methodology, and a poor early understanding of the somatotopy of the dorsal horn representation. There is sufficient evidence to conclude that following a small peripheral denervation (such as a digit amputation) there is a long-term reorganization of the mechanoreceptor representation in laminae III to V (which includes the neurons that give rise to the spinocervical tract). This reorganization is mostly rostrocaudal (Wilson, 1987), of up to 3 mm, and may derive from either physiological or local anatomical changes to widely arborizing primary afferents that, throughout some of their extent, are normally ineffective in driving second order neurons (Meyers and Snow, 1984). Larger denervations, such as cutting major nerve branches, do not produce reorganization and leave permanently deafferented regions of the dorsal horn (Brown et al., 1984; Pubols, 1984).

Where reorganization occurs in the dorsal horn in response to peripheral denervation it develops slowly and there is no immediate or medium-term (e.g. <20 days; Snow and Wilson, 1991) plasticity. Nevertheless, the size of some dorsal horn neuron receptive fields and/or their response level can be dramatically increased in the short term by manipulating descending input (Wall, 1967) or peripheral C–fibre input activity (Cook et al., 1987). From the perspective of the present review, plasticity at the level of the dorsal horn must be considered in parallel to, rather than contributing to, plasticity as seen at the cortical level. The spinocervical projecting neurons do not represent a major contribution to inputs of thalamic nuclei that project to primary somatosensory cortex. These receive a vast majority of their inputs from the medial lemniscus, fed from primary afferent terminations in the dorsal column nuclei. However a role for polymodal neurons of upper dorsal horn laminae that project via the spinothalamic tract is implicated by experiments that show peripheral C-fibre denervation (with capsaicin) effects on neural receptive fields in S1 (see Synaptic Plasticity section, below).

Dorsal coumun nuclei: a potential capacity for receptive field plasticity in the dorsal column nuclei was shown by expansion of the receptive fields of some neurons following local application of 4-aminopyridine (4-AP: increases the number of vesicles released at a synapse; Saade et al., 1982). Convincing evidence of a functional plasticity in the form of shifts to neural receptive fields at the level of the dorsal column nuclei has recently been presented (Panestos et al., 1995; Pettit and Schwark, 1993). Recording in the adult rat nucleus gracilis, Panestos et al. (1995) showed that local anaesthesia of the area of the original receptive field of a neuron leads to the immediate expression of a new receptive field (Figure.7). Similar results were reported for adult cat cuneate nucleus neurons (Pettit and Schwark, 1993), except that the responses to the ectopic receptive fields unmasked in this study did not necessarily cease when anaesthesia returned to the affected area and responses to stimulation of the original receptive field returned - as was the case in the rat gracilis (Panestos et al., 1995) and cortex of rats, flying foxes and monkeys (Byrne and Calford, 1991; Calford and Tweedale, 1990b; Calford and Tweedale, 1991a; Panestos et al., 1995). These studies confirm an earlier report of an immediate shift in receptive fields of neurons in the cat nucleus gracilis following cold block of the dorsal columns at L4 (Dostrovsky et al., 1976).

It is unclear what proportion of dorsal column nuclei neurons show short-term unmasking of new receptive fields and whether further changes occur in the representation with chronic partial denervation. Population analyses that have been attempted at various recovery times after selective dorsal root rhizotomies are compromised by the normal presence of neurons with out-of-sequence receptive fields (Dostrovsky et al., 1976; Millar et al., 1976). However, with a large denervation of cutting all dorsal roots caudal of L4, except S1 or L7, the proportion of unresponsive neurons in adult cat nucleus gracilis decreased from around 30% to around 5% after 8 months recovery (Millar et al., 1976).

Thalamus: the earliest study of short-term changes to the receptive fields of central pathway somatosensory neurons was that of Nakahama et al. (1966), recording in the cat ventrobasal thalamus (equivalent to ventroposterior lateral nucleus, vpl, in other species). This study showed that the majority of thalamic neurons have receptive fields consisting of excitatory and inhibitory (usually surround) regions. Around 1/4 of the neurons sampled showed an expansion of the receptive field or appearance of a new receptive field when the centre of the original field was temporarily anaesthetized with procaine. More recent studies have resulted in similar proportions (Nicolelis et al., 1993), far greater proportions (Rasmusson et al., 1993; Shin et al., 1995) or very small proportions (English and Calford, 1992) of thalamic neurons showing such acute changes (Table 1). However it is very difficult to compare across studies, since different species, recording techniques and anaesthetics have been used. In the study by English and Calford (1992) concurrent recordings were made in S1 where all neurons studied were seen to show receptive field plasticity. This and the expectation that the vast majority of cortical cells would have been so affected in the other studies in which a small proportion of thalamic cells showed acute plasticity indicates that, at the least, there is some unique cortical contribution to these changes. A possible problem in interpreting these studies is assessing the contribution of corticothalamic inputs. Indeed, concurrent latency shifts in the thalamic responses reported in two studies (Nicolelis et al., 1993; Shin et al., 1995) may be evidence for a cortical involvement. Alternatively, the later components may reflect an inherent longer latency associated with drive from the edges of the underlying excitatory field, and spatiotemporal profiles are needed for a full interpretation. In raccoon, total digit denervation unmasks large inhibitory receptive fields for thalamic neurons (Rasmusson et al., 1993) in line with the effect on cortical neurons in this species (Rasmusson and Turnbull, 1983). In two cases capsaicin was applied to the exposed digital nerves to block only C-fibre activity. This produced a 2 to 4 times enlargement of excitatory receptive fields located on the digit tips.

Table 1: Results from all available studies that have reported the short-term effects of denervations on responses recorded in somatosensory thalamus. SA and RA are slowly- and rapidly-adapting response categories; vpl and vpm are ventroposterior-nucleus lateral and medial divisions, respectively. From, English and Calford, 1992; Nakahama et al., 1966; Nicolelis et al., 1993; Shin et al., 1995; Rasmusson et al., 1993.

study	nucleus	recording type	denervation	observations	% new/larger fields	anaesthesia
N '65	cat vpl	single unit-RA	local-procaine[a]	36	25%	light barbiturate NM block
N '65	cat vpl	single unit-SA	local-procaine[a]	23	17%	light barbiturate NM block
E '92	cat vpl	single unit-RA	local-lignocaine[a]	8	0%	ketamine/xylazine
E '92	rat vpl	multi unit-RA	local-lignocaine[a]	9	0%	ketamine/xylazine
E '92	rat vpl	multi unit-RA	digit amputation	16	6%	ketamine/xylazine
N '93	rat vpm	discrim. su-RA	local-lidocaine	44[b]	23% (61%[c])	barbiturate
S '95	rat vpl	single unit-RA	local-lidocaine	8[b]	100%[c]	urethane
S '95	rat vpl	single unit-RA	digit-amputation	10[b]	90%[c]	urethane
R '93	raccoon vpl	multi unit-RA	digit-denervation[d]	13	76%[e]	chloralose
R '93	raccoon vpl	multi unit-RA	capsaicin to nerve	2	100%	chloralose

[a]subcutaneous injection; [b]total samples discounted by cases where lidocaine failed to block the original RF; [c]includes neurons where there was a concomitant temporal shift with the main response being at a longer latency; [d]includes both digital nerve cut and local anaesthesia; [e]in the raccoon total digit denervation unmasked inhibitory receptive fields on adjacent skin and digits as is also the case in cortex.

An intriguing aspect of the Nicolelis et al. (1993) study was that the local anaesthetic sometimes blocked responses from the original receptive field (whisker) for some neurons and not for others in the same preparation or for some response epochs and not others for the same neuron. This points to a differential block of peripheral innervation, possibly due to multiple nerve types. In vpm it is known that there is a convergence from 3 brainstem terminal nuclei of the trigeminal nerve. However, it is unlikely that any of the unmasking of new responses with peripheral denervation is due to interaction of these inputs since it has been shown that lesions of the principal nucleus do not unmask viable inputs from the other nuclei in the short term (Rhoades et al., 1987). However, 6 days after such lesions 53% of rat vpm neurons were responsive, with the majority then receiving afferent input via the interpolaris subnucleus.

Large changes in the topographic representation in thalamus have also been reported following chronic peripheral denervations (Garraghty and Kaas, 1991a). Whereas this was interpreted as indicating that thalamic or subthalamic changes may be responsible for, or largely contribute to, cortical changes, the strong interconnections of the lemniscal representations in thalamus and cortex make another explanation possible. This being that the thalamic responses may reflect cortical changes transmitted by the massive S1 to vpl projection. Alternatively, the plasticity may be a property of the combined thalamocortical circuit. In a study of the effects of different patterns of trained finger use in owl monkeys Wang et al. (1995) showed changes in the hand representation and receptive fields of neurons in cortex but not in thalamus. They concluded that the "learned" responses observed were a result of changes at the cortical level. Other evidence for there being a cortical effect in addition to any thalamic or subthalamic contribution to representational plasticity comes from short-term studies with inactivation of some corticocortical inputs to S1 (Clarey et al., 1996; see **Inhibitory plasticity**, section below). In addition, Diamond et al., (1994) report that representational changes in S1 barrel field, following whisker trimming, occur in supragranular and infragranular layers prior to changes in the thalamic-projection recipient layers.

6. Mechanisms of representational plasticity

While almost every publication that has presented data on representational plasticity in adult cortex has advanced some variant of a mechanism to account for the effect, the proposals can be grouped into three broad categories: unmasking of existing but normally unexpressed inputs, axonal sprouting, and synaptic plasticity; Before introducing these, it is worth pointing out that no author in this field has pushed for a single explanation to the exclusion of others and that these categories simply cover the range of plausible mechanisms.

Unmasking of existing but normally unexpressed inputs to cortex: Studies showing short-term plasticity, which are the main subject of this chapter, support the general concept that plasticity may result from the unmasking of normally unexpressed inputs to a cortical locus. Despite the clear demonstration of such unmasking, an explanation for how this occurs is not immediately apparent. The basis of the problem is that a peripheral denervation of a small body part leads to expansion of the receptive field of some cortical neurons onto adjacent body areas, but does not directly affect the input from these areas. As discussed above, the areas around a normal receptive field can contribute inhibitory inputs and these have the potential to mask weaker excitatory inputs. But why should denervation of a nearby body part affect this relationship? After all there are no peripheral interconnections between body areas, and primary mechanoreceptor afferents are silent when not stimulated. Therefore their denervation would, in itself, have no influence on the balance of excitation and inhibition resulting at a cortical neuron for stimulation of a nearby intact area. This problem has not been widely recognized. Two types of solution have been suggested:

1) that the denervation produces a discharge in the directly affected axons which triggers a disinhibition (Kolarik et al., 1994);
2) that there are some afferents which provide a tonic signal and that these contribute centrally to a tonic level of inhibition which is disrupted by a denervation (Calford and Tweedale, 1990a; Calford and Tweedale, 1991c).

The first of these solutions was presented after considering only those studies in which denervations were affected by nerve cuts or amputation. The proposal links short-term changes in cortex to short-term changes that have been shown in second-order polymodal neurons of the dorsal horn, where intense stimulation of peripheral C-fibres leads to expansion of the mechanoreceptor component of the receptive field of these cells (e.g. Cook et al., 1987). It is conceivable (but not shown) that some injury discharge may occur with denervation, to trigger such an event. However, such nerve cuts do not produce short term receptive field plasticity in the dorsal horn (Snow and Wilson, 1991); thus if there is a site at which nerve injury produces some release of peptides or other co-factors which may modify inhibition it must be higher in the pathway. In this laboratory we initially considered explanations of this type, but when recording from cutaneous-stimulation receptive neurons in cortex (Calford and

Tweedale, unpublished observations) or cuneate nucleus (Martin and Calford, unpublished observations) no massive increases in spontaneous rate or any other manifestation of an injury discharge were noted. In addition attempts to produce receptive field changes of single neurons in flying-fox S1, with electrical-stimulation tetani (at amplitudes above C-fibre threshold) to the median nerve were unsuccessful (Calford and Tweedale, 1991b; Calford and Tweedale, unpublished observations). Irrespective of these considerations, the explanation does not appear useful for instances in which functional denervation has been obtained with local anaesthesia (e.g. Calford and Tweedale, 1990a; Calford and Tweedale, 1991c; Panestos et al., 1995) where there is no "denervation" signal.

In consideration of a possible tonic peripheral activity that may provide a source of input to central inhibitory neurons, C-fibre activity has been blocked using the selective neurotoxin, capsaicin. The receptive fields of cutaneous-stimulation responsive single neurons in S1 of cat (Calford and Tweedale, 1991b), flying-fox (Calford and Tweedale, 1991b) and mouse (Nussbaumer and Wall, 1985), vpl of raccoon (Rasmusson et al., 1993) and cuneate nucleus of cat (Pettit and Schwark, 1996) have been shown to expand with peripheral capsaicin application. Applied subcutaneously, or directly to a peripheral nerve, capsaicin rapidly blocks C-fibre conduction with a concomitant expansion of cortical receptive fields (Figure. 8). Since mechanoreceptor A-fibres are unaffected, responses in the area of the original receptive field are maintained. Both the peripheral conduction block and the receptive field expansion exceed the period of an acute experiment. The clear implication of these experiments is that some sub-group of peripheral C-fibres provides a source of activity which ultimately drives inhibition that contributes to shaping the receptive fields of central pathway neurons. It is unclear which group of C-fibres are involved, but the implication is that they must have some level of tonic activity.

Axonal sprouting: Clearly axonal sprouting can not be the mechanism by which new receptive fields are unmasked in the short term. However, appropriate sprouting has the potential to be the mechanism involved in cases where there is a gradual "filling-in" of a deafferented

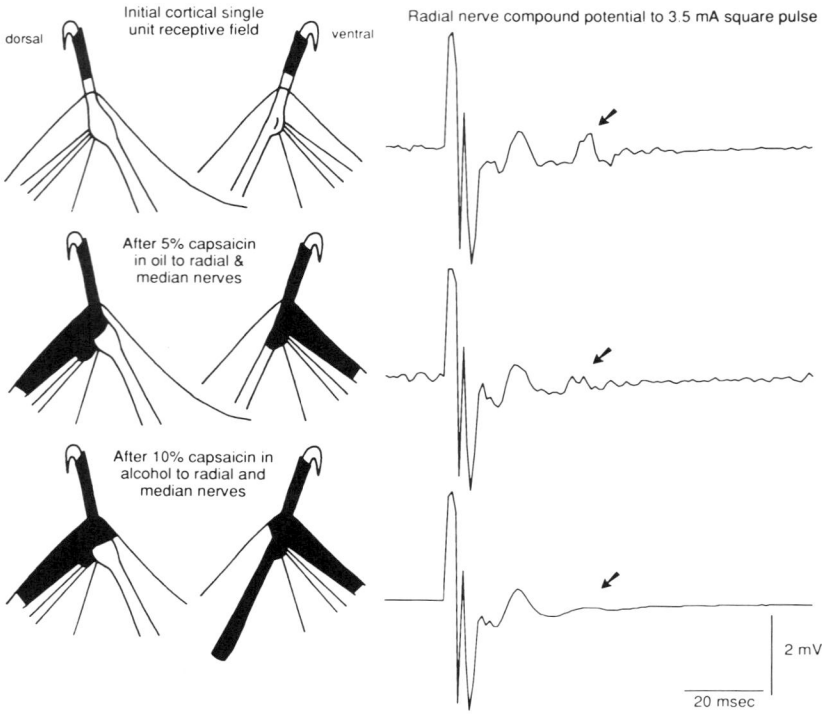

Figure 8. Expansion of the mechanoreceptor receptive field of a cortical neuron with blockage of C-fibre activity in a peripheral nerve. The left panel shows dorsal and ventral views of part of one wing of a flying fox, with the receptive field inicated in black. The right panel shows the compound action potential (recorded with hook electrodes) of the radial nerve to a square wave stimulus above C-fibre threshold. The C-fibre wave (arrowed) is diminished after application of capsaicin directly to the exposed radial and median nerves. This produces an immediate expansion of the receptive field of the cortical neuron (from Calford and Tweedale, 1991b).

area (e.g. Cusick et al., 1990; see above). Rasmusson and colleagues have looked at the possibility that sprouting of axons contributes to adult plasticity at the level of the dorsal column termination of primary afferents (Rasmusson, 1988) and the thalamocortical projection (Rasmusson and Nance, 1986) in raccoon: there was no evidence of such change.

A number of studies have shown that collaterals of primary mechanosensory afferents will sprout into denervated regions of the dorsal horn of the spinal cord (e.g. Fitzgerald et al., 1990; Florence et al., 1993; LaMotte and Kapadia, 1993; McMahon and Kett-White, 1991; Molander et al., 1988) but these axons do not sprout at their terminal target area, the dorsal column nuclei (Florence et al., 1993). It must be remembered that these axons are the central branches from cells that also send a peripheral axon to the skin which is capable of extensive regrowth if injured. Furthermore, in order for the collateral sprouting to occur, not only must neighbouring inputs be removed but the peripheral axons of these cells must be in a growth phase following direct damage. Thus, such sprouting would not be expected in most of the experimental paradigms in which central pathway representational plasticity of the somatosensory pathway has been demonstrated (i.e. following amputations or discrete nerve cuts). In addition, for reasons outlined above (see, Subcortical contributions to representational plasticity in cortex) any functional effect of sprouting within the dorsal horn may not necessarily have a major impact on the lemniscal somatosensory pathway. A recent report of axonal sprouting in the dorsal column termination in monkeys following long-term amputation (Florence et al., 1995) is not convincing. In that study, tracer injections were made into the skin above the amputation and to a matched skin region on the opposite limb. In comparing the central distributions of the transported label it would hvae been impossible to distinguish between changes brought about by central sprouting and or peripheral sprouting (within the skin above the amputation) from the severed nerves. It is possible that colateral sprouting may have occured within the innervation of the limb.

Levin and Dunn-Meynell (1993) used unilateral vibrissectomy, sparing C5, to show increases in mRNA for GAP-43 (a growth associated protein linked to axonal growth cones) in the ipsilateral trigeminal nuclei and contralateral S1 of adult rats. While suggestive of axonal sprouting in these areas, this finding requires confirmation with a more direct method for there are reasonable levels of GAP-43 expression in normal adults. The change could, thus, reflect a metabolic or activity-driven change in this noraml expression rather than a growth-related effect. Depletion of noradrenaline blocked the increase in GAP-43 mRNA. This may be consistent with explanations based on either adrenergic control of plasticity, or adrenergic effects on activity (see below). However, support for an axonal growth involvement in cortical representational plasticity comes from work on visual cortex. Darian-Smith and Gilbert (1994) reported an increased density of biocytin-labelled intrinsic projections into areas of cat primary visual cortex directly affected by discrete retinal lesions. The intrinsic projections did not appear to project further, but to have an increased density within the geometric limits of their normal projection. While the time-course of this change was not fully investigated it appeared to take many weeks to be measurable. In contrast the increase in GAP-43 message expression in rat S1 lasted less than two weeks. As with somatosensory thalamocortical projections (Rasmusson and Nance, 1986), Darian-Smith and Gilbert (1995) found no evidence for axonal sprouting of thalamocortical afferents. Given that the limits of spread of the thalamocortical projection are exceeded by demonstrated short- and/or long-term representational plasticity in both somatosensory (e.g., Diamond et al., 1994; Pons et al., 1991, Rasmusson and Nance, 1986) and visual (Darian-Smith and Gilbert, 1995; Heinen and Skaenski, 1991; Schmid et al. 1995; 1996) cortex, it is inviting to consider intrinsic projections as the primary source of the capacity for plasticity.

Synaptic plasticity: Strong support for changes in synaptic efficacy, as a mechanism for plasticity of the cortical somatosensory representation, comes from experiments in which changes in the representation were brought about by behavioural training (Recanzone et al., 1992c; Recanzone et al., 1992d; Nudo, 1996) or altered patterns of body part use (e.g., syndactyly: Allard et al., 1991). In an experimental series conducted by Recanzone and colleagues, adult owl monkeys were trained to detect a difference in the frequency of a sequence of tactile stimuli to the glabrous skin of a single finger. For most animals performance on this task improved with practice (Recanzone et al., 1992a). There were concomitant changes in the representation of the stimulated digit in cortical areas 3b and 3a. In area 3b, there was a tightening of the temporal aspects of the response to sinewave stimulation (Recanzone et al., 1992e), and an increase in the size of neural receptive fields and the total representation of the stimulated area, but not of the entire digit (Recanzone et al., 1992d). In affected regions of area 3a, which normally is dominated by neurons responsive to muscle-spindle stimulation, there was an increase in the proportion of neurons responding to cutaneous-only stimulation (Recanzone et al., 1992c). None of these changes took place with passive stimulation of the digit. These studies elegantly demonstrate that even within a highly organized topographic representation in a highly structured region of cortex (area 3b: koniocortex) there is use-dependent alteration in the adult. Furthermore they show that behaviourally-relevant changes in neuronal responses - i.e. learning - can take place at the "low" level of the primary sensory representation in cortex. Work from this same laboratory showed that changes in the cortical representation appropriate to another stimulation regimen were not present in the thalamus (Wang et al., 1995). The stimulus presentation conditions and the changes found are consistent with the changes being the result of associational potentiation at excitatory synapses. The best known example of such being the Hebbian-learning rule (Hebb, 1949). This asserts that when an activated synapse is part of a successful stimulation of a neuron, then that synapse will be altered such that it has increased efficacy. A number of real examples of synaptic potentiation conforming to the Hebbian rule have been found including the NMDA-receptor mediated long term potentiation (LTP) of excitatory amino-acid synapses in adult mammalian cortex (e.g., Artola and Singer, 1987)

Direct support for NMDA-receptor mediated LTP in adult cortical plasticity comes from two studies in which such modification was not possible: Kano et al. (1991) blocked chronic reorganization in cat S1 with administration of an NMDA-receptor antagonist, and Glazewski et al. (1996) showed that chronic use-dependent plasticity in mouse vibrissal barrel-field fails to develop in mice missing the gene for a calcium-activated second messenger known to be involved in LTP (CaMKII). LTP requires the coincident-activation of a number of intracellular events but is fundamentally dependent upon flow of calcium ions through the channel of the NMDA class of glutamate receptors. However, normal physiological levels of magnesium ions are sufficient to restrict this flow and block LTP (Artola and Singer, 1987; Nicoll, 1988). A significant cellular event such as extended depolarization (e.g., brought about by tetanic stimulation) can remove the magnesium block. Thus, experiments in which a few hours of repetitive intracortical microstimulation produced changes in the response properties of S1 neurons (Recanzone et al., 1992b; Spengler and Dinse, 1994; Dinse et al., 1993), are also consistent with an NMDA-receptor mediated LTP explanation. In contrast, the immediate unmasking of larger receptive fields in sensory cortex with peripheral (see above) or central denervation (Clarey et al., 1996) do not appear to be dependent on this mechanism. It can be argued on theoretical grounds that the conditions for Hebbian type learning are not met in partial-denervation experiments (see Calford and Tweedale, 1991a), in that the changes take place rapidly and without the necessity of appropriate stimulation. However, the strongest argument along these lines can be made on the basis of the finding that peripheral-denervation-induced plasticity is reflected (as enlarged neural receptive fields) in the representation of the mirror-image body part in cortex ipsilateral to the denervation (Calford and Tweedale, 1990b). Short-term unmasking is also rapidly reversed when temporary denervation is used (Byrne and Calford, 1991; Calford and Tweedale, 1991a; Calford and Tweedale, 1991c). In addition most of the partial-denervation unmasking experiments have been performed under ketamine anaesthesia. Ketamine is a non-competitive antagonist at NMDA-receptors, and while systemic application is not very specific, at anaesthetic levels it would be expected to block NMDA-receptor mediated LTP. In confirmation of this reasoning, Garraghty et al. (1993) have reported that the extent of representational plasticity in New World monkeys after median nerve transection is unaffected by NMDA-receptor blockade in the short-term, but that longer term (11-28 days) effects fail to develop with the blockade.

In the raccoon S1, Smits et al. (Smits et al., 1991; Zarzecki et al., 1993) have shown through in vivo intracellular recording a potentiation (faster rising phases) of the excitatory projection from the adjacent somatosensory area, the "heterogeneous zone". By virtue of its wide corticocortical projection and imprecise topography, this area has the potential to be the source of the considerable plasticity seen in the raccoon S1 following peripheral partial denervation. Loss of activity in the primary projection to a zone in S1 may allow this weaker projection to become more effective. However, the corticocortical EPSPs after chronic (22 week) d4 removal did not differ in their latencies, amplitudes, half-widths, or integrated amplitudes, but only in the rising phase (Zarzecki et al., 1993). In contrast, after d3-d4 syndactyly, corticocortical EPSPs (heterogeneous zone to S1) were more common than in normal animals; implicating this projection in perceptual-learning plasticity. Such a result could be consistent with either synpatic-potentiation or synaptic-proliferation based explanations. Similar studies have yet to be conducted with other species, but the interconnectivity of the multiple somatosensory fields and the presence of nontopographically-matched connections between them (Krubitzer et al., 1993) suggest a role for interareal corticocortical projections in adult plasticity (see Calford, 1991). Despite demonstrating significant nonsomatotopically-matched intracortical connectivity in macaque monkeys, Burton and Fabri (1995) specifically argue against this potential mechanism for adult plasticity. While it may be the case that the off-focus projections appear insufficient to account for the largest demonstration of adult plasticity, that following long-term C2-C5 dorsal rhizotomies (Pons et al., 1991; but see Lund et al., 1994, for an interpretation which halves the extent), they report sufficient divergence to account for the effects which follow from digit denervation (Calford and Tweedale, 1991c; Merzenich et al., 1984).

Disinhibition as a gate for synaptic potentiation: There is a possible direct link between short-term unmasking based plasticity and long-term learning based plasticity, through an action at the magnesium-ion block of NMDA-receptors. The disinhibition which produces the unmasking should produce conditions more favourable for LTP by allowing easier depolarization of cell membrane potentials when driven by glutamatergic excitatory inputs. As in the study with visual cortex slices by Artola and Singer (1987), in which the GABAa-receptor antagonist bicuculline was used to disinhibit, this physiological disinhibition may remove the magnesium block and allow LTP. Such a two stage process, of an initial functional plasticity (unmasking; a consequence of the wiring of the circuit), followed by an excitatory synaptic plasticity (driven by the pattern of afferent activity), can account for the observation that the initially unmasked large receptive fields shrink and a new precise topographic map emerges in the week or so following small peripheral denervations (Calford and Tweedale, 1988).

While discussing mechanisms that may be permissive for chronic-denervation triggered plasticity it is appropriate to consider the role of cholinergic inputs to cortex (for review see Dykes, 1990). Cholingeric inputs emanate from the basal forebrain region and are not activated in a stimulus-locked manner. However, this major input to sensory cortex has the potential to modulate the response of somatosensory cortex neurons to peripheral stimulation. Lamour et al. (1982) showed that most corticofugal neurons of rat S1, particularly those in layer Vb were depolarized by acetylcholine, but few supragranular cells were affected. A more even distribution of cells throughout the cortical layers were seen to show an enhanced response to somatic stimulation after iontophoretically applied acetylcholine: around 34% showed such effects with concurrent stimulation and application and 18% showing enhancement that lasted more than 5 minutes after the period of iontophoresis (Lamour et al., 1988). Similar effects (but for a higher proportion of affected cells) have been reported following release of endogenous acetylcholine by stimulation of the basal forebrain in cat (Tremblay et al., 1990). The long-lasting effects suggested a modulatory role for the basal forebrain cholinergic inputs to sensory cortex in adult representational plasticity. These suggestions seemed confirmed when it was shown that rats with basal forebrain lesions failed to show a normal degree of representational plasticity in S1 4 days after sciatic nerve section (Webster et al., 1991). However, while this is a powerful demonstration, a more complete study is required before it can be concluded that cholinergic inputs are indeed permissive for chronic-denervation triggered plasticity, since it may be that weaker inputs are unsuccessful in affecting postsynpatic cells in the absence of acetylcholine. A similar problem exists in the interpretation of a study by Levin et al. (1988), in which the potential modulatory noradrenergic inputs from the locus coeruleus were blocked and a metabolic marker for representational plasticity failed to develop. The study used a spared-vibrissa paradigm and [^{14}C]2-deoxyglucose labelling. In the Webster et al. (1991) study, one condition used reserpine to decrease effects of monoamines, such as noradrenalin. This did not block representational plasticity but did increase stimulation thresholds - an effect which compounds interpretation of the locus coeruleus effect on [^{14}C]2-deoxyglucose labelling.

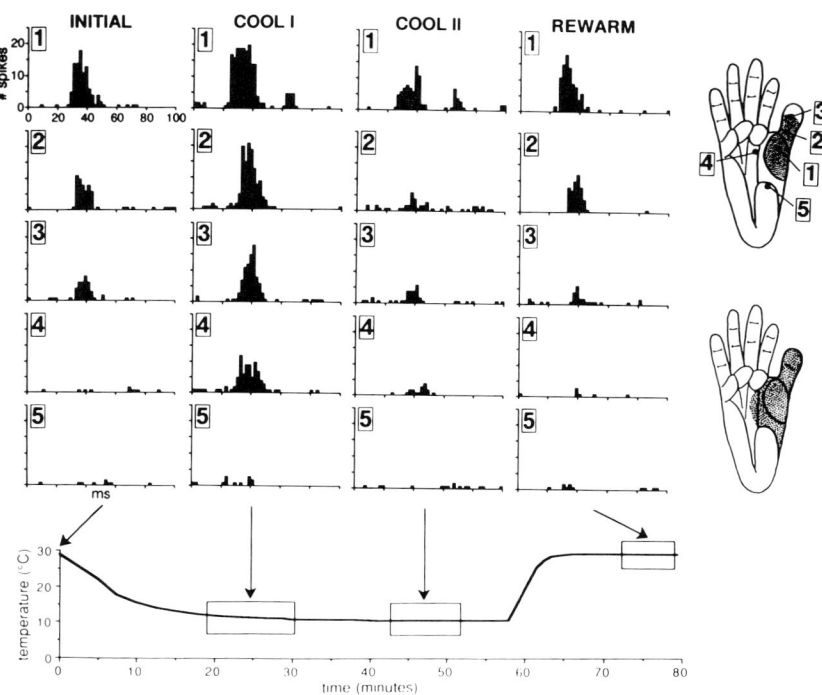

Figure 9. Poststimulus response histograms for a neuron in macaque monkey area 3b (foot representation). Each panel presents the response to stimulation at the 5 positions indicated on the drawing of the foot. The initial RF and the maximum RF obtained when the opposite-hemisphere mirror-image position was cooled are indicated with stippling. Temperature at the cooling site is indicated, and the boxes show the time windows during which the data shown were collected. An initial expansion of the RF and increase in responsivity are followed by a return to near-initial response levels even though the modification to the inputs from the opposite hemisphere is maintained (from Clarey et al., 1996).

Inhibitory plasticity: Clarey et al. (1996) reported that neurons in S1 of flying fox and macaque monkey show a rapid expansion of receptive field size and an increase in responsiveness with focal cooling of the mirror image position in the contralateral hemisphere (Figure. 9). Similar changes were noted for about 50% of neurons in S1 (area 3b) with cooling or chemical deactivation of the equivalent representation in the adjacent area 1 representation in flying foxes and marmoset monkeys (Clarey et al., 1993). These initial unmasking effects show a capacity for a rapid functional plasticity within cortex itself, brought about by the removal of a subset of the inputs to a cortical locus. The effect was readily reversible with warming of the site of deactivation. This demonstration provides a partial explanation for the finding that short-term   partial-peripheral-denervation-induced   plasticity   is transferred interhemispherically such that neural receptive fields on the mirror image body part to that directly affected also expand (Calford and Tweedale, 1990b).

The lack of a callosal interconnectivity between distal representations (Jones et al., 1978; Jones and Powell, 1969) within area 3b created a difficulty in explaining these effects. However the later results with area 1 cooling are consistent with the effect of a denervation being relayed by changes in activity in the callosal pathway connecting the area 1 representations and ipsilateral area 1 and area 3b interconnectivity. An interesting observation of the peripheral denervation interhemispheric effect (Calford and Tweedale, 1990b) was that the indirect effect was transient: that is, with small amputations the direct effect in contralateral cortex was an expansion of the receptive fields of affected neurons onto adjacent areas for the duration of the experiment, whereas the expansion of the field in ipsilateral cortex (unaffected hand) reversed after 30 to 60 minutes. Similarly the expansion of area 3b neural receptive fields with contralateral cortex cooling was transient (Figure. 9). The initial effect of the blockage was an increase in receptive size and responsiveness; this was interpreted as an unmasking or disinhibition effect due to blocking of inputs which feed onto local inhibitory neurons. After 20 minutes or so of a steady-state (in terms of manipulation) the effect reversed. Since the affected inputs are still blocked this rebound effect must involve some form of synaptic plasticity that leads to an increased local inhibition. The most parsimonious explanation for these effects is some form of increase in the efficacy of transmission at GABAa-synapses. This result points to a second type of synaptic plasticity, that may be involved in adult cortical plasticity, additional to the NMDA-receptor based LTP that is implicated in long term plasticity (see above). While it is well known from pharmacological studies that GABAergic synaptic transmission is capable of modification (e.g. the effects of benzodiazepines) such plasticity at inhibitory synapses would not conform to any associational or Hebbian rules of learning (Hebb, 1949) but would require some form of local regulation that reacts to decreased inhibition. The trigger for an increase of inhibitory efficacy may be a detection of change in local levels of activity, implying that optimal levels of transmission are actively maintained at some inhibitory synapses. A possible mechanism of regulation is an increase in the binding efficacy of GABA induced by modification of the receptor by an endogenous substance, as has been demonstrated for the GABAa-receptor in the presence of neuroactive steroids (Majewska et al., 1986; Purdy et al., 1991).

Studies on neuronal plasticity have emphasized excitatory synaptic plasticity and much effort has been aimed at elucidating the changes in efficacy of glutamatergic synapses. As discussed above, inhibition has been shown to play a role in long-term potentiation (e.g. (Artola and Singer, 1987b) However, in this role the inhibitory effect is one of a modulator or "gate" for excitatory plasticity, rather than inhibitory plasticity *per se*. With only a few exceptions (Greenough and Anderson, 1991; Kano, 1995; Marr, 1969), the concept of a modifiable inhibitory synapse has not been incorporated into theoretical models of the nervous system. However, several recent studies have provided evidence for long-term potentiation and/or depression of inhibitory synaptic transmission (Akhondzadeh and Stone, 1995; Bell et al., 1993; Kano et al., 1992; Komatsu and Iwakiri, 1993; Korn et al., 1992; Oda et al., 1995). These reports, the description of functional inhibitory plasticity in adult mammalian neocortex (Clarey et al., 1996 Figure. 9) and the description of a plausible mode of modification via the action of endogenous steroids, make a strong case for the inclusion of such plasticity into network models of learning and memory, and central nervous system functioning.

References

Adrian, E.D. (1941) Afferent discharges to the cerebral cortex from the peripheral sense organs. Journal of Physiology 100, 159-191.

Akhondzadeh, S. and Stone, T.W. (1995) Induction of a novel form of hippocampal long-term depression by muscimol: involvement of GABAa but not glutamate receptors. British Journal of Pharmacology 115, 527-33.

Allard, T., Clark, S.A., Jenkins, W.M. and Merzenich, M.M. (1991) Reorganization of somatosensory area 3b representations in adult owl monkeys after digital syndactyly. Journal of Neurophysiology 66, 1048-.

Alloway, K.D. and Burton, H. (1991) Differential effects of GABA and bicuculline on rapidly- and slowly-adapting neurons in primary somatosensory cortex of primates. Experimental Brain Research 85, 598-610.

Alloway, K.D., Rosenthal, P. and Burton, H. (1989) Quantitative measurements of receptive field changes during antagonism of GABAergic transmission in primary somatosensory cortex of cats. Experimental Brain Research 78, 514-532.

Artola, A. and Singer, W. (1987) Long-term potentiation and NMDA receptors in rat visual cortex. Nature 330, 649-652.

Basbaum, A.I. and Wall, P.D. (1976) Chronic changes in the response of cells in adult cat dorsal horn following partial deafferentation: the appearance of responding cells in a previously non-responsive region. Brain Research 116, 181-204.

Batuev, A.S., Alexandrov, A.A. and Scheynikov, N.A. (1982) Picrotoxin action on the receptive fields of the cat sensorimotor cortex neurons. Journal of Neuroscience Research 7, 49-55.

Batuev, A.S., Alexandrov, A.A., Scheynikov, N.A., Kcharazia, V.N. and Chan Chinh An (1989) The role of inhibitory processes in the formation of functional properties of neurons in vibrissal projection zone of the cat somatosensory cortex. Experimental Brain Research 76, 198-206.

Bell, C.C., Caputi, A., Grant, K. and Serrier, J. (1993) Storage of a sensory pattern by anti-Hebbian synaptic plasticity in an electric fish. Proceedings of the National Academy of Science (USA) 90, 4650-4654.

Bolz, J. and Gilbert, C.D. (1989) The role of horizontal connections in generating long receptive fields in the cat visual cortex. European Journal of Neuroscience 1, 263-268.

Brown, A.G., Fyffe, R.E.W., Noble, R. and Rowe, M.J. (1984) Effects of hind limb nerve section on lumbosacral dorsal horn neurones in the cat. Journal of Physiology (London) 354, 375-394.

Burton, H. and Fabri, M. (1995) Ipsilateral intracortical connections of physiologically defined cutaneous representations in areas 3b and 1 of macaque monkeys: projections in the vicinity of the central sulcus. Journal of Comparative Neurology 355, 508-538.

Byrne, J.A. and Calford, M.B. (1991) Short-term expansion of receptive fields in rat primary somatosensory cortex after hindpaw digit denervation. Brain Research 565, 218-224.

Calford, M.B. (1991) Curious cortical change (News and Views). Nature 352, 759-760.

Calford, M.B. and Semple, M.N. (1995) Monaural inhibition in cat auditory-cortex. Journal of Neurophysiology 73, 1876-1891.

Calford, M.B. and Tweedale, R. (1988) Immediate and chronic changes in responses of somatosensory cortex in adult flying-fox after digit amputation. Nature 332, 446-448.

Calford, M.B. and Tweedale, R. (1990a) The capacity for reorganization in adult somatosensory cortex. In: M. Rowe and L. Aitkin (Eds.), Information Processing in Mammalian Auditory and Tactile Systems, Alan R. Liss, Inc., New York, pp221-236.

Calford, M.B. and Tweedale, R. (1990b) Interhemispheric transfer of plasticity in the cerebral cortex. Science 249, 805-807.

Calford, M.B. and Tweedale, R. (1991a) Acute changes in cutaneous receptive fields in primary somatosensory cortex after digit denervation in adult flying fox. Journal of Neurophysiology 65, 178-187.

Calford, M.B. and Tweedale, R. (1991b) C-fibres provide a source of masking inhibition to primary somatosensory cortex. Proceedings of the Royal Society of London (B) 243, 269-275.

Calford, M.B. and Tweedale, R. (1991c) Immediate expansion of receptive fields of neurons in area 3b of macaque monkeys after digit denervation. Somatosensory and Motor Research 8, 249-260.

Clarey, J.C., Tweedale, R., Krubitzer, L.A. and Calford, M.B. (1993) Effect of focal cooling of area 1 on ipsilateral area 3b responses in flying foxes and marmosets. Society for Neuroscience Abstracts 19, 1568.

Clarey, J.C., Tweedale, R. and Calford, M.B. (1996) Interhemispheric modulation of somatosensory receptive fields: evidence for plasticity in primary somatosensory cortex. Cerebral Cortex 6, 196-206.

Glasby, C.M.L. and Gandevia, S.C. (1995) Afferent input from the human thumb changes the perceived size of the the thumb and lips. Proceedings of the Australian Neuroscience Society 6, 193.

Cook, A.J., Woolf, C.J., Wall, P.D. and McMahon, S.B. (1987) Dynamic receptive field plasticity in rat spinal cord dorsal horn following C-primary afferent input. Nature 325, 151-153.

Cusick, C.G., Wall, J.T., Whiting, J.H.J. and Wiley, R.G. (1990) Temporal progression of cortical reorganization following nerve injury. Brain Research 537, 355-358.

Darian-Smith, C. and Darian-Smith, I. (1993) Thalamic projections to areas 3a, 3b, and 4 in the sensorimotor cortex of the mature and infant macaque monkey. Journal of Comparative Neurology 335, 173-199.

Darian-Smith, C., Darian-Smith, I. and Cheema, S.S. (1990) Thalamic projections to sensorimotor cortex in the macaque monkey: use of multiple retrograde fluorescent tracers. Journal of Comparative Neurology 299, 17-46.

Darian-Smith, C. and Gilbert, C.D. (1994) Axonal sprouting accompanies functional reorganization in adult cat striate cortex. Nature 368, 737-740.

Daw, N.W., Fox, K., Sato, H. and Czepita, D. (1992) Critical period for monocular deprivation in the cat visual cortex. Journal of Neurophysiology 67, 197-202.

Diamond, M.E., Huang, W. and Ebner, F.F. (1994) Laminar comparison of somatosensory cortical plasticity. Science 265, 1885-1888.

Dinse, H.R., Recanzone, G.H. and Merzenich, M.M. (1993) Alterations in correlated activity parallel ICMS-induced representational plasticity. NeuroReport 5, 173-176.

Doetsch, G.S., Johnston, K.W. and Hannan Jr, C.J. (1990) Physiological changes in the somatosensory forepaw cerebral cortex of adult raccoons following lesions of a single cortical digit representation. Experimental Neurology 108, 162-175.

Dostrovsky, J.O., Millar, J. and Wall, P.D. (1976) The immediate shift of afferent drive of dorsal column nucleus cells following deafferentation: a comparison of acute and chronic deafferentation in gracile nucleus and spinal cord. Experimental Neurology 52, 480-495.

Dykes, R.W. (1990) Acetylcholine and neuronal plasticity in somatosensory cortex. In: M. Steriade and D. Biesold (Eds.), Brain Cholinergic Systems, Oxford University Press, New York, pp294-313.

Dykes, R.W. (1983) Parallel processing of somatosensory information: A theory. Brain Res. Rev. 6, 47-115.

Dykes, R.W. and Gabor, A. (1981) Magnification functions and receptive field sequences for submodality-specific bands in SI cortex of cats. Journal of Comparative Neurology 202, 597-620.

Dykes, R.W., Landry, P., Metherate, R. and Hicks, T.P. (1984) Functional role of GABA in cat primary somatosensory cortex: shaping receptive fields of cortical neurons. Journal of Neurophysiology 52, 1066-1093.

Dykes, R.W., Rasmusson, D.D. and Hoeltzell, P.B. (1980) Organization of primary somatosensory cortex in the cat. Journal of Neurophysiology 43, 1527-1546.

Elbert, T., Flor, H., Birbaumer, N., Knecht, S., Hampson, S., Larbig, W. and Taub, E. (1994) Extensive reorganization of the somatosensory cortex in adult humans after nervous-system injury. NeuroReport 5, 2593-2597.

English, P. and Calford, M. (1992) Peripheral denervation induced unmasking of large receptive fields in somatosensory cortex but not in thalamus. Proceedings of the Australian Neuroscience Society 3, 133.

Favorov, O.V. and Diamond, M.E. (1990) Demonstration of discrete place-defined columns - segregates - in the cat SI. Journal of Comparative Neurology 298, 97-112.

Favorov, O.V., Diamond, M.E. and Whitsel, B.L. (1987) Evidence for a mosaic representation of the body surface in area 3b of the somatic cortex of cat. Proceedings of the National Academy of Science (USA) 84, 6606-6610.

Fitzgerald, M., Woolf, C.J. and Shortland, P. (1990) Collateral sprouting of the central terminals of cutaneous primary afferent neurons in the rat spinal cord: pattern, morphology, and influence of targets. Journal of Comparative Neurology 300, 370-385.

Flor, H., Elbert, T., Knecht, S., Wienbruch, C., Pantev, C., Birbaumer, N., Larbig, W. and Taub, E. (1995) Phantom-limb pain as a perceptual correlate of cortical reorganization following arm amputation. Nature 375, 482-484.

Florence, S.L. and Kaas, J.H. (1995) Large scale reorganization at multiple levels of the somatosensory pathway follows therapeutic amputation of the hand in monkeys. Journal of Neuroscience 15, 8083-8095.

Florence, S.L., Garraghty, P.E., Carlson, M. and Kaas, J.H. (1993) Sprouting of peripheral nerve axons in the spinal cord of monkeys. Brain Research 601, 343-348.

Fox, K. (1992) A critical period for experience-dependent synaptic plasticity in rat barrel cortex. Journal of Neuroscience 12, 1826-1838.

Franck, J.I. (1980) Functional reorganization of cat somatic sensory-motor cortex (SM1) after selective dorsal root rhizotomies. Brain Research 186, 458-462.

Garraghty, P.E., Hanes, D.P., Florence, S.L. and Kaas, J.H. (1994) Pattern of peripheral deafferentation predicts reorganizational limits in adult primate somatosensory cortex. Somatosensory and Motor Research 11, 109-117.

Garraghty, P.E. and Kaas, J.H. (1991a) Functional reorganization in adult monkey thalamus after peripheral nerve injury. NeuroReport 2, 747-750.

Garraghty, P.E. and Kaas, J.H. (1991b) Large-scale functional reorganization in adult monkey cortex after peripheral nerve injury. Proceedings of the National Academy of Science (USA) 88, 6976-6980.

Garraghty, P.E., Pons, T.P., Sur, M. and Kaas, J.H. (1989) The arbors of axons terminating in middle cortical layers of somatosensory area 3b in owl monkeys. Somatosensory and Motor Research 6, 410-411.

Garraghty, P.E., Muja, N. and Hoard, R. (1993) NMDA receptor blockade prevents most cortical reorganization after peripheral nerve injury in adult monkeys. Society for Neuroscience Abstracts 19, 1569.

Gilbert, C.D. and Wiesel, T.N. (1979) Morphology and intracortical projections of functionally characterised neurones in the cat visual cortex. Nature 280, 120-125.

Gilbert, C.D. and Wiesel, T.N. (1989) Columnar specificity of intrinsic horizontal and corticocortical connections in cat visual cortex. Journal of Neuroscience 9, 2432-2442.

Gilbert, C.D. and Wiesel, T.N. (1992) Receptive field dynamics in adult primary visual cortex. Nature 356, 150-152.

Glazewski, S., Chen, C.-M., Silva, A. and Fox, K. (1996) Requirement for a-CaMKII in experience-dependent plasticity of the barrel cortex. Science 272, 421-423.

Greenough, W.T. and Anderson, B.J. (1991) Cerebellar synaptic plasticity. Relation to learning versus neural activity. Annals of the New York Academy of Sciences 627, 231-247.

Halligan, P.W., Marshall, J.C. and Wade, D.T. (1994) Sensory disorganization and perceptual plasticity after limb amputation - a follow-up-study. NeuroReport 5, 1341-1345.

Halligan, P.W., Marshall, J.C., Wade, D.T., Davey, J. and Morrison, D. (1993) Thumb in Cheek - Sensory reorganization and perceptual plasticity after limb amputation. NeuroReport 4, 233-236.

Hawken, M.J. and Parker, A.J. (1987) Spatial properties of neurons in the monkey striate cortex. Proceedings of the Royal Society of London (B) 231, 251-288.

Hebb, D.O. (1949) The Organization of Behavior. Wiley, New York.

Heinen, S.J. and Skaenski, A.A. (1991) Recovery of visual responses in foveal V1 neurons following bilateral foveal lesions in adult monkey. Exp. Brain Res. 83, 670-674.

Hicks, T.P. and Dykes, R.W. (1983) Receptive field size for certain neurons in primary somatosensory cortex is determined by GABA-mediated intracortical inhibition. Brain Research 274, 160-164.

Hicks, T.P., Metherate, R., Landry, P. and Dykes, R.W. (1986) Bicuculline-induced alterations of response properties in functionally identified ventroposterior thalamic neurones. Experimental Brain Research 63, 248-264.

Houser, C.R., Vaughn, J.E., Hendry, S.H.C., Jones, E.G. and Peters, A. (1984) GABA neurons in the cerebral cortex. In: E.G. Jones and A. Peters (Eds.), Cerebral Cortex, Volume 2 - Functional Properties of Cortical Cells, Plenum Press, New York, pp63-89

Hubel, D.H. and Wiesel, T.N. (1962) Receptive fields, binocular interaction and functional architecture in the cat's visual cortex. Journal of Physiology (London) 160, 106-154.

Hubel, D.H. and Wiesel, T.N. (1970) The period of susceptibility to the physiological effects of unilateral eye closure in kittens. Journal of Physiology (London) 206, 419-436.

Jain, N., Florence, S.L. and Kaas, J.H. (1995) Limits on plasticity in somatosensory cortex of adults rats: hindlimb cortex is not reactivated after dorsal column section. Journal of Neurophysiology 73, 1537-1546.

Jänig, W., Schoultz, T. and Spencer, W.A. (1977) Temporal and spatial parameters of excitation and afferent inhibition in cuneothalamic relay neurons. Journal of Neurophysiology 40, 822-835.

Jänig, W., Spencer, W.A. and Younkin, S.G. (1979) Spatial and temporal features of afferent inhibition of thalamocortical relay cells. Journal of Neurophysiology 42, 1450-1460.

Jenkins, W.M. and Merzenich, M.M. (1987) Reorganization of neocortical representations after brain injury: a neurophysiological model of the bases of recovery from stroke. In: F.J. Seil, E. Herbert and B.M. Carlson (Eds.), Progress In Brain Research, Vol. 71, Elsevier, pp249-266

Johansson, B.B. and Grabowski, M. (1994) Functional recovery after brain infarction - plasticity and neural transplantation. Brain Pathol 4, 85-95.

Jones, E.G., Coulter, J.D. and Hendry, S.H.C. (1978) Intracortical connectivity of architectonic fields in the somatic sensory, motor and parietal cortex of monkeys. Journal of Comparative Neurology 181, 291-348.

Jones, E.G. and Powell, T.P.S. (1969) Connexions of the somatic sensory cortex of the rhesus monkey. II. Contralateral cortical connexions. Brain 92, 717-730.

Juliano, S.L., Eslin, D.E. and Tommerdahl, M. (1994) Developmental regulation of plasticity in cat somatosensory cortex. Journal of Neurophysiology 72, 1706-1716.

Kaas, J.H., Krubitzer, L.A., Chino, Y.M., Langston, A.L., Polley, E.H. and Blair, N. (1990) Reorganization of retinotopic cortical maps in adult mammals after lesions of the retina. Science 248, 229-231.

Kalaska, J. and Pomeranz, B. (1979) Chronic paw denervation causes an age-dependent appearance of novel responses from forearm in "paw cortex" of kittens and adult cats. Journal of Neurophysiology 42, 618-633.

Kaneko, T. and Hicks, T.P. (1988) Baclofen and gamma-aminobutyric acid differentially suppress the cutaneous responsiveness of primary somatosensory cortical neurones. Brain Research 443, 360-366.

Kaneko, T. and Hicks, T.P. (1990) GABA(B)-related activity involved in synaptic processing of somatosensory information in S1 cortex of the anaesthetized cat. British Journal of Pharmacology 100, 689-698.

Kano, M. (1995) Plasticity of inhibitory synapses in the brain: a possible memory mechanism that has been overlooked. Journal of Neuroscience Research 21, 177-182.

Kano, M., Lino, K. and Kano, M. (1991) Functional reorganization of adult cat somatosensory cortex is dependent upon NMDA receptors. NeuroReport 2, 77-80.

Kano, M., Rexhausen, U., Dreessen, J. and Konnerth, A. (1992) Synaptic excitation produces a long-lasting rebound potentiation of inhibitory synaptic signals in cerebellar Purkinje cells. Nature 356, 601-604.

Kelahan, A.M. and Doetsch, G.S. (1984) Time dependent changes in the functional organization of somatosensory cerebral cortex following digit amputation in adult raccoons. Somatosensory Research 2, 49-81.

Kelahan, A.M., Ray, R.H., Carson, L.V., Massey, C.E. and Doetsch, G.S. (1981) Functional reorganization of adult raccoon somatosensory cerebral cortex following neonatal digit amputation. Brain Research 223, 151-159.

Kolarik, R.C., Rasey, S.K. and Wall, J.T. (1994) The consistency, extent, and locations of early-onset changes in cortical nerve dominance aggregates following injury of nerves to primate hands. Journal of Neuroscience 14, 4269-4288.

Komatsu, Y. and Iwakiri, M. (1993) Long-term modification of inhibitory synaptic transmission in developing visual cortex. NeuroReport 4, 907-910.

Korn, H., Oda, Y. and Faber, D.S. (1992) Long-term potentiation of inhibitory circuits and synapses in the central nervous system. Proceedings of the National Academy of Science (USA) 89, 440-443.

Krubitzer, L.A. and Calford, M.B. (1992) Five topographically organized fields in the somatosensory cortex of the flying fox: microelectrode maps, myeloarchitecture and cortical modules. Journal of Comparative Neurology 317, 1-30.

Krubitzer, L.A., Calford, M.B. and Schmid, L.M. (1993) Connections of somatosensory cortex in megachiropteran bats: the evolution of cortical fields in mammals. Journal of Comparative Neurology 327, 473-506.

LaMotte, C.C. and Kapadia, S.E. (1993) Deafferentation-induced terminal field expansion of myelinated saphenous afferents in the adult rat dorsal horn and the nucleus gracilis following pronase injection of the sciatic nerve. Journal of Comparative Neurology 330, 83-94.

Lamour, Y., Cutar, P. and Jobert, A. (1982) Excitatory effect of acetylcholine on different types of neurons in the first somatosensory neocortex of the rat: laminar distribution and pharmacological characteristics. Neuroscience 7, 1483-1494.

Lamour, Y., Dutar, P., Jobert, A. and Dykes, R.W. (1988) An iontophoretic study of single somatosensory neurons in rat granular cortex serving the limbs: a laminar analysis of glutamate and acetylcholine effects on receptive-field properties. Journal of Neurophysiology 60, 725-750.

Landry, P. and Deschênes, M. (1981) Intracortical arborizations and receptive fields of identified ventrobasal thalamocortical afferents to the primary somatic sensory cortex in the cat. Journal of Comparative Neurology 199, 345-371.

Laskin, S.E. and Spencer, W.A. (1979) Cutaneous masking. II. Geometry of excitatory and inhibitory receptive fields of single units in somatosensory cortex of the cat. Journal of Neurophysiology 42, 1061-1082.

LeVay, S., Wiesel, T.N. and Hubel, D.H. (1981) The postnatal development and plasticity of ocular-dominance columns in the monkey. In: F.O. Schmitt, F.G. Worden, G. Adelman and S.G. Dennis (Eds.), The Organization of Cerebral Cortex, The M.I.T. Press, Cambridge, U.S.A., pp29-45.

Levin, B.E., Craik, R.L. and Hand, P.J. (1988) The role of norepinephrine in adult rat somatosensory (SmI) cortical metabolism and plasticity. Brain Research 443, 261-271.

Levin, B.E. and Dunn-Meynell, A. (1993) Regulation of growth-associated protein 43 (GAP-43) messenger RNA associated with plastic change in the adult rat barrel receptor complex. Molecular Brain Research 18, 59-70.

Li, C.X., Waters, R.S., Oladehin, A., Johnson, E.F., McCandlish, C.A. and Dykes, R.W. (1994) Large unresponsive zones appear in cat somatosensory cortex immediately after ulnar nerve cut. Canadian Journal of Neurological Science 21, 233-247.

Lund, J.P., Sun, G.D. and Lamarre, Y. (1994) Cortical reorganization and deafferentation in adult macaques. Science 265, 546-548.

Majewska, M.D., Harrison, N.L., Schwartz, R.D., Barker, J.L. and Paul, S.M. (1986) Steroid hormone metabolites are barbiturate-like modulators of the GABA receptor. Science 232, 1004-1007.

Marr, D. (1969) A theory of cerebellar cortex. Journal of Physiology (London) 202, 437-470.

McMahon, S.B. and Kett-White, R. (1991) Sprouting of peripherally regenerating primary sensory neurones in the adult central nervous system. Journal of Comparative Neurology 304, 307-315.

Merzenich, M.M. and Kaas, J. (1980) Principles of organization of sensory-perceptual systems in mammals. In: J.M. Sprague and A.N. Epstein (Eds.), Progress in Psychobiology and Physiological Psychology, Academic Press, New York, pp1-42

Merzenich, M.M., Kaas, J.H., Wall, J., Nelson, R.J., Sur, M. and Felleman, D. (1983a) Topographic reorganization of somatosensory cortical areas 3b and 1 in adult monkeys following restricted deafferentation. Neuroscience 8, 33-55.

Merzenich, M.M., Kaas, J.H., Wall, J.T., Sur, M., Nelson, R.J. and Felleman, D.J. (1983b) Progression of change following median nerve section in the cortical representation of the hand in areas 3b and 1 in adult owl and squirrel monkeys. Neuroscience 10, 639-665.

Merzenich, M.M., Nelson, R.J., Stryker, M.P., Cynader, M.S., Schoppmann, A. and Zook, J.M. (1984) Somatosensory cortical map changes following digit amputation in adult monkeys. Journal of Comparative Neurology 224, 591-605.

Merzenich, M.M., Recanzone, G.H., Jenkins, W.M. and Grajski, K.A. (1990) Adaptive mechanisms in cortical networks underlying cortical contributions to learning and nondeclarative memory. In: Cold Spring Harbor Symposium on Quantitative Biology Vol. LV, Cold Spring Harbor Laboratory Press, pp873-887.

Merzenich, M.M., Schreiner, C., Jenkins, W. and Wang, X. (1993) Neural mechanisms underlying temporal integration, segmentation, and input sequence representation: some implications for the origin of learning disabilities. Annals of the New York Academy of Sciences 682, 1-22.

Meyers, D.E.R. and Snow, P.J. (1984) Somatotopically inappropriate projections of single hair follicle afferent fibres to the cat spinal cord. Journal of Physiology (London) 347, 59-73.

Millar, J., Basbaum, A.I. and Wall, P.D. (1976) Restructuring of the somatotopic map and appearance of abnormal neruronal activity in the gracile nucleus after partial deafferentation. Experimental Neurology 50, 658-672.

Mogilner, A., Grossman, J.A., Ribary, U., Joliot, M., Volkmann, J., Rapaport, D., Beasley, R.W. and Llinas, R.R. (1993) Somatosensory cortical plasticity in adult humans revealed by magnetoencephalography. Proceedings of the National Academy of Science (USA) 90, 3593-3597.

Molander, C., Kinnman, E. and Aldskogius, H. (1988) Expansion of spinal cord primary sensory afferent projection following combined sciatic nerve resection and saphenous nerve crush: a horseradish peroxidase study in the adult rat. Journal of Comparative Neurology 276, 436-441.

Moore, D.R. (1993) Plasticity of binaural hearing and some possible mechanisms following late-onset deprivation. Journal of the American Academy of Audiology 4, 277-283.

Morin, C. and Molotchnikoff, S. (1992) Generation of end-inhibition in striate neurons in rabbits. Brain Research Bulletin 28, 323-327.

Mountcastle, V.B. (1957) Modality and topographic properties of single neurons of cat's somatic sensory cortex. Journal of Neurophysiology 20, 408-434.

Mountcastle, V.B. and Powell, T.P. (1959) Neural mechanisms subserving cutaneous sensibility, with special reference to the role of afferent inhibition in sensory perception and discrimination. Bulletin of the Johns Hopkins Hospital 105, 210-232.

Nakahama, H., Nishioka, S. and Otsuka, T. (1966) Excitation and inhibition in ventrobasal thalamic neurons before and after cutaneous input deprivation. Progress in Brain Research 21, 180-196.

Nicolelis, M.A., Lin, R.C., Woodward, D.J. and Chapin, J.K. (1993) Induction of immediate spatiotemporal changes in thalamic networks by peripheral block of ascending cutaneous information. Nature 361, 533-536.

Nicoll, R.A. (1988) The coupling of neurotransmitter receptors to ion channels in the brain. Science 241, 545-551.

Nussbaumer, J.C. and Wall, P.D. (1985) Expansion of receptive fields in the mouse cortical barrelfield after administration of capsaicin to neonates or local application on the infraorbital nerve in adults. Brain Research 360, 1-9.

Oda, Y., Charpier, S., Murayama, Y., Suma, C. and Korn, H. (1995) Long-term potentiation of glycinergic inhibitory synaptic transmission. Journal of Neurophysiology 74, 1056-1074.

Oka, J.I. and Hicks, T.P. (1990) Benzodiazepines and synaptic processing in the spatial domain within the cat's primary somatosensory cortex. Canadian Journal of Physiology and Pharmacology 68, 1025-40.

Oka, J.I., Jang, E.K. and Hicks, T.P. (1986) Benzodiazepine receptor involvement in the control of receptive field size and responsiveness in primary somatosensory cortex. Brain Research 376, 194-198.

Panestos, F., Nuñez, A. and Avendaño, C. (1995) Local anaesthesia induces immediate receptive field changes in nucleus gracilis and cortex. NeuroReport 7, 150-152.

Pearson, J.C., Finkel, L.H. and Edelman, G.M. (1987) Plasticity in the organization of adult cerebral cortical maps: a computer simulation based on neuronal group selection. Journal of Neuroscience 7, 4209-4223.

Pettit, M.J. and Schwark, H.D. (1993) Receptive field reorganization in dorsal column nuclei during temporary denervation. Science 262, 2054-2056.

Pettit, M.J. and Schwark, H.D. (1996) Capsaicin-induced receptive field reorganization in cuneate neurons. Journal of Neurophysiology 75, 1117-1125.

Pons, T.P., Garraghty, P.E., Ommaya, K., Kaas, J.H., Taub, E. and Mishkin, M. (1991) Massive cortical reorganization after sensory deafferentation in adult macaques. Science 252, 1857-1860.

Pubols, L.M. (1984) The boundary of proximal hindlimb representation in the dorsal horn following peripheral nerve lesions in cats: a reevaluation of plasticity in the somatotopic map. Somatosensory Research 2, 19-32.

Purdy, R.H., Morrow, A.L., Moore, P.H.J. and Paul, S.M. (1991) Stress-induced elevations of γ-aminobutyric acid type A receptor-active steroids in the rat brain. Proceedings of the National Academy of Science (USA) 88, 4553-4557.

Purves, D., Riddle, D.R. and LaMantia, A-S. (1992) Iterated patterns of brain activity (or how the cortex gets its spots). Trends in Neurosciences 15, 362-368.

Rajan, R., Irvine, D.R.F., Wise, L.Z. and Heil, P. (1993) Effect of unilateral partial cochlear lesions in adult cats on the representation of lesioned and unlesioned cochleas in primary auditory cortex. Journal of Comparative Neurology 338, 17-49.

Ramachandran, V.S., Stewart, M. and Rogers-Ramachandran, D.C. (1992) Perceptual correlates of massive cortical reorganization. NeuroReport 3, 583-586.

Rasmusson, D.D. (1982) Reorganization of raccoon somatosensory cortex following removal of the fifth digit. Journal of Comparative Neurology 205, 313-326.

Rasmusson, D.D. (1988) Projection patterns of digit afferents to the cuneate nucleus in the raccoon before and after partial deafferentation. Journal of Comparative Neurology 277, 549-556.

Rasmusson, D.D., Louw, D.F. and Northgrave, S.A. (1993) The immediate effects of peripheral denervation on inhibitory mechanisms in the somatosensory thalamus. Somatosensory and Motor Research 10, 69-80.

Rasmusson, D.D. and Nance, D.M. (1986) Non-overlapping thalamocortical projections for separate forepaw digits before and after cortical reorganization in the raccoon. Brain Research Bulletin 16, 399-406.

Rasmusson, D.D. and Turnbull, B.G. (1983) Immediate effects of digit amputation on S1 cortex in the raccoon: unmasking of inhibitory fields. Brain Research 288, 368-370.

Recanzone, G.H., Jenkins, W.M., Hradek, G.T. and Merzenich, M.M. (1992a) Progressive improvement in discriminative abilities in adult owl monkeys performing a tactile frequency discrimination task. Journal of Neurophysiology 67, 1015-1030.

Recanzone, G.H., Merzenich, M.M. and Dinse, H.R. (1992b) Expansion of the cortical representation of a specific skin field in primary somatosensory cortex by intracortical microstimulation. Cerebral Cortex 2, 181-196.

Recanzone, G.H., Merzenich, M.M. and Jenkins, W.M. (1992c) Frequency discrimination training engaging a restricted skin surface results in an emergence of a cutaneous response zone in cortical area 3a. Journal of Neurophysiology 67, 1057-1070.

Recanzone, G.H., Merzenich, M.M., Jenkins, W.M., Grajski, K.A. and Dinse, H.R. (1992d) Topographic reorganization of the hand representation in cortical area 3b of owl monkeys trained in a frequency-discrimination task. Journal of Neurophysiology 67, 1031-1056.

Recanzone, G.H., Merzenich, M.M. and Schreiner, C.E. (1992e) Changes in the distributed temporal response properties of SI cortical neurons reflect improvements in performance on a temporally based tactile discrimination task. Journal of Neurophysiology 67, 1071-1091.

Rhoades, R.W., Belford, G.R. and Killackey, H.P. (1987) Receptive-field properties of rat ventral posterior medial neurons before and after selective kainic acid lesions of the trigeminal brain stem complex. Journal of Neurophysiology 57, 1577-600.

Roberts, W.A., Eaton, S.A. and Salt, T.E. (1992) Widely distributed GABA-mediated afferent inhibition processes within the ventrobasal thalamus of rat and their possible relevance to pathological pain states and somatotopic plasticity. Experimental Brain Research 89, 363-372.

Robertson, D. and Irvine, D.R.F. (1989) Plasticity of frequency organization in auditory cortex of guinea pigs with partial unilateral deafness. Journal of Comparative Neurology 282, 456-471.

Rodieck, R.W. (1965) Quantitative analysis of cat retinal ganglion cell response to visual stimuli. Vision Research 5, 583-601.

Roe, A.W., Pallas, S.L., Hahm, J.O. and Sur, M. (1990) A map of visual space induced in primary auditory cortex. Science 250, 818-820.

Saade, N.E., Banna, N.R., Khoury, A., Jabbur, S.J. and Wall, P.D. (1982) Cutaneous receptive field alterations induced by 4-aminopyridine. Brain Research 232, 177-180.

Schmid, L.M., Rosa, M.G.P., Ambler, J.S. and Calford, M.B. (1996) Visuotopic reorganization in the primary visual cortex of adult cats following monocular and binocular retinal lesions. Cerebral Cortex. 6, 388-405.

Schmid, L.M., Rosa, M.G.P. and Calford, M.B. (1995) Retinal-Detachment Induces Massive Immediate Reorganization in Visual-Cortex. NeuroReport 6, 1349-1353.

Shin, H-C., Park, S., Son, J. and Sohn, J-H. (1995) Responses from new receptive fields of VPL neurones following deafferentation. NeuroReport 7, 33-36.

Smith, D.C., Spear, P.D. and Kratz, K.E. (1978) Role of visual experience in postcritical-period reversal of effects of monocular deprivation in cat striate cortex. Journal of Comparative Neurology 178, 313-328.

Smits, E., Gordon, D.C., Witte, S., Rasmusson, D.D. and Zarzecki, P. (1991) Synaptic potentials evoked by convergent somatosensory and corticocortical inputs in raccoon somatosensory cortex: substrates for plasticity. Journal of Neurophysiology 66, 688-695.

Snow, P.J., Nudo, R.J., Rivers, W., Jenkins, W.M. and Merzenich, M.M. (1988) Somatotopically inappropriate projections from thalamocortical neurons to the SI cortex of the cat demonstrated by the use of intracortical microstimulation. Somatosensory Research 5, 349-372.

Snow, P.J. and Wilson, P. (1991) Plasticity in the somatosensory system of developing and mature mammals - the effects of injury to the central and peripheral nervous system. Progress in Sensory Physiology Vol. 11, Springer-Verlag, Berlin,

Spengler, F. and Dinse, H.R. (1994) Reversible Relocation of Representational Boundaries of Adult-Rats by Intracortical Microstimulation. NeuroReport 5, 949-953.

Spinelli, D.N. and Jensen, F.E. (1979) Plasticity: the mirror of experience. Science 203, 75-78.

Tremblay, N., Warren, R. and Dykes, R.W. (1988) The effects of strychnine on neurons in cat somatosensory cortex and its interaction with the inhibitory amino acids, glycine, taurine and beta-alanine. Neuroscience 26, 745-62.

Tremblay, N., Warren, R.A. and Dykes, R.W. (1990) Electrophysiological studies of acetylcholine and the role of the basal forebrain in the somatosensory cortex of the cat. II. Cortical neurons excited by somatic stimuli. Journal of Neurophysiology 64, 1212-1222.

Turnbull, B.G. and Rasmusson, D.D. (1990) Acute effects of total or partial digit denervation on raccoon somatosensory cortex. Somatosensory and Motor Research 7, 365-389.

Turnbull, B.G. and Rasmusson, D.D. (1991) Chronic effects of total or partial digit denervation on raccoon somatosensory cortex. Somatosensory and Motor Research 8, 201-213.

Waite, P.M.E. (1984) Rearrangement of neuronal responses in the trigeminal system of the rat following peripheral nerve section. Journal of Physiology (London) 352, 425-445.

Wall, J.T. (1988) Development and maintenance of somatotopic maps of the skin: a mosaic hypothesis based on peripheral and central contiguities. Brain Behavior and Evolution. 31, 252-268.

Wall, J.T. and Cusick, C.G. (1984) Cutaneous responsiveness in primary somatosensory (S-I) hindpaw cortex before and after partial hindpaw deafferentation in adult rats. Journal of Neuroscience 4, 1499-1515.

Wall, J.T. and Cusick, C.G. (1986) The representation of peripheral nerve inputs in the S-1 hindpaw cortex of rats raised with incompletely innervated hindpaws. Journal of Neuroscience 6, 1129-1147.

Wall, P.D. (1967) The laminar organization of dorsal horn and effects of descending inpulses. Journal of Physiology (London) 188, 403-423.

Wall, P.D. and Egger, M.D. (1971) Formation of new connexions in adult rat brains after partial deafferentation. Nature 232, 542-545.

Wang, X., Merzenich, M.M., Sameshima, K. and Jenkins, W.M. (1995) Remodelling of hand representation in adult cortex determined by timing of tactile stimulation. Nature 378, 71-75.

Webster, H.H., Hanisch, U.K., Dykes, R.W. and Biesold, D. (1991) Basal forebrain lesions with or without reserpine injection inhibit cortical reorganization in rat hindpaw primary somatosensory cortex following sciatic nerve section. Somatosensory and Motor Research 8, 327-346.

Wilson, P. (1987) Absence of mediolateral reorganization of dorsal horn somatotopy after peripheral deafferentation in the cat. Experimental Neurology 95, 432-447.

Woolsey, C.N., Marshall, W.H. and Bard, P. (1942) Representation of cutaneous tactile sensibility of the monkey as indicated by evoked potentials. Bulletin of the Johns Hopkins Hospital 70, 399-441.

Zarzecki, P., Witte, S., Smits, E., Gordon, D.C., Kirchberger, P. and Rasmusson, D.D. (1993) Synaptic mechanisms of cortical representational plasticity: somatosensory and corticocortical EPSPs in reorganized raccoon SI cortex. Journal of Neurophysiology 69, 1422-1432.

SUBJECT INDEX